Thermal Analysis of Materials

Thermal Analysis of Materials

Editors

Sergey V. Ushakov
Shmuel Hayun

MDPI • Basel • Beijing • Wuhan • Barcelona • Belgrade • Manchester • Tokyo • Cluj • Tianjin

Editors
Sergey V. Ushakov
Arizona State University,
Tempe, AZ, USA

Shmuel Hayun
Ben-Gurion University of the Negev,
Beer-Sheva, Israel

Editorial Office
MDPI
St. Alban-Anlage 66
4052 Basel, Switzerland

This is a reprint of articles from the Special Issue published online in the open access journal *Materials* (ISSN 1996-1944) (available at: https://www.mdpi.com/journal/materials/special_issues/Thermal_Analysis).

For citation purposes, cite each article independently as indicated on the article page online and as indicated below:

LastName, A.A.; LastName, B.B.; LastName, C.C. Article Title. *Journal Name* **Year**, *Volume Number*, Page Range.

ISBN 978-3-0365-6846-1 (Hbk)
ISBN 978-3-0365-6847-8 (PDF)

Cover image courtesy of Sergey V. Ushakov and Shmuel Hayun.

© 2023 by the authors. Articles in this book are Open Access and distributed under the Creative Commons Attribution (CC BY) license, which allows users to download, copy and build upon published articles, as long as the author and publisher are properly credited, which ensures maximum dissemination and a wider impact of our publications.
The book as a whole is distributed by MDPI under the terms and conditions of the Creative Commons license CC BY-NC-ND.

Contents

Sergey V. Ushakov and Shmuel Hayun
Special Issue: Thermal Analysis of Materials
Reprinted from: *Materials* **2021**, *14*, 4923, doi:10.3390/ma14174923 1

Zhi Wang, Ruichao Wei, Xuehui Wang, Junjiang He and Jian Wang
Pyrolysis and Combustion of Polyvinyl Chloride (PVC) Sheath for New and Aged Cables via Thermogravimetric Analysis-Fourier Transform Infrared (TG-FTIR) and Calorimeter
Reprinted from: *Materials* **2018**, *11*, 1997, doi:10.3390/ma11101997 3

Feng Wang, Dong-Sheng Qian, Peng Xiao and Song Deng
Accelerating Cementite Precipitation during the Non-Isothermal Process by Applying Tensile Stress in GCr15 Bearing Steel
Reprinted from: *Materials* **2018**, *11*, 2403, doi:10.3390/ma11122403 19

Shuhua Liu, Qiaoling Li and Xinyi Zhao
Hydration Kinetics of Composite Cementitious Materials Containing Copper Tailing Powder and Graphene Oxide
Reprinted from: *Materials* **2018**, *11*, 2499, doi:10.3390/ma11122499 31

Hannes Fröck, Michael Reich, Benjamin Milkereit and Olaf Kessler
Scanning Rate Extension of Conventional DSCs through Indirect Measurements
Reprinted from: *Materials* **2019**, *12*, 1085, doi:10.3390/ma12071085 47

Kyungju Kim, Dasom Kim, Kwangjae Park, Myunghoon Cho, Seungchan Cho and Hansang Kwon
Effect of Intermetallic Compounds on the Thermal and Mechanical Properties of Al–Cu Composite Materials Fabricated by Spark Plasma Sintering
Reprinted from: *Materials* **2019**, *12*, 1546, doi:10.3390/ma12091546 69

Luís Felipe dos Santos Carollo, Ana Lúcia Fernandes de Lima e Silva and Sandro Metrevelle Marcondes de Lima e Silva
A Different Approach to Estimate Temperature-Dependent Thermal Properties of Metallic Materials
Reprinted from: *Materials* **2019**, *12*, 2579, doi:10.3390/ma12162579 83

Yuri Kirshon, Shir Ben Shalom, Moran Emuna, Yaron Greenberg, Joonho Lee, Guy Makov and Eyal Yahel
Thermophysical Measurements in Liquid Alloys and Phase Diagram Studies
Reprinted from: *Materials* **2019**, *12*, 3999, doi:10.3390/ma12233999 99

Weipeng Luo, Shuai Xue, Cun Zhao, Meng Zhang and Guoxi Li
Robust Interferometry for Testing Thermal Expansion of Dual-Material Lattices
Reprinted from: *Materials* **2020**, *13*, 313, doi:10.3390/ma13020313 119

Barak Ratzker, Sergey Kalabukhov and Nachum Frage
Spark Plasma Sintering Apparatus Used for High-temperature Compressive Creep Tests
Reprinted from: *Materials* **2020**, *13*, 396, doi:10.3390/ma13020396 133

Sergey V. Ushakov, Shmuel Hayun, Weiping Gong and Alexandra Navrotsky
Thermal Analysis of High Entropy Rare Earth Oxides
Reprinted from: *Materials* **2020**, *13*, 3141, doi:10.3390/ma13143141 145

Yuval Mordekovitz, Yael Shoval, Natali Froumin and Shmuel Hayun
Effect of Structure and Composition of Non-Stoichiometry Magnesium Aluminate Spinel on Water Adsorption
Reprinted from: *Materials* **2020**, *13*, 3195, doi:10.3390/ma13143195 165

Iga Szpunar, Ragnar Strandbakke, Magnus Helgerud Sørby, Sebastian Lech Wachowski, Maria Balaguer, Mateusz Tarach, et al.
High-Temperature Structural and Electrical Properties of BaLnCo$_2$O$_6$ Positrodes
Reprinted from: *Materials* **2020**, *13*, 4044, doi:10.3390/ma13184044 179

Angelika Plota and Anna Masek
Lifetime Prediction Methods for Degradable Polymeric Materials—A Short Review
Reprinted from: *Materials* **2020**, *13*, 4507, doi:10.3390/ma13204507 197

Amalie Gunnarshaug, Maria Monika Metallinou and Torgrim Log
Study of Industrial Grade Thermal Insulation at Elevated Temperatures
Reprinted from: *Materials* **2020**, *13*, 4613, doi:10.3390/ma13204613 223

Bartłomiej Rogalewicz, Agnieszka Czylkowska, Piotr Anielak and Paweł Samulkiewicz
Investigation and Possibilities of Reuse of Carbon Dioxide Absorbent Used in Anesthesiology
Reprinted from: *Materials* **2020**, *13*, 5052, doi:10.3390/ma13215052 239

Krzysztof Labus
Comparison of the Properties of Natural Sorbents for the Calcium Looping Process
Reprinted from: *Materials* **2021**, *14*, 548, doi:10.3390/ma14030548 249

Sergey V. Ushakov, Jonas Niessen, Dante G. Quirinale, Robert Prieler, Alexandra Navrotsky and Rainer Telle
Measurements of Density of Liquid Oxides with an Aero-Acoustic Levitator
Reprinted from: *Materials* **2021**, *14*, 822, doi:10.3390/ma14040822 269

Editorial

Special Issue: Thermal Analysis of Materials

Sergey V. Ushakov [1,*] and Shmuel Hayun [2,*]

[1] School of Molecular Sciences and Center for Materials of the Universe, Arizona State University, Tempe, AZ 85287, USA
[2] Department of Materials Engineering at the Ben-Gurion University of the Negev, Beer-Sheva 84105, Israel
* Correspondence: sushakov@asu.edu (S.V.U.); hayuns@bgu.ac.il (S.H.)

The measurement of any physical property as a function of temperature brings the method used into the realm of thermal analysis. This makes "Thermal Analysis of Materials" a broad interdisciplinary subject. This Special Issue combines contributions solicited by editors that reflect their scientific interests and networks, with general related submissions received over two years. It contains 16 articles and 1 short review submitted by authors from 10 countries.

In several papers, well-established techniques of differential thermal analysis, thermogravimetry, and gas adsorption calorimetry are applied to research related to environmental problems, such as capturing CO_2 from industrial processes by calcium looping [1]; the utilization of copper tailings wastes as an addition to Portland cement [2]; the reuse of soda lime CO_2 absorbent in spacecrafts, submarines, anesthetics, and diving apparatuses [3]; and the passive fire protection offered by inorganic material-based insulation [4].

Another set of papers deals with the development of new or application-specific approaches, namely laser interferometry techniques for high precision measurements on non-standard samples, such as composite truss structures [5]; measurements of high temperature compressive creep in spark plasma sintering apparatuses [6]; measurements of the density of liquid oxides with aero-acoustic levitators [7]; and the simultaneous estimation of thermal conductivity and heat capacity in metal alloys [8]. The range of materials covered includes polymers [9,10]; natural complex inorganic materials, such as natural sorbents [1] and industrial-grade thermal insulation [4]; single phase ceramic materials, such as spinel $MgAl_2O_4$ [11], rare earth sesquioxides [12], and double perovskite cobaltites [13]; metals, alloys, and intermetallic compounds [8,14–16].

The use of thermal analysis techniques to elucidate phase transformation kinetics can be found in papers on the hydration of cement [2] and in the study of microstructural evolutions in high-carbon and chromium-bearing steel [15]. A review by Plota and Masek [10] is devoted to the extrapolation of kinetic data on polymers' degradation for lifetime predictions. A paper by Frock et al. [16] provides a new method to study kinetics at heating rates between 5 and 1000 K per second. These heating rates are relevant to welding and laser melting processing but too fast for conventional differential scanning calorimeters and too slow for fast-scanning chip calorimetry. Examples of the applications of thermal analysis and calorimetry techniques to obtain thermodynamic data vary from measurements of enthalpies of water adsorption on defect spinel surfaces [11] to phase transformation enthalpies and entropies above 2000 °C in rare earth oxides [12].

Thermal analysis contributes to the calculation of phase diagrams (CALPHAD), both through experimental phase diagrams and by providing thermodynamic data for optimizations. Krishon et al. [17] extend the CALPHAD approach by calculating pressure dependence of several binary metal phase diagrams based on the temperature dependence of sound velocity and density data in liquid alloys. Thus, this Special Issue provides a sampling of the current use, diversity, and ongoing developments of techniques and approaches in the thermal analysis of materials.

Conflicts of Interest: The authors declare no conflict of interest.

References

1. Labus, K. Comparison of the Properties of Natural Sorbents for the Calcium Looping Process. *Materials* **2021**, *14*, 548. [CrossRef] [PubMed]
2. Liu, S.; Li, Q.; Zhao, X. Hydration Kinetics of Composite Cementitious Materials Containing Copper Tailing Powder and Graphene Oxide. *Materials* **2018**, *11*, 2499. [CrossRef] [PubMed]
3. Rogalewicz, B.; Czylkowska, A.; Anielak, P.; Samulkiewicz, P. Investigation and Possibilities of Reuse of Carbon Dioxide Absorbent Used in Anesthesiology. *Materials* **2020**, *13*, 5052. [CrossRef] [PubMed]
4. Gunnarshaug, A.; Metallinou, M.M.; Log, T. Study of Industrial Grade Thermal Insulation at Elevated Temperatures. *Materials* **2020**, *13*, 4613. [CrossRef] [PubMed]
5. Luo, W.; Xue, S.; Zhao, C.; Zhang, M.; Li, G. Robust Interferometry for Testing Thermal Expansion of Dual-Material Lattices. *Materials* **2020**, *13*, 313. [CrossRef] [PubMed]
6. Ratzker, B.; Kalabukhov, S.; Frage, N. Spark Plasma Sintering Apparatus Used for High-temperature Compressive Creep Tests. *Materials* **2020**, *13*, 396. [CrossRef] [PubMed]
7. Ushakov, S.V.; Niessen, J.; Quirinale, D.G.; Prieler, R.; Navrotsky, A.; Telle, R. Measurements of Density of Liquid Oxides with an Aero-Acoustic Levitator. *Materials* **2021**, *14*, 822. [CrossRef] [PubMed]
8. Dos Santos Carollo, L.F.; de Lima e Silva, A.L.F.; de Lima e Silva, S.M.M. A Different Approach to Estimate Temperature-Dependent Thermal Properties of Metallic Materials. *Materials* **2019**, *12*, 2579. [CrossRef] [PubMed]
9. Wang, Z.; Wei, R.; Wang, X.; He, J.; Wang, J. Pyrolysis and Combustion of Polyvinyl Chloride (PVC) Sheath for New and Aged Cables via Thermogravimetric Analysis-Fourier Transform Infrared (TG-FTIR) and Calorimeter. *Materials* **2018**, *11*, 1997. [CrossRef] [PubMed]
10. Plota, A.; Masek, A. Lifetime Prediction Methods for Degradable Polymeric Materials—A Short Review. *Materials* **2020**, *13*, 4507. [CrossRef] [PubMed]
11. Mordekovitz, Y.; Shoval, Y.; Froumin, N.; Hayun, S. Effect of Structure and Composition of Non-Stoichiometry Magnesium Aluminate Spinel on Water Adsorption. *Materials* **2020**, *13*, 3195. [CrossRef] [PubMed]
12. Ushakov, S.V.; Hayun, S.; Gong, W.; Navrotsky, A. Thermal Analysis of High Entropy Rare Earth Oxides. *Materials* **2020**, *13*, 3141. [CrossRef] [PubMed]
13. Szpunar, I.; Strandbakke, R.; Sørby, M.H.; Wachowski, S.L.; Balaguer, M.; Tarach, M.; Serra, J.M.; Witkowska, A.; Dzik, E.; Norby, T.; et al. High-Temperature Structural and Electrical Properties of BaLnCo$_2$O$_6$ Positrodes. *Materials* **2020**, *13*, 4044. [CrossRef] [PubMed]
14. Kim, K.; Kim, D.; Park, K.; Cho, M.; Cho, S.; Kwon, H. Effect of Intermetallic Compounds on the Thermal and Mechanical Properties of Al–Cu Composite Materials Fabricated by Spark Plasma Sintering. *Materials* **2019**, *12*, 1546. [CrossRef] [PubMed]
15. Wang, F.; Qian, D.-S.; Xiao, P.; Deng, S. Accelerating Cementite Precipitation during the Non-Isothermal Process by Applying Tensile Stress in GCr15 Bearing Steel. *Materials* **2018**, *11*, 2403. [CrossRef] [PubMed]
16. Fröck, H.; Reich, M.; Milkereit, B.; Kessler, O. Scanning Rate Extension of Conventional DSCs through Indirect Measurements. *Materials* **2019**, *12*, 1085. [CrossRef] [PubMed]
17. Kirshon, Y.; Ben Shalom, S.; Emuna, M.; Greenberg, Y.; Lee, J.; Makov, G.; Yahel, E. Thermophysical Measurements in Liquid Alloys and Phase Diagram Studies. *Materials* **2019**, *12*, 3999. [CrossRef] [PubMed]

Article

Pyrolysis and Combustion of Polyvinyl Chloride (PVC) Sheath for New and Aged Cables via Thermogravimetric Analysis-Fourier Transform Infrared (TG-FTIR) and Calorimeter

Zhi Wang, Ruichao Wei, Xuehui Wang, Junjiang He and Jian Wang *

State Key Laboratory of Fire Science, University of Science and Technology of China, Hefei 230026, China; ustc14wz@mail.ustc.edu.cn (Z.W.); rcwei@mail.ustc.edu.cn (R.W.); wxuehui@mail.ustc.edu.cn (X.W.); hjj0513@mail.ustc.edu.cn (J.H.)
* Correspondence: wangj@ustc.edu.cn; Tel.: +86-551-6360-6463

Received: 9 August 2018; Accepted: 12 October 2018; Published: 16 October 2018

Abstract: To fill the shortages in the knowledge of the pyrolysis and combustion properties of new and aged polyvinyl chloride (PVC) sheaths, several experiments were performed by thermogravimetric analysis (TG), Fourier transform infrared (FTIR), microscale combustion calorimetry (MCC), and cone calorimetry. The results show that the onset temperature of pyrolysis for an aged sheath shifts to higher temperatures. The value of the main derivative thermogravimetric analysis (DTG) peak of an aged sheath is greater than that of a new one. The mass of the final remaining residue for an aged sheath is also greater than that of a new one. The gas that is released by an aged sheath is later but faster than that of a new one. The results also show that, when compared with a new sheath, the heat release rate (HRR) is lower for an aged one. The total heat release (THR) of aged sheath is reduced by 16.9–18.5% compared to a new one. In addition, the cone calorimetry experiments illustrate that the ignition occurrence of an aged sheath is later than that of a new one under different incident heat fluxes. This work indicates that an aged sheath generally pyrolyzes and it combusts more weakly and incompletely.

Keywords: aged cable; pyrolysis; TG-FTIR; combustion; calorimeter

1. Introduction

Due to the outstanding electric insulation, prominent mechanical properties, high chemical resistance, natural flame retardant effect, ease of processing, efficient recyclability, etc., Polyvinyl chloride (PVC), one of the most extensively used plastics, has been widely applied in the wire and cable industry as the main constituent of insulation and sheathing [1–3]. However, PVC plastic that is commercially applied in cable sheathing is considerably flammable, even when treated by stabilizer, lubricant, plasticizer, and flame retardant [1]. The fire statistics illustrate that cable faults are among the most common causes of electrical fires, which involves the new and aged cables [4]. The aging degradation will lead to changes in initial properties as a result of simultaneous chemical and physical processes, causing changes in chemical composition and structure of materials [5]. There must be some differences of fire protection properties between the new and aged cables [2,4]. The outer PVC sheath is recognized as the main combustible part of a cable, and the investigation of its pyrolysis and combustion behavior is key to the study of fire properties of cables [4,6]. It will be aged firstly and fiercely by used for a long time. Limited systematic work has focused on comparing the pyrolysis and combustion properties between new and aged cable sheaths. Besides, as cables of the type used in this study are mostly used indoors, the temperature is considered as the most important factor to make them aged during the long-term service [2,5]. Thus, the thermal aging is supposed to simulate the natural indoors aging process.

Many studies have been conducted on the thermal degradation characteristics and combustion properties of typical PVC cable [7–15]. Benes et al. [16] used the thermogravimetric analysis (TG) coupled to mass spectrometry and Fourier transform infrared (FTIR) spectroscopy to study the thermal degradation of PVC cable under different atmospheres. It was proposed that the pyrolysis of PVC backbone is accompanied by the release of HCl, H_2O, CO_2 and benzene. Gao [17] explored the pyrolysis characteristics of insulative PVC materials applied for fire retardant cable. Wang et al. [10] performed several TG experiments coupled with FTIR analysis to determine the pyrolysis behavior of PVC sheath of flame-retarded cables. They proposed that the pyrolysis process for PVC sheath could be divided into two regions and the amount of six components were detected. Courty et al. [18] employed two tests method for the characterization of PVC/PVC cable pyrolysis and flammability. Fernandez-Pello et al. [19] studied the fire performance of seven types of complex cables, including PVC cable that is commonly used in electrical installations and focused on the ignition and flame spread. Andersson et al. [20] carried out both small and large scale fire experiments with PVC sheathed cable with PVC insulation around the individual wires under well-ventilated and vitiated conditions. McGrattan et al. [21,22] investigated the cable heat release, ignition, and spread in tray installations during a fire, corresponding to the PVC cable. Grayson et al. [23] studied the fire performance of several types of electric cables containing PVC cable to design improved standard testing methods to determine the fire property of cables. Matala et al. [1] investigated the effects of the modelling decisions and parameter estimation methods on the pyrolysis modelling of two PVC cables. It should be noted that all above works focus on the fresh PVC cables. Certainly, there are some studied on the aged PVC cable [2–5,24–27]. Quennehen et al. [24] analyzed the two sets of single core cables with PVC insulation to determine the aging mechanism that is responsible for this decrease of electrical properties. Jakubowicz et al. [5] studied the effects of accelerated and natural ageing on plasticized PVC cable and concluded that the accelerated ageing did not significantly affect the tensile properties of the insulation materials. Yu et al. [12] summarized thermal degradation of PVC waste. Emanuelsson et al. [2] studied the effect of accelerated aging on the fire performance of building wires involving one PVC-based cable and one flame-retarded polyolefin-based cable using cone calorimetry and FTIR. Wang et al. [28] performed experiments to estimate the fire characteristics of new and aged building wires using a cone calorimeter. Xie et al. [4] employed TG, FTIR, and microscale combustion calorimetry (MCC) to investigate the fire protection properties of PVC sheaths for new and old cables: the old one was taken from an old building's electric power system, which had been in use for more than ten years; the new one represented a typical PVC cable manufactured at present. The new and old cables have significant differences in the compositions and structures due to the different commercial companies made at different time. Whereas, the current work is an integral study to compare the pyrolysis and combustion behaviors of the same cable sheath at different thermal aging degrees.

In this study, one flame-retardant PVC cable with different thermal aging degrees was adopted. The PVC sheath part removed from the cable was prepared to the follow-up tests. TG experiments were carried out to study the pyrolysis properties of PVC sheaths of new and aged cables with different heating rates (5, 10, 20, 30 and 40 K min^{-1}) in nitrogen atmosphere. The onset temperatures of pyrolysis, mass loss, mass loss rate, and residue mass were recorded. Meanwhile, the gaseous release during the thermal degradation process was analyzed by TG coupled with FTIR spectroscopy. The combustion characteristics including heat release rate and total heat release were experimentally analyzed by MCC. In addition, a cone calorimeter was applied to investigate the time to ignition of new and aged cable sheaths. Finally, a comparison of the pyrolysis and combustion properties between new and aged cable sheaths was made and discussed.

2. Experimental

2.1. Sample Preparation

The sheath sample used in the present study was obtained from a flame-retardant PVC cable (ZR-RVV) that was provided by Jiangsu Xinchangfeng Cable Co., Ltd. in Wuxi, China. The main components of the sheath were PVC, antimony trioxide (Sb_2O_3), plasticizers, etc. Flame retardant accounts for about 7 wt %. The flame-retardant PVC cable was subjected to different degrees of thermal aging treatment to obtain the aged sheath sample. A rough elemental analysis of the sample was conducted, and the measured results can be observed in Table 1. It can be found that the contents of carbon and chlorine are decreased while the content of oxygen is increased with thermal aging degree. That could be ascribed to the dehydrochlorination and oxidation of the material during the thermal aging [24,29]. Varied sample preparation methods were applied to the tests. For the microscale TG and MCC experiment, the samples were milled to less than 0.5 mm. For the bench-scale cone calorimetry experiment, the cable was cut into pieces 10 cm long with 10 mm thickness. Prior to testing, all of the samples underwent a drying procedure at 80 °C for approximately 1 h to remove moisture.

Table 1. Elemental analysis of new and aged cable sheaths.

Sample	C (%)	H (%)	O (%)	N (%)	Cl (%)	Pb	Sb	Ca
New	33.17	4.16	19.66	0.07	40.67	0.32	1.7	0.25
30 days	30.61	3.63	20.43	0.08	37.38	3.91	1.62	2.34
60 days	28.21	3.24	22.24	0.06	34.61	5.39	1.56	4.69

2.2. Thermal Aging

Thermal aging of PVC sheaths was performed in a convection oven (GHX-100L, Hefei Anke Environmental Test Equipment Co., Ltd. in Hefei, China) in air at 100 ± 0.1 °C for 30 days and 60 days, respectively. After thermal aging, the samples presented a more pronounced brown color, as shown in Figure 1. Severe discoloration that occurred is due to the formation of chromophore groups [5]. The mass loss of PVC sheaths was monitored in the aging process using a precise balance, and the mass loss ratios were 10.2% for 30 days and 11.8% for 60 days when compared with the new sheath. Besides, the PVC sheath was also rigidized with prolonging the aging time. However, it is beyond the scope of the current study.

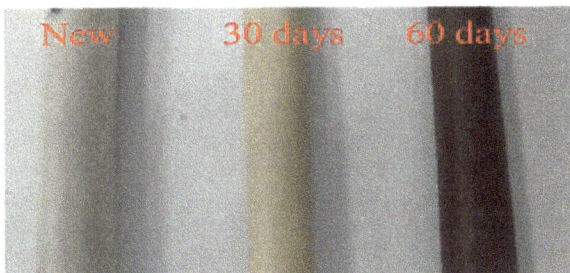

Figure 1. Digital photographs of polyvinyl chloride (PVC) sheaths before and after thermal aging.

2.3. TG-FTIR Measurements

The tests were conducted by a TA Instruments SDT Q600 thermal analyzer (TA Instruments in New Castle, PA, USA) coupled with FTIR spectroscopy (PerkinElmer in Waltham, MA, USA). The TG test ranged from 303 to 1073 K with a gas flowrate of 50 mL min^{-1} in pure nitrogen atmosphere. Five typical heating rates 5, 10, 20, 30 and 40 K min^{-1} were applied. The sample particles were placed into an alumina crucible with a mass of approximately 5 ± 0.2 mg in each test. A flow cell with a recommended temperature of 280 °C was applied to the connection between the TG and FTIR spectrometer to prevent the condensation of gaseous products of the cable sheath. The spectral range was set as 4000–450 cm^{-1} with a scan frequency of eight times.

2.4. Calorimeter Measurements

The MCC tests of the samples were carried out using a microscale combustion calorimeter (MCC-2, Govmark in New York, NY, USA) based on ASTM D7309 [30] (Method A). A linear heating rate of 60 K min^{-1} was applied in all tests. The temperature of the combustor was set at 1173 K. The heat release of the PVC sheaths of the new and aged cables was measured using the oxygen consumption principle. The heat release rate (HRR) as a function of time and sample temperature was experimentally analyzed and compared.

In general, the ignition of a cable essentially depends on the properties of the polymeric sheath. To obtain detailed information of the ignition risk of PVC sheathing, the time to ignition (TTI) of new and aged cables was measured based on a bench-scale cone calorimeter by Fire Testing Technology. Typical incident heat fluxes of 25, 35, 50, and 75 kW m^{-2} were selected for the ignition tests.

3. Results and Discussion

3.1. Thermogravimetric Analysis

Figures 2 and 3 describe the curves of TG and derivative thermogravimetric analysis (DTG) for PVC sheaths of new and aged cables under heating rates of 5, 10, 20, 30 and 40 K min^{-1} in nitrogen atmosphere. It is noted that all TG curves for any sample show similar trends when ignoring the heating rates; the same result is obtained for the DTG curves. The thermal degradation process of the PVC sheath is generally defined as a two-step process, which has been reported and discussed by Zhu et al. [31] and Xie et al. [4]. In the first step, the dehydrochlorination of PVC polymer occurs accompanied by the formation of conjugated double bonds; the second step includes the continuous degradation of PVC polymer with cracking and pyrolysis, resulting in the formation of low hydrocarbons with linear or cyclic structures. Essentially, some weak points, including C–H, C–C, and C–Cl in PVC chains relate to the degradation of PVC polymer [32]. The pyrolysis residue has a conjugated polyene structure with cis- and trans-arrangements [10]. In the present TG and DTG profiles, three significant pyrolysis stages involving two strong peaks and one weak peak can be seen, as shown in Figures 2 and 3, whether in the new or aged cable sheath. The first DTG peak relates to the rapid volatilization and removal of HCl. The cleavage of the conjugated polyene chain and chain structural reconstruction in PVC and additives contribute to the second peak. The third peak may be attributed to the sluggish degradation of remaining residues. In addition, a displacement of TG curves and an increase of DTG peaks with heating rates are observed, as described by previous researchers [32–35].

The TG and DTG curves corresponding to new and aged PVC sheaths at different heating rates (5, 10, 20, 30 and 40 K min^{-1}) in nitrogen atmosphere are depicted in Figures 4 and 5, respectively. Figure 4 illustrates that there is no difference in the TG curves of new and aged PVC sheaths in the initial pyrolysis stage. However, the TG curve of the new cable sheath decreases faster than that of the aged one with increasing sample temperature. The mass of lost residue from the new cable sheath is significantly less than that of the aged one. Namely, the new cable sheath experienced a more sufficient pyrolysis process than the aged one. In addition, the lower the heating rate, the less the residue mass. Overall, the pyrolysis difference between new and aged cable sheaths is clearly visible, regardless of the heating rates. Figure 5 shows that the onset temperature, which is related to the ignition time of polymer in fire, shifts slightly upward for the aged cable sheath at all heating rates. This indicates that the aged cable sheath pyrolyzes later than the new one. The values of DTG peaks for the aged cable sheath are higher than those of the new one. Table 2 lists the typical pyrolysis parameters of PVC sheaths of new and aged cables in nitrogen atmosphere with different heating rates. Note that the changes of chemical composition, structure, and additives in polymer after long-term thermal aging are the main reason for the pyrolysis difference of PVC sheaths between aged and new cables. However, it can be noted that the T_{onset} and DTG_{peak} for the aged sheath of 60 days is lower than that of 30 days, which may be attributed to the complex factors, including the dehydrochlorination, consumption of stabilizers and the aggravated chain scission and cross-linking.

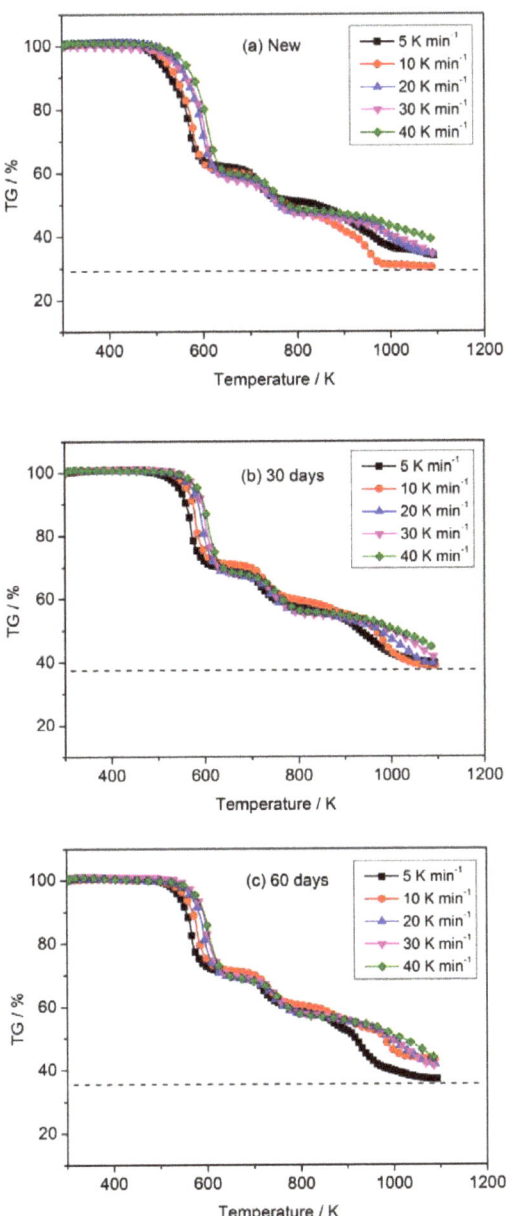

Figure 2. Mass loss (thermogravimetric analysis (TG)) curves of PVC sheaths at different heating rates in nitrogen atmosphere: (**a**) new; (**b**) aged 30 days and (**c**) aged 60 days.

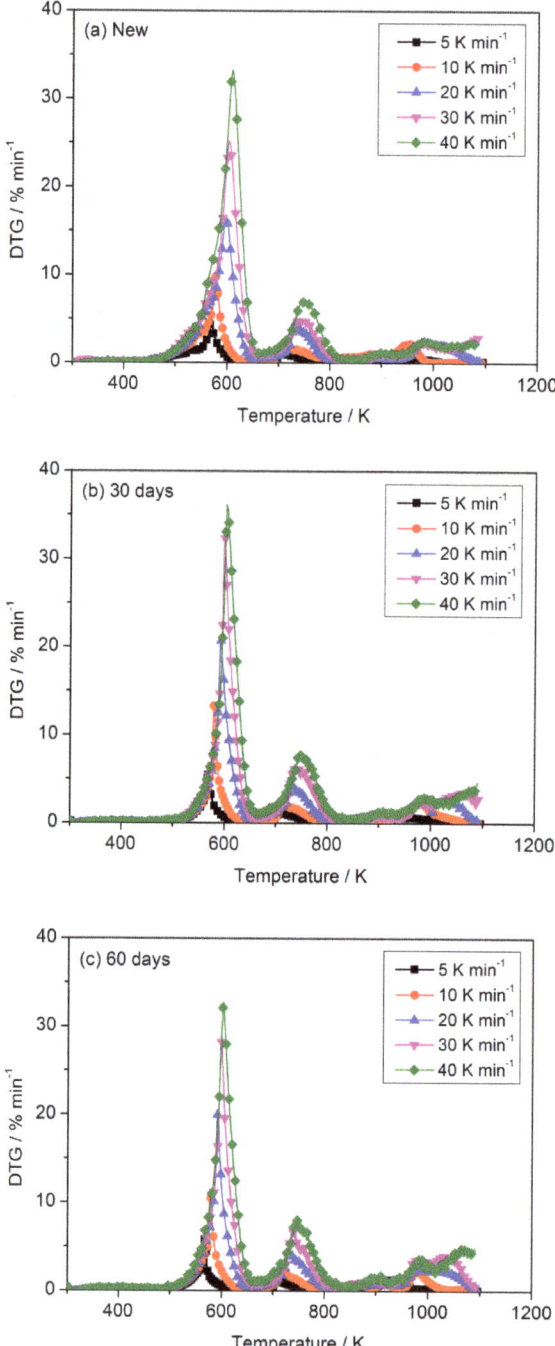

Figure 3. Mass loss rate (derivative thermogravimetric analysis (DTG)) curves of PVC sheath at different heating rates in nitrogen atmosphere: (**a**) new; (**b**) aged 30 days and (**c**) aged 60 days.

In addition, the effect of the flame-retardant on the thermal degradation of PVC sheath was also analyzed. As shown in Figures 4 and 5, the T_{onset} of new PVC sheath is lower than that of the pure PVC raw material when the flame retardant is added to the feedstock. However, the new PVC sheath exhibits a lower mass loss rate of while gives a more final residue, as compared with the pure PVC raw material. The whole pyrolysis process has been prolonged after the flame retardant added. Meanwhile, there is an obvious decrease of the main DTG_{peak} for pure PVC raw material. The results are primarily attributed to the formation of volatile $SbCl_3$ and $SbOCl$ between the antimony trioxide and PVC during the thermal degradation. Besides, it should be noted the losses of both chlorine and antimony content are clearly seen after thermal aging from the element analysis, which is agreement with the previous works [15,36]. They also proposed that $SbCl_3$ production was possible by reaction of HCl with Sb_2O_3 without involving the high-temperature disproportionation reactions of intermediate SbOCl. Namely, the higher T_{onset}, larger DTG_{peak} and less HCl gas emission may be explained by this point, to some extent. The detailed study of this point is beyond the scope of the current work, taking the experimental design, thermal aging period, etc. into account. However, the relevant research will be conducted in the next plan.

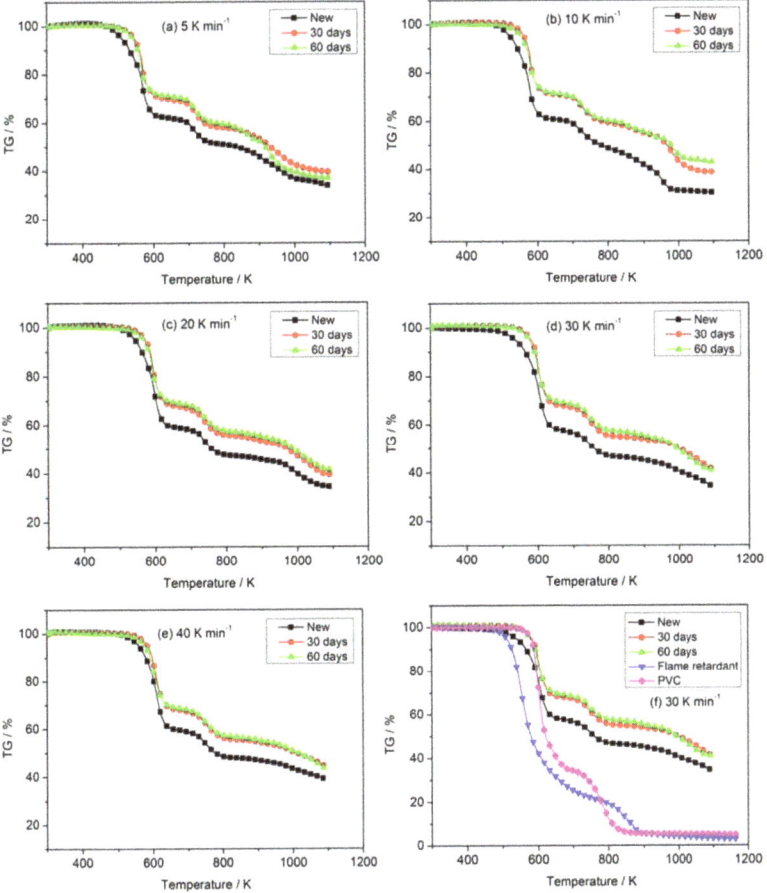

Figure 4. Comparison of TG of PVC sheaths for new and aged cables at different heating rates in nitrogen atmosphere. (**a**) 5 K min^{-1}; (**b**) 10 K min^{-1}; (**c**) 20 K min^{-1}; (**d**) 30 K min^{-1}; (**e**) 40 K min^{-1} and (**f**) 30 K min^{-1} involving raw materials.

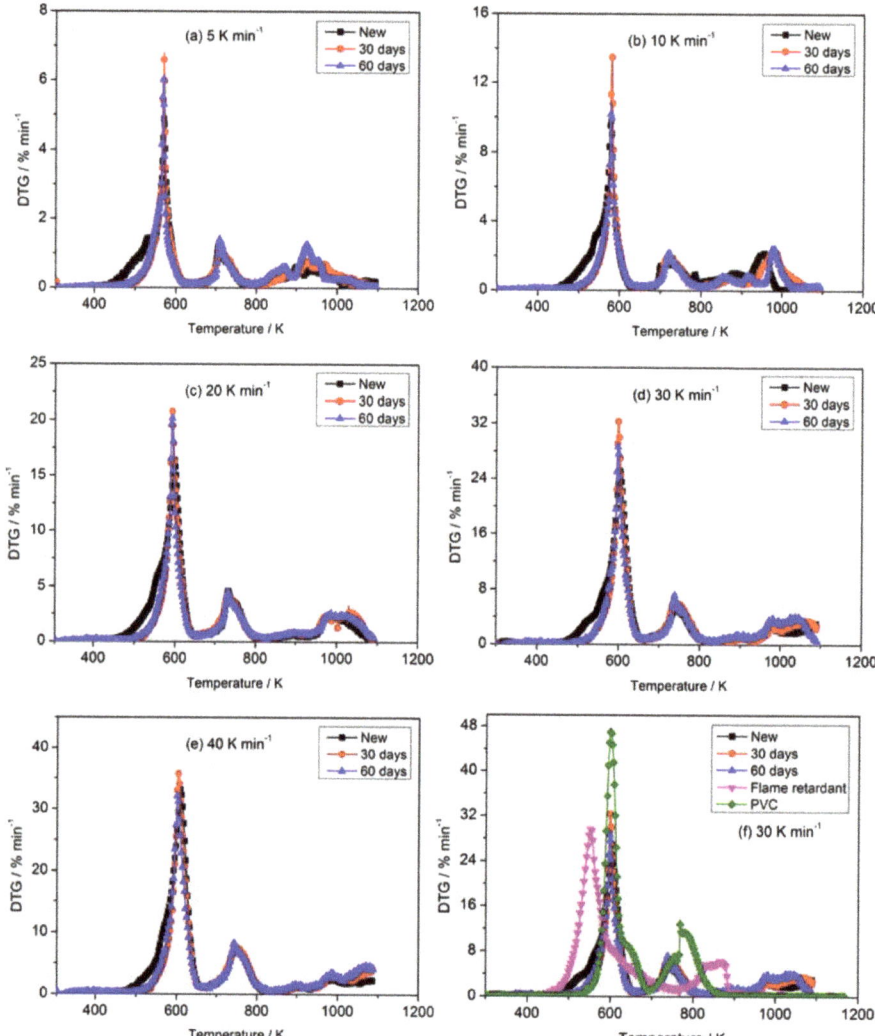

Figure 5. Comparison of DTG of PVC sheaths for new and aged cables at different heating rates in nitrogen atmosphere. (**a**) 5 K min^{-1}; (**b**) 10 K min^{-1}; (**c**) 20 K min^{-1}; (**d**) 30 K min^{-1}; (**e**) 40 K min^{-1} and (**f**) 30 K min^{-1} involving raw materials.

Table 2. Pyrolysis parameters of PVC sheaths for new and aged cables in nitrogen atmosphere with different heating rates.

Sample	Heating Rate (K min^{-1})	T_{onset} (K)	DTG_{peak} (% min^{-1})	Residue$_{mass}$ (%)
New	5	510.41	4.88	33.72
	10	522.36	9.91	30.17
	20	543.28	16.46	34.47
	30	552.96	25.18	34.47
	40	559.12	33.22	39.07
30 days	5	545.84	6.69	39.63
	10	561.77	13.58	38.70
	20	572.44	20.70	39.42
	30	580.89	32.68	41.50
	40	584.08	35.97	44.36
60 days	5	540.75	6.29	36.91
	10	552.54	10.78	42.87
	20	566.09	20.44	41.50
	30	576.14	28.77	41.11
	40	580.24	32.75	43.42

3.2. FTIR Analysis

FTIR spectroscopy is an effective method to analyze the evolved gas products of materials in the pyrolysis process. Figure 6 gives the three-dimensional (3D) surface for the FTIR spectra of gases that are released by new and aged cable sheaths. It is clearly shown that the temperature range of gases released is well consistent with the evolution of DTG, regardless of the third relatively weak peak. In addition, gases are produced from the new cable sheath much earlier than the aged one. However, the kinds and concentrations of gases released in pyrolysis between new and aged cable sheaths have little difference. Table 3 presents typical gases released from PVC sheaths in nitrogen atmosphere at a heating rate of 30 K min^{-1} at different times. It suggests that the main products including HCl, CO_2, and H_2O during the whole pyrolysis process are similar for new and aged cable sheaths. However, more dangerous products, such as alkene, benzene, styrene, etc., are not involved in current work. The spectrum band of 2400–2260 cm^{-1} is associated with CO_2. HCl can be observed in the band range of 3100–2600 cm^{-1}, and the absorption band of H_2O corresponds to 1800–1300 cm^{-1} and 4000–3500 cm^{-1}. The spectrum band in the range of 700–550 cm^{-1} is attributed to the stretching vibrations of C–Cl. The results show that the gases from the aged cable sheath are released later, but more quickly, than those by the new one, especially HCl gas.

Table 3. Gases released from PVC sheaths in nitrogen atmosphere at a heating rate of 30 K min^{-1} with different times.

Sample	Gas	Absorbance			
		8.64 min	10.83 min	16.80 min	30.14 min
New	HCl	7.26×10^{-3}	1.00×10^{-2}	4.33×10^{-3}	1.40×10^{-3}
	CO_2	1.07×10^{-2}	6.49×10^{-2}	3.59×10^{-2}	1.81×10^{-2}
	H_2O	2.95×10^{-3}	5.50×10^{-3}	6.57×10^{-4}	5.47×10^{-4}
30 days	HCl	7.17×10^{-4}	1.70×10^{-3}	3.59×10^{-3}	2.96×10^{-4}
	CO_2	1.29×10^{-2}	5.24×10^{-2}	2.79×10^{-2}	1.27×10^{-2}
	H_2O	1.31×10^{-5}	1.25×10^{-3}	7.66×10^{-5}	1.54×10^{-5}
60 days	HCl	3.99×10^{-4}	1.79×10^{-3}	5.20×10^{-3}	1.18×10^{-3}
	CO_2	1.42×10^{-2}	5.23×10^{-2}	2.26×10^{-2}	7.08×10^{-3}
	H_2O	4.71×10^{-3}	6.35×10^{-3}	3.74×10^{-3}	9.39×10^{-4}

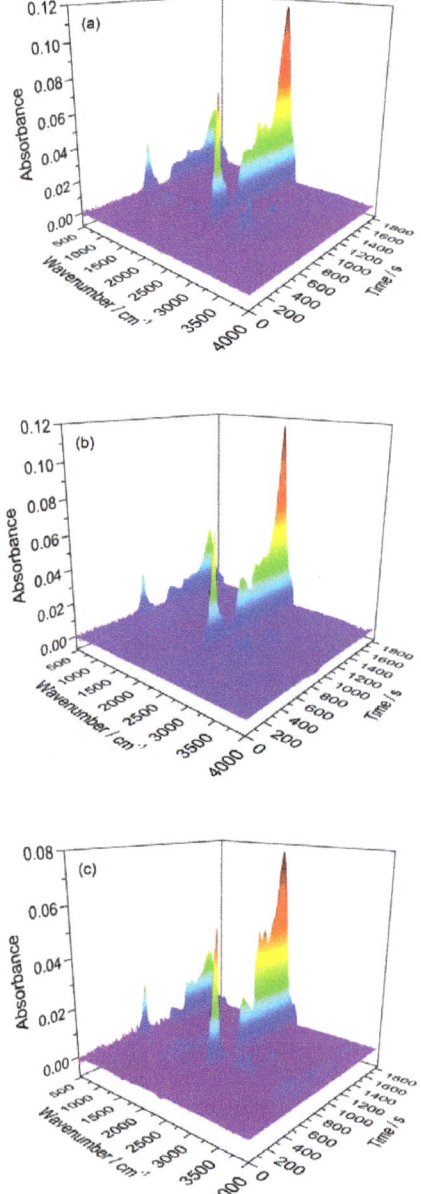

Figure 6. Typical three-dimensional thermogravimetric analysis-Fourier transform infrared (3D TG-FTIR) spectrogram of PVC sheath at a heating rate of 30 K min^{-1}: (**a**) new; (**b**) aged 30 days; and, (**c**) aged 60 days.

3.3. MCC Analysis

The HRR curves of PVC sheaths for both new and aged cables are shown in Figure 7. Two peaks of heat release rate (PHRRs) can be observed for new and aged sheaths. The value of the first HRR peak for the new cable sheath is greater than that of the aged one, and the value of the second HRR peak has a slight difference. The first HRR peak is produced by the combustion of pyrolysis gases, and the second one is formed by the char oxidation. The result indicates that the faster and more combustible gases were released for aged cable sheath during MCC tests, which can be supported by the measurements of FTIR. The temperature corresponding to two HRR peaks (TPHRRs) laterally shifts upward, and the duration from onset decomposition to PHRR of the aged cable sheath is longer than that of the new one. That means that the decomposition and combustion of aged cable sheath would be slightly weak due to the change of composition and structure. Figure 8 gives the integral HRR (total heat release (THR)) of PVC sheaths for new and aged cables. THR is related not only to the HRR peak, but also to the detailed pyrolysis process. Compared with the aged cable sheath, the THR value of the new cable sheath is relatively high. Table 4 presents the MCC data of PVC sheaths of new and aged cables. Where HRC is the capacity of heat release, PHRR is the maximum value of two peaks, TPHRR is the temperature corresponding to the first peak of HRR, and THR is defined as the value at the final temperature. It implies that the new cable sheath would burn more fiercely and amply, accompanied by more heat release, which is in good agreement with the above TG analysis.

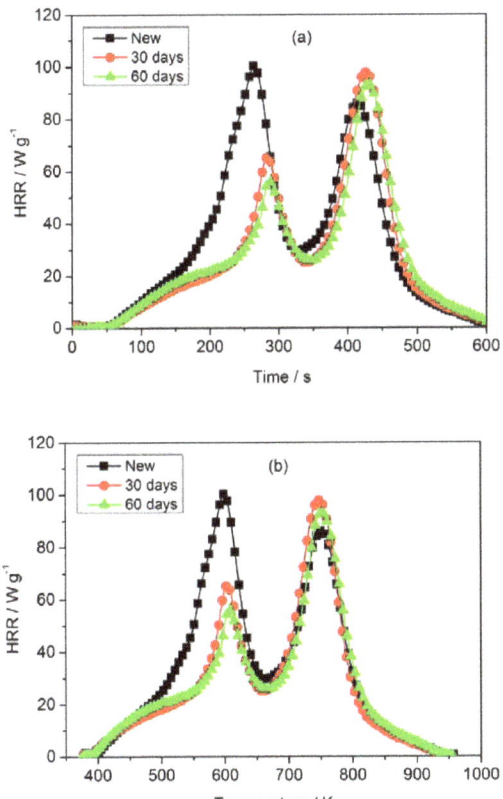

Figure 7. Heat release rate of PVC sheaths for new and aged cables: (**a**) heat release rate (HRR) vs. time and (**b**) HRR vs. temperature.

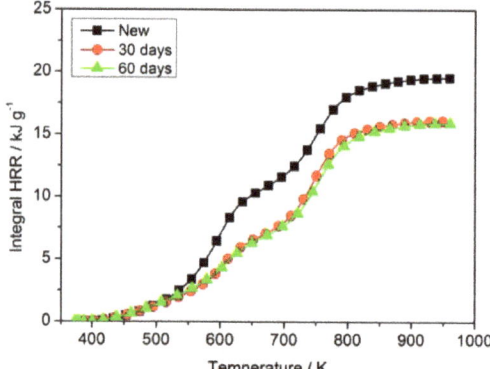

Figure 8. Integral HRR (total heat release (THR)) of PVC sheaths for new and aged cables.

Table 4. MCC data of new and aged cable sheaths.

Sample	HRC (J g^{-1} K^{-1})	PHRR (W g^{-1})	TPHRR (°C)	THR (kJ g^{-1})
New	100	100.1	324.2	19.5
30 days	104	98.0	471.4	16.2
60 days	97	93.1	476.8	15.9

3.4. Cone Calorimetry Analysis

Figure 9 shows the time to ignition (TTI) of new and aged cable sheaths under the varied incident heat fluxes ranging from 25 kW m^{-2} to 75 kW m^{-2}. The TTI value of the new cable sheath is greater than that of the aged one at different incident heat fluxes. The TTI difference between new and aged cable sheaths decreases with the increasing incident heat flux. The aged cable sheath has a maximal ignition time 93% greater than that of the new cable sheath at 25 kW m^{-2}. This result is consistent with the onset temperature and thermal degradation process analysis of pyrolysis. It suggests again that the new cable sheath has a higher ignition risk.

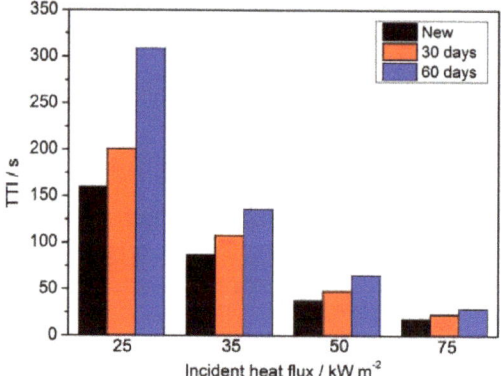

Figure 9. Time to ignition (TTI) of PVC sheaths for new and aged cables with different incident heat fluxes.

4. Conclusions

In the current study, several experiments using TG-FTIR, MCC, and cone calorimetry were employed for PVC sheaths of new and aged cables. The results show that the pyrolysis behavior between aged cable sheaths and new cable sheaths with varied heating rates is markedly different in nitrogen atmosphere. The onset temperature of mass loss for the aged cable sheath is greater than that of the new one, regardless of the heating rates. In addition, the mass of the pyrolysis residue of the aged cable sheath is slightly greater than that of the new cable sheath. This indicates that the new cable sheath starts to pyrolyze more easily and completely than the aged one. It is also concluded that there is a main DTG peak for new and aged cable sheaths under all conditions. The value of the main DTG peak of the aged cable sheath is clearer than that of the new one. The evolved gas that was measured by FTIR spectra illustrates that the aged cable sheath releases pyrolysis gases slightly later but more quickly than the new cable sheath. The results also show that the values of PHRR and THR for the aged cable sheath are clearly less than those of the new one. However, the duration from onset decomposition to PHRR and time to ignition for the aged cable sheath are significantly greater than those of the new cable sheath. It must be noted that the difference of pyrolysis and combustion between the aged sheath with 30 days and aged sheath with 60 days is slight, which may indicate that there is a critical stage during the thermal aging. The pyrolysis and combustion properties of materials change slightly when the materials is aged in a long enough period. Generally, the pyrolysis and combustion properties depend on the material itself under the same condition. In consequence, the modification of chemical composition, chain structure, and additives might be deduced to be the reason for different pyrolysis and combustion properties of the new and aged cable sheaths, eventually resulting in the change of flammability characteristics. Whereas, the currently available evidence is insufficient for this deduction and more research is needed. This work adds to the understanding of the difference in pyrolysis and combustion performances between new and aged cable sheaths. Finally, pyrolysis and combustion of waste plastic allow the obtainment of valuable chemicals, hydrocarbon compounds, combustible, gases, and energy. Knowledge of the pyrolysis mechanisms and combustion properties of typical aged cable sheath will benefit the recycling of plastic waste and energy conversion, which deserves further examination in future study.

Author Contributions: Conceptualization, Z.W. and J.W.; Methodology, Z.W., J.H and J.W.; Investigation, Z.W., R.W. and X.W.; Data Curation, Z.W. and J.H.; Writing-Original Draft Preparation, Z.W.; Writing-Review & Editing, Z.W., R.W., X.W. and J.W.; Supervision, J.W.; Project Administration, J.W.

Funding: This work was supported by the National Key R&D Program of China (No. 2016YFC0802500). The authors gratefully acknowledge this support.

Conflicts of Interest: The authors declare no conflict of interest.

References

1. Matala, A.; Hostikka, S. Pyrolysis modelling of PVC cable materials. *Fire Saf. Sci.* **2011**, *10*, 917–930. [CrossRef]
2. Emanuelsson, V.; Simonson, M.; Gevert, T. The effect of accelerated ageing of building wires. *Fire Mater.* **2007**, *31*, 311–326. [CrossRef]
3. Brebu, M.; Vasile, C.; Antonie, S.R.; Chiriac, M.; Precup, M.; Yang, J.; Roy, C. Study of the natural ageing of PVC insulation for electrical cables. *Polym. Degrad. Stab.* **2000**, *67*, 209–221. [CrossRef]
4. Xie, Q.; Zhang, H.; Tong, L. Experimental study on the fire protection properties of PVC sheath for old and new cables. *J. Hazard. Mater.* **2010**, *179*, 373–381. [CrossRef] [PubMed]
5. Jakubowicz, I.; Yarahmadi, N.; Gevert, T. Effects of accelerated and natural ageing on plasticized polyvinyl chloride (PVC). *Polym. Degrad. Stab.* **1999**, *66*, 415–421. [CrossRef]
6. Meinier, R.; Sonnier, R.; Zavaleta, P.; Suard, S.; Ferry, L. Fire behavior of halogen-free flame retardant electrical cables with the cone calorimeter. *J. Hazard. Mater.* **2018**, *342*, 306–316. [CrossRef] [PubMed]
7. Ma, R.-H.; Lin, Y.-C.; Kuo, C.-P. The study of thermal pyrolysis mechanisms for chloro organic compounds in electric cable and medical wastes. *J. Anal. Appl. Pyrolysis* **2006**, *75*, 245–251. [CrossRef]
8. Henrist, C.; Rulmont, A.; Cloots, R.; Gilbert, B.; Bernard, A.; Beyer, G. Toward the understanding of the thermal degradation of commercially available fire-resistant cable. *Mater. Lett.* **2000**, *46*, 160–168. [CrossRef]

9. Cheng, J.Q.; Jia, W.N.; Shu, Z.J.; Yang, S.S.; Chen, N. Experimental research on rules of releasing of the halogen gas of PVC cable under hydrogenation condition. *Fire Sci. Technol.* **2007**, *26*, 383.
10. Wang, C.; Liu, H.; Zhang, J.; Yang, S.; Zhang, Z.; Zhao, W. Thermal Degradation of Flame-retarded high-voltage Cable Sheath and Insulation via TG-FTIR. *J. Anal. Appl. Pyrolysis* **2018**, *134*, 167–175. [CrossRef]
11. Huggett, C.; Levin, B.C. Toxicity of the pyrolysis and combustion products of poly(vinyl chlorides): A literature assessment. *Fire Mater.* **1987**, *11*, 131–142. [CrossRef]
12. Yu, J.; Sun, L.; Ma, C.; Qiao, Y.; Yao, H. Thermal degradation of PVC: A review. *Waste Manag.* **2016**, *48*, 300–314. [CrossRef] [PubMed]
13. Arlman, E. Thermal and oxidative decomposition of polyvinyl chloride. *J. Polym. Sci.* **1954**, *12*, 547–558. [CrossRef]
14. Hjertberg, T.; Sörvik, E.M. On the influence of HCl on the thermal degradation of poly(vinyl chloride). *J. Appl. Polym. Sci.* **1978**, *22*, 2415–2426. [CrossRef]
15. Vahabi, H.; Sonnier, R.; Ferry, L. Effects of ageing on the fire behaviour of flame-retarded polymers: A review. *Polym. Int.* **2015**, *64*, 313–328. [CrossRef]
16. Beneš, M.; Milanov, N.; Matuschek, G.; Kettrup, A.; Plaček, V.; Balek, V. Thermal degradation of PVC cable insulation studied by simultaneous TG-FTIR and TG-EGA methods. *J. Therm. Anal. Calorim.* **2004**, *78*, 621–630. [CrossRef]
17. Gao, Y. Study on pyrolysis characteristics of insulative PVC material used for fire retardant cable. *Eng. Plast. Appl.* **2007**, *35*, 44–47.
18. Courty, L.; Garo, J. External heating of electrical cables and auto-ignition investigation. *J. Hazard. Mater.* **2017**, *321*, 528–536. [CrossRef] [PubMed]
19. Fernandez-Pello, A.; Hasegawa, H.; Staggs, K.; Lipska-Quinn, A.; Alvares, N. A study of the fire performance of electrical cables. *Fire Saf. Sci.* **1991**, *3*, 237–247. [CrossRef]
20. Andersson, P.; Rosell, L.; Simonson, M.; Emanuelsson, V. Small and Large Scale Fire Experiments with Electric Cables under Well-Ventilated and Vitiated Conditions. *Fire Technol.* **2004**, *40*, 247–262. [CrossRef]
21. McGrattan, K.B.; Lock, A.J.; Marsh, N.D.; Nyden, M.R. Cable Heat Release, Ignition, and Spread in Tray Installations during Fire (CHRISTIFIRE): Phase 1-Horizontal Trays. Available online: https://www.nist.gov/publications/cable-heat-release-ignition-and-spread-tray-installations-during-fire-christifire-0 (accessed on 11 July 2012).
22. McGrattan, K.B.; Bareham, S.D. *Cable Heat Release, Ignition, and Spread in Tray Installations during Fire (CHRISTIFIRE) Phase 2: Vertical Shafts and Corridors*; NUREG/CR-7010; United States Nuclear Regulatory Commission (U.S.NCR): Washington, DC, USA, 2013; Volume 2.
23. Grayson, S.; Van Hees, P.; Green, A.M.; Breulet, H.; Vercellotti, U. Assessing the fire performance of electric cables (FIPEC). *Fire Mater.* **2001**, *25*, 49–60. [CrossRef]
24. Quennehen, P.; Royaud, I.; Seytre, G.; Gain, O.; Rain, R.; Espilit, T.; François, S. Determination of the aging mechanism of single core cables with PVC insulation. *Polym. Degrad. Stab.* **2015**, *119*, 96–104. [CrossRef]
25. Ekelund, M.; Edin, H.; Gedde, U.W. Long-term performance of poly(vinyl chloride) cables. Part 1: Mechanical and electrical performances. *Polym. Degrad. Stab.* **2007**, *92*, 617–629. [CrossRef]
26. Ekelund, M.; Azhdar, B.; Hedenqvist, M.S.; Gedde, U.W. Long-term performance of poly (vinyl chloride) cables, Part 2: Migration of plasticizer. *Polym. Degrad. Stab.* **2008**, *93*, 1704–1710. [CrossRef]
27. Gumargalieva, K.; Ivanov, V.; Zaikov, G.; Moiseev, J.V.; Pokholok, T. Problems of ageing and stabilization of poly (vinyl chloride). *Polym. Degrad. Stab.* **1996**, *52*, 73–79. [CrossRef]
28. Wang, Z.; Wang, J. An Experimental Study on the Fire Characteristics of New and Aged Building Wires Using a Cone Calorimeter. Available online: https://link.springer.com/article/10.1007/s10973-018-7626-8 (accessed on 14 August 2018).
29. Pimentel Real, L.E.; Ferraria, A.M.; do Rego, A.M.B. The influence of weathering conditions on the properties of poly(vinyl chloride) for outdoor applications. An analytical study using surface analysis techniques. *Polym. Test.* **2007**, *26*, 77–87. [CrossRef]
30. ASTM D7309-13. *Standard Test Method for Determining Flammability Characteristics of Plastics and Other Solid Materials Using Microscale Combustion Calorimetry*; ASTM International: West Conshohocken, PA, USA, 2013.
31. Zhu, H.; Jiang, X.; Yan, J.; Chi, Y.; Cen, K. TG-FTIR analysis of PVC thermal degradation and HCl removal. *J. Anal. Appl. Pyrolysis* **2008**, *82*, 1–9. [CrossRef]

32. Conesa, J.A.; Marcilla, A.; Font, R.; Caballero, J.A. Thermogravimetric studies on the thermal decomposition of polyethylene. *J. Anal. Appl. Pyrolysis* **1996**, *36*, 1–15. [CrossRef]
33. Park, J.W.; Oh, S.C.; Lee, H.P.; Kim, H.T.; Yoo, K.O. A kinetic analysis of thermal degradation of polymers using a dynamic method. *Polym. Degrad. Stab.* **2000**, *67*, 535–540. [CrossRef]
34. Yang, J.; Miranda, R.; Roy, C. Using the DTG curve fitting method to determine the apparent kinetic parameters of thermal decomposition of polymers. *Polym. Degrad. Stab.* **2001**, *73*, 455–461. [CrossRef]
35. Aboulkas, A.; El Bouadili, A. Thermal degradation behaviors of polyethylene and polypropylene. Part I: Pyrolysis kinetics and mechanisms. *Energy Convers. Manag.* **2010**, *51*, 1363–1369. [CrossRef]
36. Clough, R.L. Aging effects on fire-retardant additives in polymers. *J. Polym. Sci. Polym. Chem. Ed.* **1983**, *21*, 767–780. [CrossRef]

© 2018 by the authors. Licensee MDPI, Basel, Switzerland. This article is an open access article distributed under the terms and conditions of the Creative Commons Attribution (CC BY) license (http://creativecommons.org/licenses/by/4.0/).

Article

Accelerating Cementite Precipitation during the Non-Isothermal Process by Applying Tensile Stress in GCr15 Bearing Steel

Feng Wang [1,2], Dong-Sheng Qian [1,2,*], Peng Xiao [1,2] and Song Deng [2]

1. School of Materials Science and Engineering, Wuhan University of Technology, Wuhan 430070, China; wangfengwhut@163.com (F.W.); whutxiaopeng@163.com (P.X.)
2. Hubei Key Laboratory of Advanced Technology for Automotive Components, Wuhan 430070, China; guoheng0722@126.com
* Correspondence: qiands@whut.edu.cn; Tel.: +86-27-8716-8391

Received: 7 November 2018; Accepted: 26 November 2018; Published: 28 November 2018

Abstract: In this work, the non-isothermal process of GCr15 bearing steel after quenching and tempering (QT) under different tensile stress (0, 20, 40 MPa) was investigated by kinetic analysis and microstructural observation. The Kissinger method and differential isoconversional method were employed to assess the kinetic parameters of the microstructural evolution during the non-isothermal process with and without applied stress. It is found that the activation energy of retained austenite decomposition slightly increases from 109.4 kJ/mol to 121.5 kJ/mol with the increase of tensile stress. However, the activation energy of cementite precipitation decreases from 179.4 kJ/mol to 94.7 kJ/mol, proving that tensile stress could reduce the energy barrier of cementite precipitation. In addition, the microstructural observation based on scanning and transmission electron microscopy (SEM and TEM) shows that more cementite has formed for the specimens with the applied tensile stress, whereas there is still a large number of ε carbides existing in the specimens without stress. The results of X-ray diffraction (XRD) also verify that carbon in martensite diffuses more and participates in the formation of cementite under the applied tensile stress, which thus are in good agreement with the kinetic analysis. The mechanisms for the differences in cementite precipitation behaviors may lie in the acceleration of carbon atoms migration and the reduction of the nucleation barrier by applying tensile stress.

Keywords: GCr15 bearing steel; cementite precipitation; stress filed; kinetic analysis

1. Introduction

GCr15 bearing steel, with high carbon and chromium concentrations, is widely used for the rolling elements in bearing. It is routinely treated by spheroidising, quenching and tempering (QT) to obtain a microstructure comprised of spherical cementite, small amount of retained austenite and tempered martensite. Generally, bearing needs to be served under high speed and high load conditions. It has been reported [1,2] that the service temperature of ordinary bearing may reach 200 °C, while the aerospace bearing can be serviced up to 300 °C. Moreover, deterioration of bearing performance may occur during the cyclic contact loading. Microstructural development under the temperature and stress field is the crucial factor influencing the service life of the bearing steel [2,3]. Therefore, clarifying the microstructural evolution caused by the temperature and stress field to understand the mechanism of performance degradation is always an attractive topic.

Many researchers have used transmission electron microscope [4], resistivity [5], thermal analysis [6,7] and other methods to study the microstructural evolution of bearing steel during tempering or aging. There are also some studies focusing on the effect of stress on the transformation of retained austenite [8,9] and carbide precipitation [10] without an applied temperature field. Nevertheless, these studies do not consider the coupling effect of the stress field and temperature field on the microstructural transformation behaviors. Therefore, it is necessary to study the effect of stress on microstructures development during the non-isothermal process. Until now, some studies have been concentrated on the effect of stress on phase transformation behaviors, such as isothermal ferrite and pearlite transformations [11], austenite decomposition [12], martensitic phase transformations [13], bainitic phase transformation [14] and so on. However, to our best knowledge, there is still no research concerning on the microstructural transformation, and especially the cementite precipitation behaviors of GCr15 bearing steel during the non-isothermal process under applied stress.

As the non-isothermal process is close to the high-temperature tempering process for GCr15 bearing steel, the microstructure evolution during tempering should be clarified first. According to the theory of tempering, the process of tempering can be divided into several stages with the increase of temperature [15]: (1) Stage 0, 0~80 °C, the migration of carbon atoms to dislocation and defects [16]; (2) Stage I, 100~200 °C, the precipitation of ε/η transition carbides [17]; (3) Stage II, 250~350 °C, the decomposition of retained austenite [18]; (4) Stage III, 300~400 °C, the conversion of the transition carbide into cementite [19].

It is well known that the finished bearing products have gone through Stage 0 and Stage I after the traditional QT treatment, so the aim of the present work is mainly to investigate the microstructural evolution (Stage II and Stage III) during the higher tempering temperature under different tensile stresses. The kinetics parameters of microstructural development were compared and analyzed considering the effect of applied tensile stress. At the same time, the impact of applied stress on microstructural evolution during the non-isothermal process was analyzed based on microstructural analysis. In particular, the mechanism of tensile stress on cementite precipitation was discussed.

2. Experimental and Theory

The material used in the current study was GCr15 bearing steel, with the nominal composition as presented in Table 1. The experimental materials were austenitized at 860 °C for 15 min and followed by quenching in oil at 60 °C for 5 min. Then they were tempered at 160 °C for 2 h. After the QT heat treatment, the steel was designed to be treated under the couple action of the stress and temperature field.

Table 1. The chemical composition of GCr15 bearing steel (wt %).

C	Cr	Mn	Si	P	S	Fe
0.960	1.430	0.350	0.270	0.012	0.002	Bal.

In order to observe the microstructural evolution during the non-isothermal process under different tensile stresses, the specimens with a size of φ 10 mm × 105 mm were subjected to a continuous tensile stress (0, 20 and 40 MPa) and heated from room temperature to 300 °C (considering the limit service temperature of bearing and the temperature range of stage II and stage III) at the heating rate of 10 °C/min by means of a Gleeble 3500 thermo-mechanical simulator. The microstructure of these specimens was examined by a field-emission scanning electron microscope (FESEM, FEI Quanta 450, Hillsboro, OR, USA) and transmission electron microscopy (TEM, JEOL-2100F, Akishima, Japan). The specimens for FESEM were etched in an alcohol solution containing 4% nitric acid (volume fraction) for 10 s. The specimens for TEM were prepared by mechanically polishing and then electro-polishing in a twin-jet polisher (Struers, TenuPol-5, Ballerup, Denmark) using a solution of 10% perchloric acid and 90% acetic acid. Moreover, X-ray diffraction (XRD) data were recorded on a D/MAX-RB diffraction analyser (Rigaku, Tokyo, Japan) at 12 kW.

Activation energy, as an important kinetic parameter, can be used to evaluate the difficulty of phase transformation. The Kissinger method [20,21] and isoconversional method [22,23] are the most widely used methods to calculate activation energy, and they are usually combined with thermal analysis. In this study, for analysing the influence of tensile stress on the kinetic of microstructural evolution during the non-isothermal process, a group of specimens was heated to 500 °C under a 40 MPa tensile stress at different heating rate of 10, 15 and 20 °C/min with the aim of calculating the activation energy. It was noted that the thermal expansion curves were recorded by measuring the variation in diameter during the non-isothermal process. Considering the activation energy determined from the thermal expansion curves and heat flow curves (both as functions of the heating rate) is really close [7,24], the kinetic parameters of microstructural evolution without stress was obtained by differential scanning calorimetry (DSC) experiments using a STA449F3 thermal analyzer (Netzsch, Selb, Germany) as a comparison. The specimens for DSC were cut into ϕ 4 mm × 0.6 mm and then heated from ambient temperature to 500 °C at different heating rates of 5, 10, 15 °C/min, respectively. Here, pure aluminum disks were used as the reference material, and the baseline was determined by performing a rerun at the same heating rate.

The effective activation energies of different stages during the non-isothermal process were calculated by using the Kissinger equation based on the fact that the peak temperature depends on the heating rate:

$$\ln\left(T_P^2/\varnothing\right) = Q/RT_P + \text{const} \tag{1}$$

where T_P is the peak temperature, $\varnothing$ is the heating rate, and R is the gas constant (R = 8.314 kJ/mol). By plotting $\ln\left(T_P^2/\varnothing\right)$ as a function of $1/T_P$, the activation energy can be obtained.

The differential isoconversional method was also employed to obtain the kinetic parameters of microstructural evolution during the non-isothermal process under different tensile stress. In this method, the reaction rate can be assumed to be a function of temperature ($k(T)$) and converted fraction ($f(\alpha)$), which is expressed as:

$$d\alpha/dt = k(T)f(\alpha) \tag{2}$$

where α and T is the converted fraction and temperature, respectively. $k(T)$ can be obtained by the Arrhenius equation as:

$$k(T) = A \cdot \exp(-Q/RT) \tag{3}$$

where A is the pre-exponential factor and Q is the activation energy. By combining Equations (2) and (3) and taking the equation to a logarithm, then the differential isoconversional method can be proposed, as follows:

$$\ln(d\alpha/dt) = \ln[A \cdot f(\alpha)] - Q/RT \tag{4}$$

For the non-isothermal process with a constant heating rate $\varnothing$, Equation (4) can be expressed as:

$$\ln[\varnothing \cdot (d\alpha/dt)] = \ln[A \cdot f(\alpha)] - Q/RT \tag{5}$$

By plotting $\ln[\varnothing \cdot (d\alpha/dt)]$ as a function of $1/T$, the slope indicates a value of Q/R. Then the curve of the activation energy varying with the converted fraction can be further obtained.

3. Results

The thermal expansion curves of continuous heating up to 500 °C at the heating rate of 10 °C/min under different tensile stress is shown in Figure 1. It can be found that the relative dimensional change first increases with increasing temperature linearly. When the temperature exceeds 200 °C, the slopes between temperature and relative dimensional change begins to increase and then decline. With the temperature further increasing to 400 °C, the slopes no longer change. According to the theory of tempering, the specimens may have gone through Stage 0 and Stage I, thus the slopes remained unchanged when temperature was below 200 °C. However, it is known that Stage II leads to expansion and Stage III decreases the dimension. Therefore, the variation of slopes during 200~400 °C is closely related to the microstructural evolution in Stage II and Stage III.

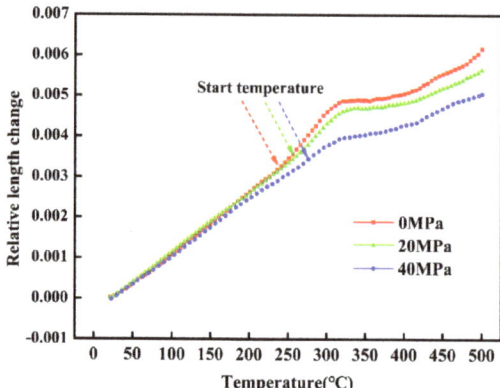

Figure 1. The thermal expansion curves during the non-isothermal process under different tensile stress.

Based on the above thermal expansion curves (Figure 1), the starting and ending temperatures of slope change was obtained by the tangents method [25] as shown in Table 2. In fact, according to the theory of tempering, the starting temperatures of slope change can represent the beginning of retained austenite decomposition (Stage II), while the ending temperature of slope change can reflect the complete formation of cementite (Stage III). The results listed in Table 2 clearly show that the starting temperature of retained austenite decomposition increases from 238 °C to 276 °C when the applied tensile stress increases from 0 MPa to 40 MPa. This may be due to the increase of stability of retained austenite induced by applied mechanical stress. However, it can be further found that the ending temperature of cementite formation obviously decreases from 474 °C to 418 °C with the increase of tensile stress, which may indicate a premature formation of cementite.

Table 2. The starting and ending temperatures of slope change under different tensile stress.

Tensile Stress	0 MPa	20 MPa	40 MPa
Starting temperature (Stage II)	238 °C	257 °C	276 °C
Ending temperature (Stage III)	474 °C	421 °C	418 °C

Furthermore, the activation energy of microstructural evolution without stress and with 40 MPa tensile stress were obtained by DSC testing and thermo-mechanical simulator experiments, respectively. Figure 2a illustrates the representative DSC curves at heating rate of 5 °C/min, where it can be seen that the first run curve (solid line) contains a significant heat flow peak in the range of 200~400 °C. Meanwhile, the rerun curve (dash line) had no visible peaks, which was quite necessary to act as a baseline to remove the influence of experimental instruments and environment.

By subtracting the baseline and dividing by the mass of the specimens, the revised DSC curve was obtained as shown in Figure 2b. It is obvious that no heat flow peak can be observed when the temperature is less than 200 °C, indicating that QT-treated specimens have gone through Stage 0 and Stage I. In the subsequent non-isothermal process, two heat flow peaks appeared in the range of 200~260 °C and 300~400 °C, which corresponds to Stage II and Stage III, respectively. Furthermore, by fitting the revised DSC curve, the curves of retained austenite decomposition (Stage II) and carbide transformation (Stage III) were both obtained, as shown in Figure 2b.

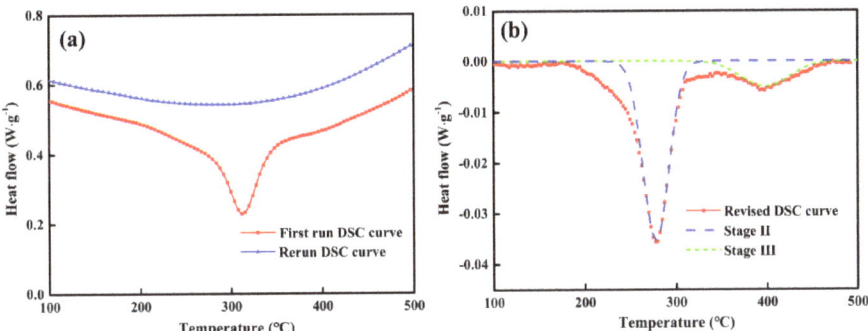

Figure 2. (a) The differential scanning calorimetry (DSC) curves at heating rate of 5 °C/min; (b) revised DSC curves after subtracting baseline.

Accordingly, the peak temperatures of Stage II and Stage III at different heating rates were listed in Table 3. As the heating rate increases, their peak temperatures gradually shift higher, indicating that higher heating rate will delay the transformation of Stage II and Stage III. According to the Kissinger method, the Kissinger straight lines of two stages can be obtained respectively, as shown in Figure 3. As a result, the activation energy of retained austenite decomposition was calculated to be 109.4 kJ/mol, and the activation energy of cementite precipitation was 179.4 kJ/mol, which is consistent with the report in literature [7].

Table 3. The peak temperatures of Stage II and Stage III without applied stress at different heating rates.

Heating Rates	Peak Temperatures	
	Stage II	Stage III
5 °C/min	278 °C	344 °C
10 °C/min	291 °C	353 °C
15 °C/min	309 °C	368 °C

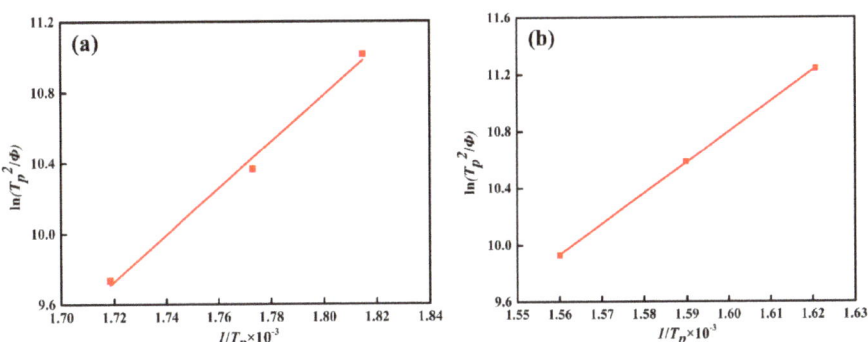

Figure 3. Kissinger analysis ($\ln(T_P^2/\varnothing)$ versus $1/T_P \times 10^{-3}$) for the determination of the individual activation energy of Stage II (a) and Stage III (b) without applied stress.

Figure 4a indicates the thermal expansion curves under 40 MPa tensile stress at different heating rate. As the most common method for analyzing the thermal expansion curves, the leverage theorem was used in this study and the transformed fraction curves were shown in Figure 4b. It can be observed that higher heating rate will result in an increase of the temperature of phase transformation, which is well consistent with the previous DSC results. In addition, the median temperatures (transformed fraction = 0.5) were listed in Table 4 as well, which was selected to calculate the activation energy by means of the Kissinger method. Consequently, the activation energy obtained by leverage law was 102.1 kJ/mol. However, considering that the transformed fraction curves calculated by the leverage law could not distinguish Stage II and Stage III, a follow-up calculation was thus performed using the isoconversional method.

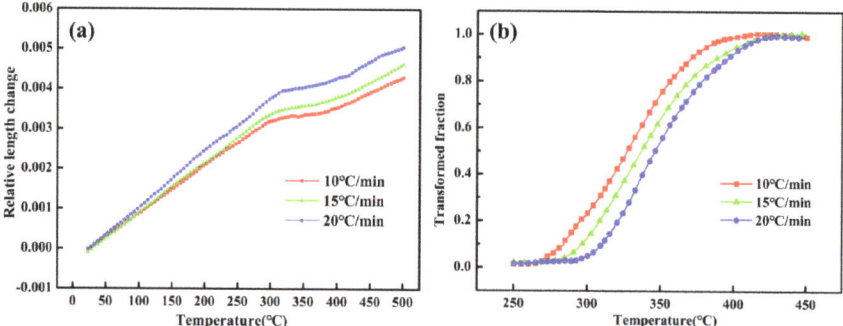

Figure 4. The thermal expansion curves (**a**) and transformed fraction curves (**b**) under 40 MPa tensile stress at a heating rate of 10, 15, 20 °C/min.

Table 4. The median temperature under 40 MPa at different heating rates.

Heat Rate	10 °C/min	15 °C/min	20 °C/min
Median temperature	327 °C	337 °C	346 °C

The isoconversional method can effectively reflect the change of activation energy during the whole process, which can be considered as a more accurate supplement for the Kissinger method. A detailed activation energy and modified pre-exponential factor ($\ln[A \cdot f(\alpha)]$) with respect to different transformed fraction were estimated using differential isoconversional method, as illustrated in Figure 5. It can be found that the activation energy first increases to 121.5 kJ/mol and then decreases to 72.8 kJ/mol with the increasing transformed fraction. During the early stage ($0.2 \leq \alpha \leq 0.4$) and later stage ($0.6 \leq \alpha \leq 0.7$), there are two platforms whose average activation energy are 121.5 kJ/mol and 94.7 kJ/mol, respectively. In fact, the platforms can be interpreted as different stages as reported by Wang et al. [26]. In the present work, according to the reaction sequence of Stage II and Stage III and kinetics analysis by the Kissinger method, the two platforms can be interpreted as the two stages (retained austenite decomposition and cementite precipitation). Meanwhile, the average value of these two stages was 108.1 kJ/mol, which was close to the activation energy obtained by Kissinger method (102.1 kJ/mol) as before. Compared with the activation energy without stress, it was found that the activation energies of Stage II and Stage III changed significantly. Under the applied tensile stress, the activation energy of the decomposition of retained austenite slightly increase to 121.5 kJ/mol, while the cementite precipitation was significantly accelerated due to the decreased activation energy of cementite precipitation (from 179.4 kJ/mol to 94.7 kJ/mol). Moreover, the results show that the variation of modified pre-exponential factor ($\ln[A \cdot f(\alpha)]$) presented similar trends to those of the activation energies.

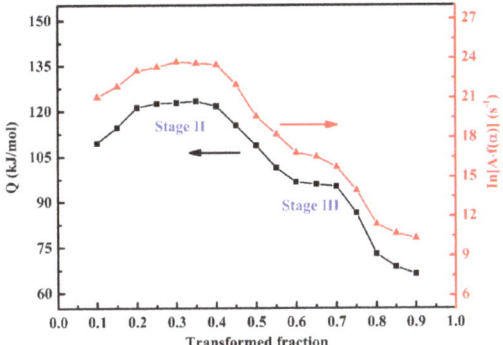

Figure 5. The activation energy and modified pre-exponential factor with respect to the transformed fraction.

To observe the microstructural evolution during the non-isothermal process, the QT treated specimens were heated to 300 °C and then air cooled to ambient temperature for microstructural analysis. Figure 6a shows the FESEM microstructure of the QT treated specimens which comprises undissolved spherical cementite, retained austenite and tempered martensite. Figure 6b–d illustrates the microstructure of the specimens after heating to 300 °C with different tensile stress, where it can be seen the matrix is mainly consisted of undissolved spherical carbides and precipitated needle-like cementite. In addition, the region highlighted by the yellow oval is poor of the precipitated cementite, indicating that there is still a large amount of cementite not formed for the specimens without tensile stress. However, more cementite has formed for the specimens with increasing tensile stress, which thus proves that the applied tensile stress can accelerate the formation of cementite efficiently.

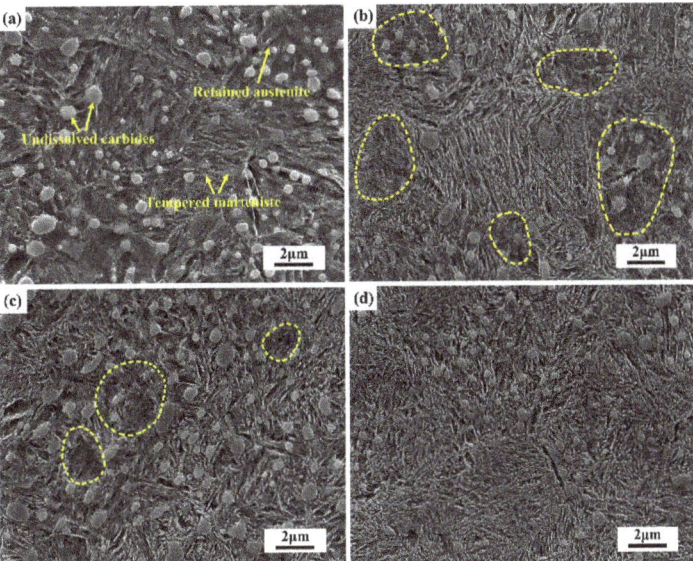

Figure 6. Typical field-emission scanning electron microscope (FESEM) microstructure of the (**a**) quenching and tempering (QT) treated specimens, and the QT treated specimens after heating to 300 °C under tensile stress of (**b**) 0 MPa, (**c**) 20 MPa and (**d**) 40 MPa.

Figure 7 illustrates the TEM micrographs and corresponding dark field of the microstructure of the specimen with and without 40 MPa stress, in which typical morphologies of carbides are observed. As shown in Figure 7a, some nano-size particles were found for the specimens without stress. These particles were identified using selected-area electron diffraction (SAED) to be ε-carbides. For the specimens with 40 MPa stress (Figure 7b), the needle-like precipitates presented in the matrix were identified to be θ-carbides (cementite), which is consistent with the SEM results. It can be inferred that a large number of ε-carbides still exist in the specimens without applied stress, and they have not been transformed into stable cementite at 300 °C. However, numerous stable θ-carbides (cementite) have formed for the specimens with 40 MPa stress, demonstrating again that the applied stress can accelerate the formation of cementite during the non-isothermal process.

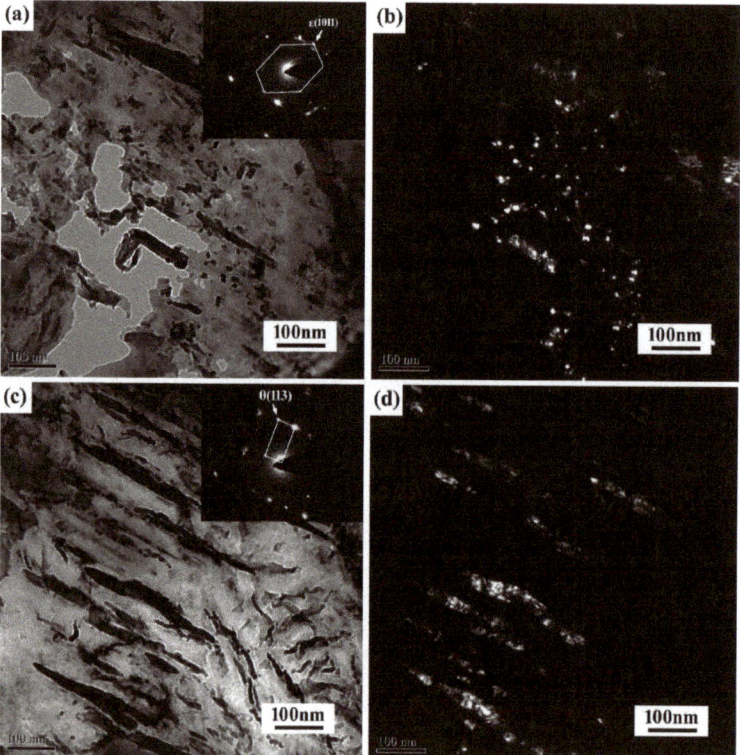

Figure 7. The presence of ε-carbides and θ-carbides for the specimens (**a**) without stress and (**c**) with 40 MPa, respectively. Where (**b,d**) are the corresponding dark field of (**a,c**), respectively.

The results of the XRD diffraction pattern (Figure 8) clearly show that the retained austenite has been completely decomposed after heating to 300 °C, regardless of the applied tensile stress. In addition, according to the observation of the (110)α diffraction peak (as seen in the insert), it can be seen that the martensite diffraction peak shifts to a higher angle with the increase of applied tensile stress. It has been reported that the diffraction peak information of martensite is closely related to the carbon content in martensite [27]. In the present work, the higher diffraction angle of (110) martensite indicates the lower carbon content of tempered martensite for the specimens subjected to tensile stress [28]. Therefore, it can be inferred that interstitial carbon atoms in martensite diffuse more and participate in the formation of cementite under the applied tensile stress.

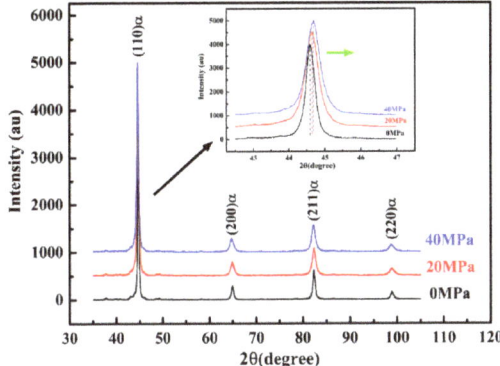

Figure 8. The X-ray diffraction (XRD) pattern of the specimens after being heated to 300 °C under different tensile stress.

4. Discussion

The mechanism of cementite formation was mainly determined by the diffusion of carbon atoms in the long range or short range. Chen [23] and Li [29] have studied the diffusion behavior of interstitial atoms in α-Fe under the strain field and they find that with the increase of tensile stress, the diffusion barrier decreases and atom diffusion gradually becomes easier. In this study, when external tensile stress is applied, the cubic crystal produces a weak elastic deformation along the tensile direction, then the diffusion barrier and atomic transition distance changes, which affects the diffusion rate of the carbon atoms. As a consequence, the applied tensile stress will exert a significant influence on the migration of carbon atoms and the formation of cementite.

It is reported by Kim [30] that the extra lattice energy arose from the presence of defect (including dislocations, grain boundaries, twins, etc.) in the matrix could lead to a reduction of nucleation energy barrier of cementite. In the present work, according to the first law of thermodynamics, the change of internal energy of the lattice can be expressed as follows:

$$dU = dQ - dW \tag{6}$$

where dQ and dW is the heat absorption and energy dissipation of the lattice, respectively. Assuming that the elastic strain of lattice (Δl) occurs along the tensile direction when tensile stress (f) is applied, the heat absorption and energy dissipation are obtained based on the second law of thermodynamics [31]:

$$dQ = TdS \tag{7}$$

$$dW = PdV - f\Delta l \tag{8}$$

where T and S is the temperature and surface area of the lattice, respectively; P and dV is the pressure and the volume change, respectively. By substituting Equations (7) and (8) into Equation (6), the modified internal energy change in the lattice (Equation (6)) can be expressed as:

$$dU = TdS - PdV + f\Delta l \tag{9}$$

For crystals, $PdV = 0$, So Equation (9) can be simplified as:

$$dU = TdS + f\Delta l \tag{10}$$

It should be noted that for the non-isothermal process, the lattice must be in the endothermic state, so the value of heat absorption will be positive ($TdS > 0$). Moreover, since the value of tensile stress is positive ($f > 0$), the change of internal energy in the lattice (dU) will increase compared with those without applied tensile stress under the same temperature. Therefore, it can be inferred that the applied tensile stress favors the increase of internal energy in the lattice, thereby leading to a reduction of the nucleation of the energy barrier for the crystal core with a same size. This is also the reason for the obvious decrease in activation energy of cementite formation when the tensile stress is applied.

5. Conclusions

The kinetic parameters of the microstructural evolution during the non-isothermal process with and without tensile stress were investigated using Kissinger and isoconversional methods. The observation of microstructure was conducted by means of scanning and transmission electron microscopy and X-ray diffraction after heating to 300 °C. The corresponding conclusions drawn are as follows:

(1) The activation energy of retained austenite decomposition slightly increases from 109.4 kJ/mol to 121.5 kJ/mol with the increase of tensile stress, which indicates the applied tensile stress is in favor of the stabilization of retained austenite. Additionally, the applied tensile stress not only lowers the end temperature of cementite formation, but also leads to a decrease of the activation energy of cementite precipitation from 179.4 kJ/mol to 94.7 kJ/mol, thereby proving that tensile stress can reduce the energy barrier of cementite precipitation.

(2) The microstructural observation shows that more cementite has formed for the specimens with the applied tensile stress, whereas there is still a large number of ε carbides existing in the specimens without stress. The results of XRD also verified that carbon in martensite diffuses more and participates in the formation of cementite under the applied tensile stress. Therefore, these results are in good agreement with the kinetic analysis, which together proves that the applied tensile stress can accelerate the precipitation of cementite during the non-isothermal process.

Author Contributions: Conceptualization, F.W. and D.-S.Q.; Methodology, P.X.; Software, P.X.; Validation, F.W. and D.-S.Q.; Formal Analysis, S.D. and F.W.; Investigation, F.W.; Resources, D.-S.Q.; Data Curation, S.D.; Writing—Original Draft Preparation, P.X.; Writing—Review & Editing, F.W.; Visualization, F.W.; Supervision, F.W.; Project Administration, D.-S.Q.; Funding Acquisition, D.-S.Q.

Funding: The authors would like to thank the National Natural Science Foundation of China (No. 51575414 and No. 51605354) and the Innovative Research Team Development Program of Ministry of Education of China (No. IRT13087) for a grant.

Acknowledgments: Grateful acknowledgement is made to my supervisor Lin Hua, who gave me considerable help by means of suggestion, comments and criticism. His encouragement and unwavering support has sustained me through frustration and depression.

Conflicts of Interest: The authors declare that they have no conflict of interest.

References

1. Shevchenko, R.P. *Lubricant Requirements for High Temperature Bearings*; Technical report; Society of Automotive Engineers: Warrendale, PA, USA, 1966.
2. Bhadeshia, H.K.D.H. Steels for bearings. *Prog. Mater. Sci.* **2012**, *57*, 268–435. [CrossRef]
3. Sidoroff, C.; Perez, M.; Dierickx, P.; Girodin, D. Advantages and shortcomings of retained austenite in bearing steels: A review. In *Bearing Steel Technologies: 10th Volume, Advances in Steel Technologies for Rolling Bearings*; ASTM International: West Conshohocken, PA, USA, 2015.
4. Jung, M.; Lee, S.J.; Lee, Y.K. Microstructural and dilatational changes during tempering and tempering kinetics in martensitic medium-carbon steel. *Metall. Mater. Trans. A* **2009**, *40*, 551–559. [CrossRef]
5. Jung, J.G.; Jung, M.; Kang, S.; Lee, Y.K. Precipitation behaviors of carbides and Cu during continuous heating for tempering in Cu-bearing medium C martensitic steel. *J. Mater. Sci.* **2014**, *49*, 2204–2212. [CrossRef]
6. Leiva, J.A.V.; Morales, E.V.; Villar-Cociña, E.; Donis-Díaz, C.A.; Bott, I.S. Kinetic parameters during the tempering of low-alloy steel through the non-isothermal dilatometry. *J. Mater. Sci.* **2010**, *45*, 418–428. [CrossRef]
7. Preciado, M.; Pellizzari, M. Influence of deep cryogenic treatment on the thermal decomposition of Fe–C martensite. *J. Mater. Sci.* **2014**, *49*, 8183–8191. [CrossRef]

8. Qiao, X.; Han, L.Z.; Zhang, W.M.; Gu, J.F. Nano-indentation investigation on the mechanical stability of individual austenite in high-carbon steel. *Mater. Charact.* **2015**, *110*, 86–93. [CrossRef]
9. Creuziger, A.; Foecke, T. Transformation potential predictions for the stress-induced austenite to martensite transformation in steel. *Acta Metal.* **2010**, *58*, 85–91. [CrossRef]
10. Fu, H.; Galindo-Nava, E.I.; Rivera-Díaz-del-Castillo, P.E.J. Modelling and characterisation of stress-induced carbide precipitation in bearing steels under rolling contact fatigue. *Acta Metal.* **2017**, *128*, 176–187. [CrossRef]
11. Ye, J.S.; Chang, H.B.; Hsu, T.Y. A kinetics model of isothermal ferrite and pearlite transformations under applied stress. *ISIJ Int.* **2017**, *44*, 1079–1085. [CrossRef]
12. Jimenez-Melero, E.; Blondé, R.; Sherif, M.Y.; Honkimäki, V.; van Dijk, N.H. Time-dependent synchrotron X-ray diffraction on the austenite decomposition kinetics in SAE 52100 bearing steel at elevated temperatures under tensile stress. *Acta Metal.* **2013**, *61*, 1154–1166. [CrossRef]
13. Zhao, H.Z.; Seok-jae, L.; Young-kook, L.; Liu, X.H.; Wang, G.D. Effects of applied stresses on martensite transformation in AISI4340 steel. *J. Iron Steel Res. Int.* **2007**, *14*, 63–67. [CrossRef]
14. Xu, Z.Y. Effect of stress on bainitic transformation in steel. *Acta Metall. Sin.* **2004**, *40*, 113–119.
15. Speich, G.R.; Leslie, W.C. Tempering of steel. *Metall. Trans.* **1972**, *3*, 1043–1054. [CrossRef]
16. Taylor, K.A.; Chang, L.; Olson, G.B.; Smith, G.D.W.; Cohen, M.; Vander Sande, J.B. Spinodal decomposition during aging of Fe-Ni-C martensites. *Metall. Trans. A* **1989**, *20*, 2717–2737. [CrossRef]
17. Barrow, A.T.W.; Kang, J.H.; Rivera-Díaz-del-Castillo, P.E.J. The ε→η→θ transition in 100Cr6 and its effect on mechanical properties. *Acta Metal.* **2012**, *60*, 2805–2815.
18. Morra, P.V.; Böttger, A.J.; Mittemeijer, E.J. Decomposition of iron-based martensite. A kinetic analysis by means of differential scanning calorimetry and dilatometry. *J. Therm. Anal. Calorim.* **2001**, *64*, 905–914. [CrossRef]
19. Perez, M.; Sidoroff, C.; Vincent, A.; Esnouf, C. Microstructural evolution of martensitic 100Cr6 bearing steel during tempering: From thermoelectric power measurements to the prediction of dimensional changes. *Acta Metal.* **2009**, *57*, 3170–3181. [CrossRef]
20. Budrugeac, P.; Segal, E. Applicability of the Kissinger equation in thermal analysis. *J. Therm. Anal. Calorim.* **2007**, *88*, 703–707. [CrossRef]
21. Wang, X.; Han, L.; Gu, J.F. Precipitation kinetics and yield strength model for NZ30K-Mg alloy. *Acta Metall. Sin.* **2014**, *50*, 355–360.
22. Liu, Q.L.; Qian, D.S.; Hua, L. Transformation from non-isothermal to isothermal tempering of steel based on isoconversional method. *J. Mater. Sci.* **2018**, *53*, 2774–2784. [CrossRef]
23. Chen, Z.; Kioussis, N.; Ghoniem, N.; Seif, D. Strain-field effects on the formation and migration energies of self-interstitials in α-Fe from first principles. *Phys. Rev. B* **2010**, *81*, 094102. [CrossRef]
24. Liu, C.; Mittemeijer, E.J. The tempering of iron-nitrogen martensite: Dilatometric and calorimetric analysis. *Metall. Trans. A* **1990**, *21*, 13–26.
25. Liu, Y.C.; Sommer, F.; Mittemeijer, E.J. Abnormal austenite–ferrite transformation behaviour in substitutional Fe-based alloys. *Acta Metal.* **2003**, *51*, 507–519. [CrossRef]
26. Wang, X.N.; Han, L.Z.; Gu, J.F. Precipitation kinetics of NZ30K-Mg alloys based on electrical resistivity measurement. *Trans. Nonferr. Met. Soc. China* **2014**, *24*, 1690–1697. [CrossRef]
27. Rammo, N.N.; Abdulah, O.G. A model for the prediction of lattice parameters of iron–carbon austenite and martensite. *J. Alloy. Compd.* **2006**, *420*, 117–120. [CrossRef]
28. Liu, X.; Zhong, F.; Zhang, J.X. Lattice-parameter variation with carbon content of martensite. I. X-ray-diffraction experimental study. *Phys. Rev. B* **1995**, *52*, 9970–9978.
29. Li, W.Y.; Zhang, Y.; Zhou, H.B.; Jin, S.; Lu, G.H. Stress effects on stability and diffusion of H in W: A first-principles study. *Nucl. Instrum. Methods Phys. Res. Sect. B* **2011**, *269*, 1731–1734. [CrossRef]
30. Kim, B.; Celada, C.; Martín, D.S.; Sourmail, T.; Rivera-Díaz-del-Castillo, P.E.J. The effect of silicon on the nanoprecipitation of cementite. *Acta Metal.* **2013**, *61*, 6983–6992. [CrossRef]
31. Roh, H.S. Internal energy transfer theory for thermodynamic non-equilibrium, quasi-equilibrium, and equilibrium. *Int. J. Heat Mass Transf.* **2014**, *78*, 778–795. [CrossRef]

 © 2018 by the authors. Licensee MDPI, Basel, Switzerland. This article is an open access article distributed under the terms and conditions of the Creative Commons Attribution (CC BY) license (http://creativecommons.org/licenses/by/4.0/).

Article

Hydration Kinetics of Composite Cementitious Materials Containing Copper Tailing Powder and Graphene Oxide

Shuhua Liu *, Qiaoling Li and Xinyi Zhao

State Key Laboratory of Water Resources and Hydropower Engineering Science, Wuhan University, Wuhan 430072, China; qllee@whu.edu.cn (Q.L.); xinyizhao@whu.edu.cn (X.Z.)
* Correspondence: shliu@whu.edu.cn; Tel.: +86-27-6877-2233

Received: 19 October 2018; Accepted: 6 December 2018; Published: 8 December 2018

Abstract: The hydration heat evolution curves of composite cementitious materials containing copper tailing powder (CT) and graphene oxide (GO) with different contents are measured and analyzed in this paper. The hydration rate and total hydration heat of the composite cementitious materials decrease with the increase of CT dosage, but improve with the increase of CT fineness and GO dosage. The hydration process of the cementitious systems undergoes three periods, namely nucleation and crystal growth (NG), phase boundary reaction (I), and diffusion (D), which can be simulated well using the Krstulovic–Dabic model. The hydration rates of the three controlling processes of the composite cementitious system decrease with the increase of CT content, but improve slightly with the increase of CT fineness. GO enhances the controlling effect of the NG process of the cementitious systems with or without CT, thus promotes the early hydration as a whole.

Keywords: composite cementitious materials; copper tailing powder; graphene oxide; hydration kinetics

1. Introduction

Copper tailings are waste materials generated during the purification of precious copper from the copper ores, and 128 t copper tailings will be left over per 1 t refined copper [1,2]. According to the United States Geological Survey Bureau, two trillion tons of copper tailings were produced worldwide in 2011 [3]. The cumulative copper tailings in China exceeded 3 trillion tons by 2014; however, only about 8.2% are recycled [3–5]. The rest are mainly disposed of in the tailing pond, which results in a lot of adverse effects. For example, the heavy metal elements in copper tailings poison the surrounding soil and water. Furthermore, as copper tailings are definitely solid gravel, the tailing pond is prone to destabilize and collapse in the case of earthquakes and flooding [6]. In addition, the construction cost and operation expense of a tailings depot are very high. As a result, the utilization of copper tailings in concrete as either fine aggregate or supplementary cementitious material is not only conducive to the resources recovery and utilization but also greatly reduces the environmental pollution, geological disasters, and other problems caused by copper tailings.

Some studies regarding the use of copper tailings as cement substitution have been conducted, and the results show that the pozzolanic activity of copper tailing powder is fairly low and its optimal replacement ratio is 5% without strength and durability reduction; however, the replacement ratio can reach up to 30–50% in mass concrete with consideration of the temperature control [2,7,8]. Moreover, it has been reported that graphene oxide (GO) has reinforcing and toughening effects on the cement paste [9–13]. The improvement of the mechanism of GO can be concluded using the following reasons. The first is called the template effect, in which GO can regulate the morphology of the hydration products of cement to form flower-shaped microcrystals with uniform shape and uniform distribution [9–11]. The second is the nucleation effect that GO lamellae can provide nucleation sites for nucleation and crystallization of calcium silicate hydrates (C-S-H), which promotes the crystallization, nucleation, and growth of C-S-H, thereby accelerating the early hydration [12]. In addition, the interface enhancement effect between GO and C-S-H is proven to be fairly high, which may also contribute to the strength and toughness of the paste [13].

As a result, copper tailing powder (CT) shows some adverse effects on early hydration and strength of the cement-based materials, but GO can improve their microstructure and properties. What will happen when the positive and the negative elements interact? How does GO promote the hydration mechanism of a cement-CT composite system? We suppose that GO also has reinforcing and toughening effects on the composite cementitious materials containing CT. In order to investigate the effect of CT on the early hydration mechanism of cement, as well as the enhancing effect of GO on the hydration of cement and cement-CT binder, the hydration heat evolution rate and cumulative hydration heat of composite cementitious materials containing CT and GO at different contents are measured at 25 °C by an isothermal calorimeter. Based on the thermodynamic data, the hydration kinetics of the composite cementitious materials containing CT and GO are investigated in detail using the Krstulovic–Dabic model to reveal the effects of CT and GO on the early hydration.

2. Materials and Methods

2.1. Raw Materials

Ordinary Portland cement (P.O 42.5) and CT supplied by China Construction Mining Corporation are used in the experiment. Their chemical compositions, determined by the X-ray fluorescence (XRF Axios FAST, Malvern Panalytical Ltd., Royston, UK), are listed in Table 1. CT contains a great deal of CaO, SiO_2, and Al_2O_3. While the total content of SiO_2, Al_2O_3, and Fe_2O_3 is 53.03%, which is less than 70% of that put forward by ASTM C618-15 Standard Specification for Coal Fly Ash and Raw or Calcined Pozzolan for use in concrete [14]. Its XRD pattern is displayed in Figure 1; the XRD patters shows that SiO_2 exists mainly in the form of andradite instead of active SiO_2. In general, the CT used in this study has a low pozzolanic activity. GO dispersion is produced by Shanxi Institute of Coal Chemistry, Chinese Academy of Sciences, Taiyuan, China. It has a GO content of 4 mg/mL and its morphology, size (about 1 μm) and thickness (about 1 nm) are characterized by the Atomic Force Microscope (AFM Tosca™ 400, Anton Paar Shanghai Trading Co. Ltd., Shanghai, China) as shown in Figure 2.

Table 1. The main chemical compositions of cement and copper tailing powder (CT)/Mass, %.

Compositions	SiO_2	Al_2O_3	CaO	Fe_2O_3	MgO	SO_3	TiO_2	K_2O	Na_2O	P_2O_5	CuO
Cement	21.25	2.91	63.09	3.24	0.68	3.36	0.31	1.12	0.31	0.17	-
CT	39.15	5.49	31.76	8.39	5.37	1.21	0.21	0.64	1.32	0.11	0.08

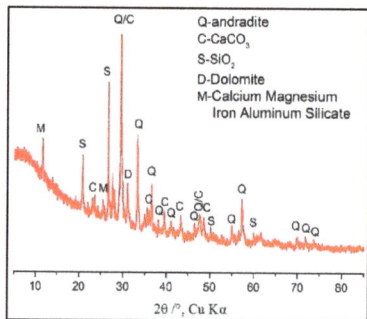

Figure 1. XRD pattern of CT.

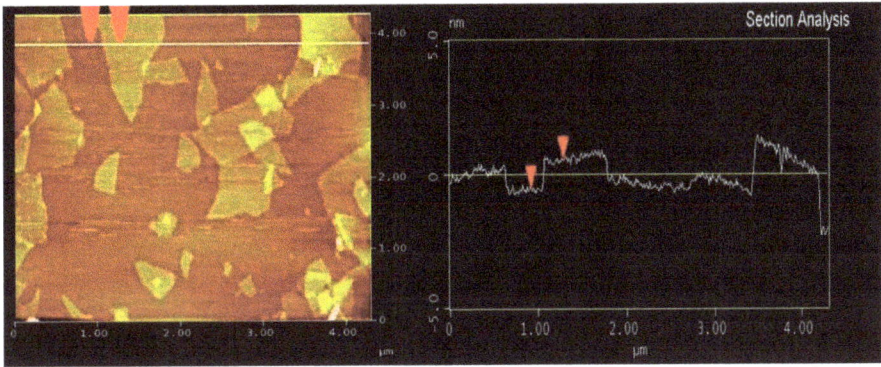

Figure 2. AFM image of graphene oxide (GO).

2.2. Testing Methods

The copper tailings were dried, first, at 60 °C and filtered through a 1 mm-square-hole sieve in order. Then, they were ground for 30 min and 60 min, respectively, in a small insulative ball mill (SM 500, Daoxu Machinery Factory, Shangyu, China). The ball mill has a speed of 48 r/min, with a loading capacity of 5 kg of sample. The powder morphology was investigated using scanning electron microscopy (JSM-5610LV, JSM Ltd., Tokyo, Japan). As shown in Figure 3, CT particles ground for 30 min and 60 min show irregular blocky, granular, and clastic shape. The particle size distribution of cement and CT, as shown in Figure 4, was measured by a laser particle size analyzer (Master size 2000, Malvern Instruments Ltd., Worcestershire, UK), ranging from 0.1 to 1000 μm. The specific surface area was calculated by software included in the laser particle size analyzer. The specific surface area of CT ground for 30 min and 60 min is 380 m^2/kg and 690 m^2/kg, respectively. The specific surface area of cement is 440 m^2/kg. The average size of CT ground for 30 min and 60 min is 6.79 μm and 3.47 μm, respectively, and the average size of cement is 5.67 μm, which is similar to their specific surface area.

Figure 3. Particle morphology of CT after being ground for (**a**) 30 min; and (**b**) 60 min.

Two paste systems, I and II, prepared for hydration heat determination are listed in Tables 2 and 3, respectively. The hydration heat evolution rate and total hydration heat emission of the samples were measured by an isothermal calorimeter (TAM Air, TA Instruments Inc., New Castle, DE, USA) at 25 °C within 72 h, with a temperature fluctuation of less than ±0.02 °C.

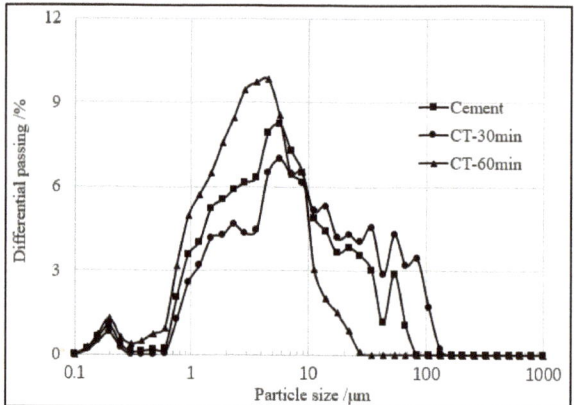

Figure 4. Particle size distribution curves of cement and CT.

Table 2. Mixture proportions I (samples cement-CT system (CTs))/Mass, %.

Sample	Grinding Time	CT	W/C
CT3-15	30 min	15%	0.4
CT3-30	30 min	30%	0.4
CT3-45	30 min	45%	0.4
CT6-15	60 min	15%	0.4
CT6-30	60 min	30%	0.4
CT6-45	60 min	45%	0.4
C	/	0%	0.4

Table 3. Mixture proportions II (samples cement-GO system (CGs) and samples cement-GO-CT system (GCTs))/Mass, %.

	Sample	GO Content	Cement	CT	W/C
CGs	C-0	0	100%	0	0.4
	C-1	0.01%	100%	0	0.4
	C-2	0.02%	100%	0	0.4
	C-3	0.03%	100%	0	0.4
GCTs	CT*-0	0	70%	30%	0.4
	CT*-1	0.01%	70%	30%	0.4
	CT*-2	0.02%	70%	30%	0.4
	CT*-3	0.03%	70%	30%	0.4

Note: Taking CT6-30 in mixture proportions I for example, 6 indicates that the grinding time of CT is 60 min, 30 represents that the mixing content of CT is 30%; For CT*-2 in mixture proportions II, CT* is the simplified name of CT6-30, 2 stands for 0.02% dosage of graphene oxide (GO).

3. Results and Discussion

3.1. Characteristics of Hydration Heat Evolution

Figure 5 shows the hydration heat evolution rate and total hydration heat of a cement-CT composite system (CTs) within 72 h. The acceleration period and the time at which the induction period of CTs ended are shown in Figure 6. The rate of the second exothermic peak and the total hydration heat evolution at different hydration times determined from heat evolution curves are listed in Table 4.

Table 4. Characteristic values of hydration heat evolution curves of samples at 25 °C.

Sample	Rate of the Second Heat Emission Peak q_{max} (J/g·h)	Total Heat Release (J/g)				Heat Release Per g of Cement (J)			
		12 h	48 h	60 h	72 h	12 h	48 h	60 h	72 h
CT3-15	8.341	44.7	193.2	210.0	222.7	52.6	227.3	247.1	262.0
CT3-30	7.044	41.0	160.2	171.6	179.9	58.6	228.9	245.1	257.0
CT3-45	5.529	34.9	130.0	139.0	145.4	63.5	236.4	252.7	264.4
CT6-15	8.541	48.2	200.6	217.5	229.8	56.7	236.0	255.9	270.4
CT6-30	7.203	43.7	166.2	179.4	189.3	62.4	237.4	256.3	270.4
CT6-45	5.768	37.6	132.0	141.4	148.3	68.4	240.0	257.1	269.6
C	9.479	51.6	217.3	235.3	249.7	51.6	217.3	235.3	249.7

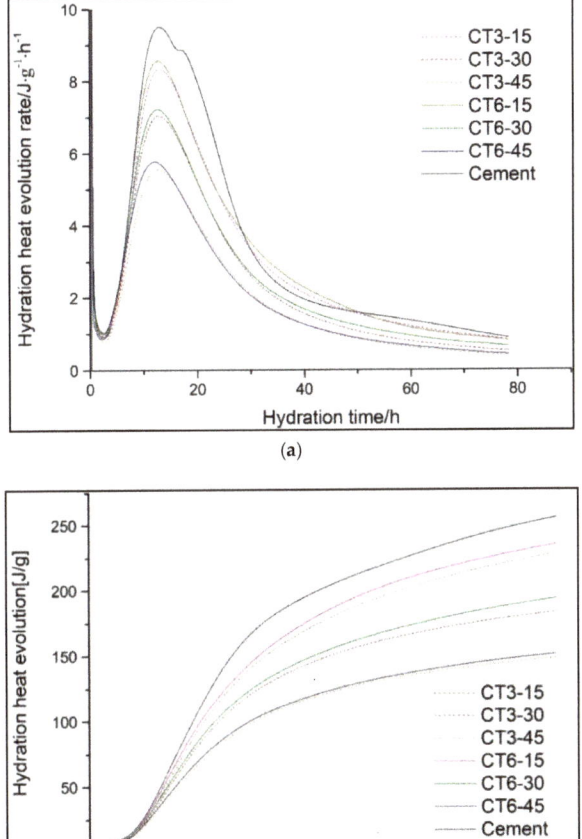

Figure 5. (a) Hydration heat evolution rate and (b) total hydration heat release of sample CTs.

As shown in Figure 5a, during the first few minutes after mixing the binder with water, a sharp exothermic peak corresponding to the first peak occurs in the curves and is attributed to the quick dissolution of cement and the quick formation of ettringite [15]. Then, the first peak declines dramatically and goes into the induction period. Figure 6 indicates that there is little difference among all the samples during the induction period. The duration of the induction of cement lasts for

1.94 h and ends at about 3.43 h. With the increase in CT content, the duration of the induction period is prolonged from 2.21 to 3.1 h. It may be caused because the cement content of the cementitious system decreases with the increase of CT content. Therefore, the dissolved Ca^{2+} concentration reduces and the time when Ca^{2+} reaches supersaturation extends, which leads to the ending time of the induction period being slightly prolonged [16–18].

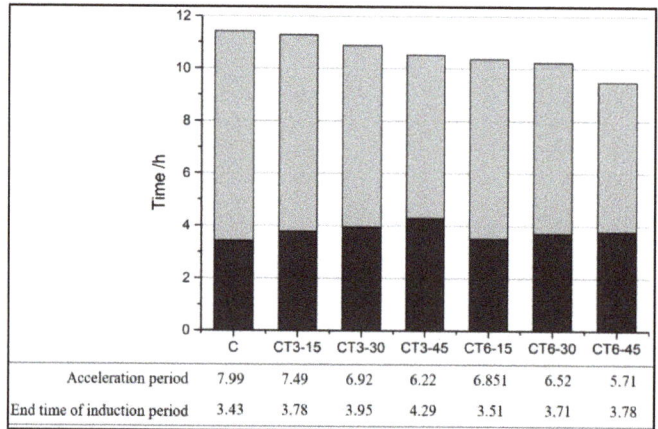

Figure 6. The acceleration period and end time of induction period.

After the acceleration period, comes the strong hydration reaction of C_3S and fast formation of C-S-H and $Ca(OH)_2$ [19]. Owing to the sufficient reactant and water supply, the reaction during the acceleration period proceeds very quickly to form hydration products around the unreacted particles, which in turn postpones the hydration reaction. When the accelerating effect is equal to the delaying effect, the maximum value of the second exothermic peak is achieved. As a whole, the peak value improves and the duration time of the induction period is prolonged with the increase of CT content. In addition, the time at which the second exothermic peak is observed for all samples is slightly different and the hydration release curves tend to narrow with the increase in CT content, which corresponds to a low total hydration heat evolution. The low hydration activity and hydration degree of CT in the early stage, which mainly plays the filling role, is negligible and leads to the decrease in the overall quantity of reactants. Then, the reaction goes into the deceleration period and stable period, and the hydration reactions are much more steady and controlled by diffusion.

Moreover, it is interesting that the heat release per g of cement of the cement-CT binder clearly improves with the increase of CT content and strengthens slightly with the increase of CT fineness, as shown in Table 4. The actual water–cement ratio increases with the increase of the CT content, thus there is more water involved in the cement hydration, and the hydration rate of the cement at an early stage is accelerated even though the total hydration rate is delayed. CT particles, especially for the fine particles, can act as nucleation sites for cement hydration and accelerate its hydration. Similar results have been found for other mineral admixtures in our previous studies [20,21] time taken to observe the second exothermic peak decreases with the increase of the specific surface area. It indicates that the high CT fineness can accelerate the early hydration of composite cementitious materials.

Figure 5b shows that the total hydration heat evolution of a composite system containing CT is evidently lower than that of pure cement. Moreover, the total hydration heat evolution reduces with the increase of CT content but improves with the increase of CT fineness. As shown in Table 4, the heat emission at 72 h decreases by at least 8%, from 249.72 to 229.75 J/g, after incorporating 15% CT and decreases no less than 33%, from 222.73 to 148.30 J/g, when the CT content increases from 15 to 45%.

Figure 7 displays the hydration heat evolution rate and total hydration heat release of a cement-GO-CT composite system (GCTs). The second exothermic peak is slightly quickened as well as the value increases slightly with the GO content increasing from 0.01 to 0.03% in both pure cement and cement-CT composite system containing 30% CT. GO lamellar can serve as nucleation sites for C-S-H to nucleate and crystalize, which promotes the hydration of the cementitious system. As shown in Figure 7b, the total hydration heat also increases slightly with the increase of GO content.

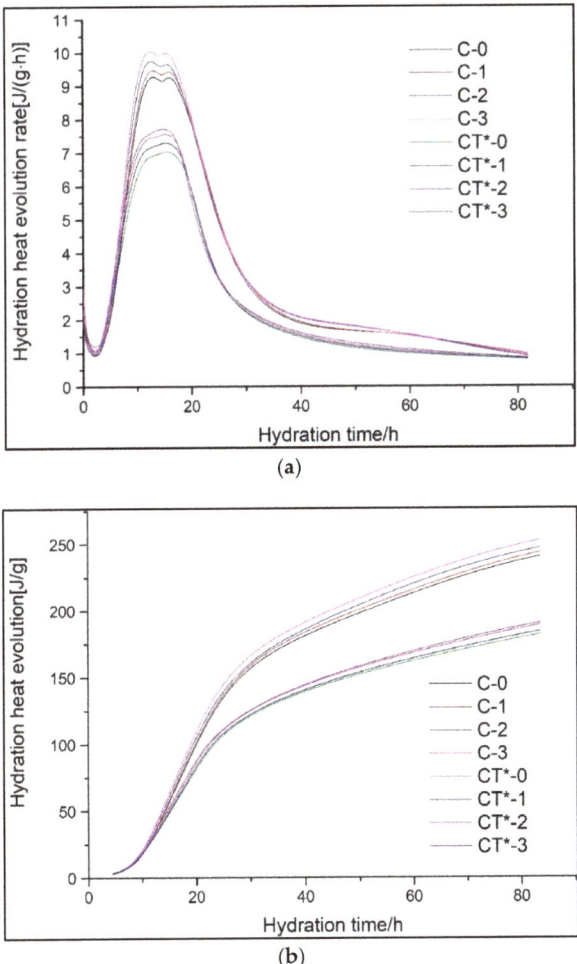

Figure 7. (a) Hydration heat evolution rate and (b) total hydration heat release of samples CGs and GCTs.

3.2. Hydration Process Simulation

3.2.1. Hydration Kinetic Model

A hydration kinetic model is often used to analyze the influence of various factors on the reaction rate and reaction direction during the hydration process, in order to reveal its control mechanism [22]. The Krstulovic–Dabic model assumes that three basic processes take place during the early hydration of the cement-based materials, namely nucleation and crystal growth (NG), phase boundary reaction (I), and diffusion (D) [23]. The three control processes may occur simultaneously, and can also occur alone or in pairs, but the hydration rate of the overall process depends on the one which reacts slowest. That is, the slowest reaction controls the reaction rate and mechanism at an early stage [24]. The Krstulovic–Dabic model also gives the basic kinetic equations to describe the three dominating processes of these three control processes as follows:

$$\text{NG}: \quad [-\ln(1-\alpha)]^{1/n} = K_1(t-t_0) = K_1'(t-t_0) \tag{1}$$

$$\text{I}: \quad [1-(1-\alpha)^{1/3}]^1 = K_2 R^{-1}(t-t_0) = K_2'(t-t_0) \tag{2}$$

$$\text{D}: \quad [1-(1-\alpha)^{1/3}]^2 = K_3 R^{-1}(t-t_0) = K_3'(t-t_0) \tag{3}$$

where α is hydration degree; $K_1(K_1')$, $K_2(K_2')$, $K_3(K_3')$ is the reaction rate constant corresponding to the hydration processes NG, I, and D; t is hydration time; t_0 is the time when the induction period ends; R is the ideal gas constant; n is the crystal growth index that reflects the geometrical crystal growth, $1 \leq n \leq 2$ [25].

When α is differentiated with respect to t in the equations above, the hydration rate of each process is obtained as follows:

$$\text{NG}: \quad d\alpha/dt = F_1(\alpha) = K_1'(1-\alpha)[-\ln(1-\alpha)]^{\frac{n-1}{n}} \tag{4}$$

$$\text{I}: \quad d\alpha/dt = F_2(\alpha) = 3K_2'(1-\alpha)^{2/3} \tag{5}$$

$$\text{D}: \quad d\alpha/dt = F_3(\alpha) = 3[K_3'(1-\alpha)^{\frac{2}{3}}]/2[1-(1-\alpha)^{\frac{1}{3}}] \tag{6}$$

where $F_1(\alpha)$, $F_2(\alpha)$, and $F_3(\alpha)$ represent the hydration processes NG, I, and D, respectively.

Based on the total hydration emission $Q(t)$ and the rate of hydration evolution dQ/dt obtained by isothermal conduction calorimetry, the hydration degree α and hydration rate $d\alpha/dt$ required for the kinetic simulation of hydration process are determined by the following equations [26]:

$$\alpha(t) = \frac{Q(t)}{Q_{max}} \tag{7}$$

$$\frac{d\alpha}{dt} = \frac{dQ}{dt} \cdot \frac{1}{Q_{max}} \tag{8}$$

$$\frac{1}{Q} = \frac{1}{Q_{max}} + \frac{t_{50}}{Q_{max} \cdot t} \tag{9}$$

where this newly defined $Q(t)$ is the heat released from the end of the induction period. Due to the fact that the dissolution progress is so fast that it is not always possible to be detected. Meanwhile, the induction period is also thought to make a small contribution to the total heat. Therefore, this simulation is conducted from the beginning of the second peak, namely the ending of the induction period [27]; Q_{max} is the total hydration heat when the reaction has completely finished and is obtained by using the Knudsen extrapolation Equation (9) to linearly fit the hydration heat evolution curves [28]: where t_{50} is the time required for half of Q_{max}.

3.2.2. Hydration Process Simulation of CTs

In order to obtain the hydration kinetic equations of CTs, the Q_{max} and hydration degree $\alpha(t)$ is determined by Equations (7) and (9), respectively and successively at first, as shown in Figure 8. Then, the kinetic parameters K_1' and n during the nucleation and crystal growth (NG) process could be calculated by substituting $\alpha(t)$ into Equation (1) and linearly fitting the double logarithmic curve of $\ln[-\ln(1-\alpha)]$ vs. $\ln(t-t_0)$, as shown in Figure 9. K_2' for I process and K_3' for the D process can also be derived by plugging $\alpha(t)$ into Equations (2) and (3) and fitting the double logarithmic curve $\ln[1-(1-\alpha)^{1/3}]$ vs. $\ln(t-t_0)$ linearly. Finally, the hydration kinetic expressions, $F_1(\alpha)$, $F_2(\alpha)$, and $F_3(\alpha)$, characterizing the hydration rate of the NG, I, and D processes as a function of the hydration degree α, are acquired. The relationships of $F_1(\alpha)$, $F_2(\alpha)$, $F_3(\alpha)$ and $d\alpha/dt$ with α are also shown in Figure 10. The intersection point α_1 is the turning point from NG to I, and α_2 is the turning points from I to D.

From Figure 10, it can be seen that the practical hydration curves of $d\alpha/dt$ can be segmentally simulated by the theoretical curves, i.e., $F_1(\alpha)$, $F_2(\alpha)$, and $F_3(\alpha)$. It means that the hydration kinetic model could basically simulate the hydration process of all samples and the hydration undergoes three processes, namely NG, I, and D, in order. The hydration of the composite system is controlled by multiple reaction mechanisms instead of a single one.

Cement reacts with water primarily, and $Ca(OH)_2$ becomes supersaturated in a few minutes and then the hydrates grow from a fixed number of nuclei [29]. For the cement–CT binder, CT particles can also serve as nucleation sites for hydrates in addition to cement grains. After that, the hydration products grow rapidly on the limited number of nuclei [30,31]. As a result, the NG process dominates the hydration process. Owing to the continued replenishment of Ca^{2+} dissolved from unhydrated particles to supply the hydration reaction, the supersaturation state of Ca^{2+} remains constant. The reaction mainly occurs at the boundary between the solid hydrates and the liquids. At that time, the phase-boundary-controlled I process plays the leading role [32]. As the hydration process continues, the hydration products increase. In the meantime, a large amount of water is consumed. The ability of water and ions to reach the surface of the unhydrated particles through the hydrated layer also becomes difficult. Therefore, the diffusion process, D, becomes the dominant process.

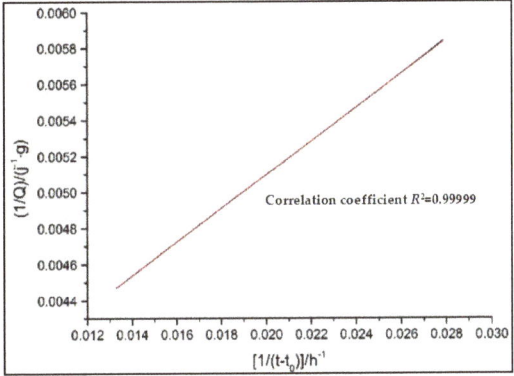

Figure 8. Determination of maximum hydration emission heat Q_{max} from linear regression.

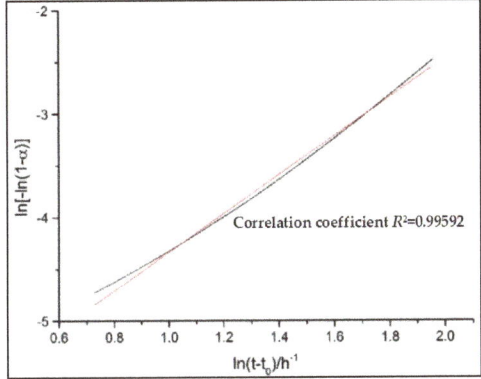

Figure 9. Determination of kinetic factors (n and K_1') of nucleation and crystal growth (NG) progress from linear regression.

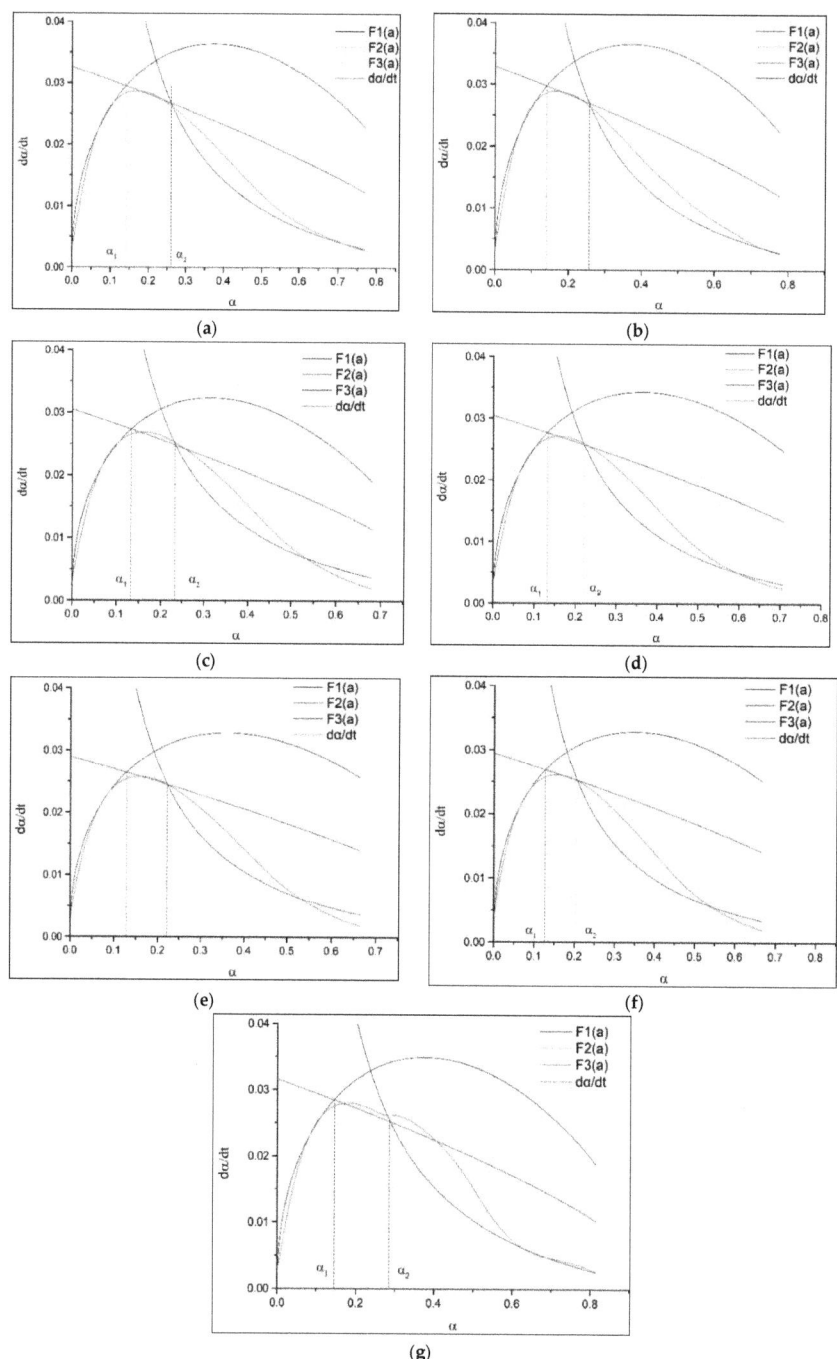

Figure 10. Hydration rate curves for samples CTs: (**a**) CT3-15; (**b**) CT6-15; (**c**) CT3-30; (**d**) CT6-30; (**e**) CT3-45; (**f**) CT6-45; (**g**) Cement.

The kinetic parameters of the hydration process of sample CTs are listed in Table 5. The value of reaction order n reduces with the increase of CT dosage while improving with the increase of CT fineness, which indicates that high fineness may affect the crystal growth geometry. The rate of chemical reaction during the NG process is very fast. The rate of the NG process is about 4–5 times the rate of the I process, and about 20 times the rate of the D process. The hydration reaction during the NG process is an autocatalytic reaction. The continuous growth of the crystal nucleus leads to the increase of their boundaries, which in turn accelerates the hydration reaction of the NG process. Therefore, the hydration reaction during the NG process is fairly fast. However, the hydration reaction during the I process is mainly controlled by the ion concentration, crystal area, and growth space for hydrates. In comparison with the NG process, a lot of reactants and water are consumed and a large amount of hydration products are formed during the I process, leading to the ion concentration to decline and the growth space for hydration products to narrow. Thus, the hydration rate of the I process is much lower than that of the NG process. With the development of hydration, the hydration reaction proceeds to the D process. During this process, massive dense C-S-H is formed due to the hydration of cement and the pozzolanic reaction of CT, which wraps on the surface of unreacted particles and makes the ion immobility difficult. As a result, the hydration rate during the D process is even lower than that during the I process.

Table 5. Kinetic parameters of the hydration process of sample CTs.

Sample	N	k_1'	k_2'	k_3'	α_1	α_2	$\alpha_2-\alpha_1$	Q_{max}'
Cement	1.8843	0.04425	0.01053	0.00226	0.1456	0.2860	0.1404	305.983
CT3-15	1.8521	0.04414	0.01088	0.00208	0.1412	0.2601	0.1189	290.475
CT3-30	1.8188	0.04260	0.01016	0.00169	0.1312	0.2340	0.1028	261.525
CT3-45	1.7522	0.04080	0.00966	0.00155	0.1280	0.2208	0.0928	215.114
CT6-15	1.8788	0.04488	0.01097	0.00206	0.1404	0.2558	0.1154	295.598
CT6-30	1.8401	0.04236	0.01014	0.00220	0.1318	0.2204	0.0886	265.947
CT6-45	1.7988	0.04136	0.00982	0.00145	0.1280	0.2055	0.0775	220.486

As shown in Table 5, k_1' decreases with the increase of CT dosage, implying that CT affects the nucleation and growth of hydrates. During the NG process, the reaction amount of CT is usually considered negligible because of its low pozzolanic activity [32]. With the increase of CT content, the cement dosage decreases, leading to the decrease of the pH value and the solubility of amorphous silicon as well as the growth rate of the crystal nucleus. Therefore, the growth rate during the NG process reduces with the increase of CT content. For the I process, the trend of the hydration rate, k_2', is consistent with that of k_1'. The reaction rate of CT is much lower than that of cement. The amount of CT accounts for a large proportion of the increase in CT content, which leads to the reduction of the k_2' value. For the D process, a similar trend of k_3' is observed. A large amount of hydration products have been formed and the reaction becomes stable at this stage. On the one hand, the higher the content of cement is, the more intense the preceding hydration reaction and the denser the hydration products will be; therefore, the more difficult the ion mobility will be. Thus, the hydration reaction of the D process is weakened. On the other hand, a higher content of cement will greatly strengthen the overall hydration reaction of the cementitious system. When the positive effect leading to the improvement of the cement dosage and decrease of the CT content outweighs the negative, the whole reaction rate during the D process will decrease with the incorporation of CT.

It is also noted that the hydration duration for both the NG process and I process is shortened with the increase of CT content, indicating that the hydration reaction of the composite binder containing CT transforms from the NG to I process and from the I to D process at a lower hydration degree with the increase of CT content. The replacement of cement by CT increases the effective water to cement ratio and provides more space for hydration products during the early hydration stage. Meanwhile, CT with high fineness can also act as nucleation sites for hydration products during the

NG process [19]. Therefore, the controlling effect of the NG and I process is strengthened, which leads to a sharp exothermic rate and narrow hydration duration.

3.2.3. Hydration Process Simulation of Composite Cementitious System Containing GO

Figure 11 shows the simulated and practical hydration exothermic curves of composite cementitious materials containing GO and CT. It is observed that curves, $F_1(1)$, $F_2(2)$, and $F_3(3)$, simulate the experimental hydration curves well, which validates that the hydration of both the cement-GO system (CGs) and cement-GO-CT system (GCTs) has a complicated process with a multiple reaction mechanism. In particular, the hydration process of CGs and GCTs after the induction period are the NG, I, and D processes in turn. Moreover, the hydration mechanism of the composite systems with GO is similar to that without it (C-0, CT*-0).

The kinetic parameters of the hydration process of composite cementitious materials are given in Table 6. The duration of the NG process is shortened while the I process is prolonged. For the CG system, the value of α_1 decreases from 0.1483 to 0.1377 while $\alpha_2 - \alpha_1$ increases from 0.2702 to 0.2833 while GO content increases from 0 to 0.03%. With regard to the GCTs system, the value of α_1 decreases from 0.1375 to 0.1306, while $\alpha_2 - \alpha_1$ increases from 0.2262 to 0.2413 along with the increase in the GO content from 0 to 0.03%. It illustrates that GO promotes the nucleation and crystal growth process by acting as nucleation sites, and leads to a higher reaction rate but a shorter duration of the NG process. This also be confirmed by the fact that k'_1 increases with the GO content. However, the hydration process during the I process is prolonged with the increase of GO dosage. It may be because GO improves the overall quantity of nucleation sites for crystal growth. In this case, the boundaries between the growing crystals and the solutions may also increase, leading to a longer duration to finish the I process. It also can be observed that the values of α_1 and $\alpha_2 - \alpha_1$ of GCTs are even less than those of CGs. It is also confirmed that the replacement of CT leads to a short hydration duration.

Overall, based on the data in Tables 5 and 6, the hydration rates, k'_1, k'_2, and k'_3 of the three controlling processes of the composite cementitious system decrease with the increase of CT content, but improve slightly with the increase of CT fineness. Although CT exerts an adverse effect on the early hydration, this can be slightly compensated for by the increase of CT fineness and can be overturned by the incorporation of GO. GO evidently accelerates the hydration of composite materials with the fact that k'_1, k'_2, and k'_3 gradually increase with GO dosage. Additionally, with the increase of CT dosage as well as its fineness, the NG and I process are gradually shortened. GO enhances the controlling effect of the NG process of the cementitious systems with or without CT, thus promoting the early hydration.

Table 6. Kinetic parameters of hydration process of sample CGs and CTGs.

Sample		N	k'_1	k'_2	k'_3	α_1	α_2	$\alpha_2 - \alpha_1$	Q'_{max}
CGs	C-0	1.806	0.04520	0.01137	0.00228	0.1483	0.2702	0.1219	306.112
	C-1	1.878	0.04599	0.01149	0.00233	0.1456	0.2742	0.1286	309.310
	C-2	1.881	0.04600	0.01151	0.00236	0.1404	0.2794	0.1390	312.495
	C-3	1.894	0.04705	0.01158	0.00241	0.1377	0.2833	0.1456	314.794
GCTs	CT*-0	1.621	0.04060	0.01005	0.00165	0.1375	0.2262	0.0887	251.889
	CT*-1	1.700	0.04182	0.01017	0.00172	0.1341	0.2333	0.0992	256.410
	CT*-2	1.717	0.04235	0.01023	0.00179	0.1318	0.2401	0.1083	263.850
	CT*-3	1.725	0.04251	0.01025	0.00181	0.1306	0.2413	0.1107	267.380

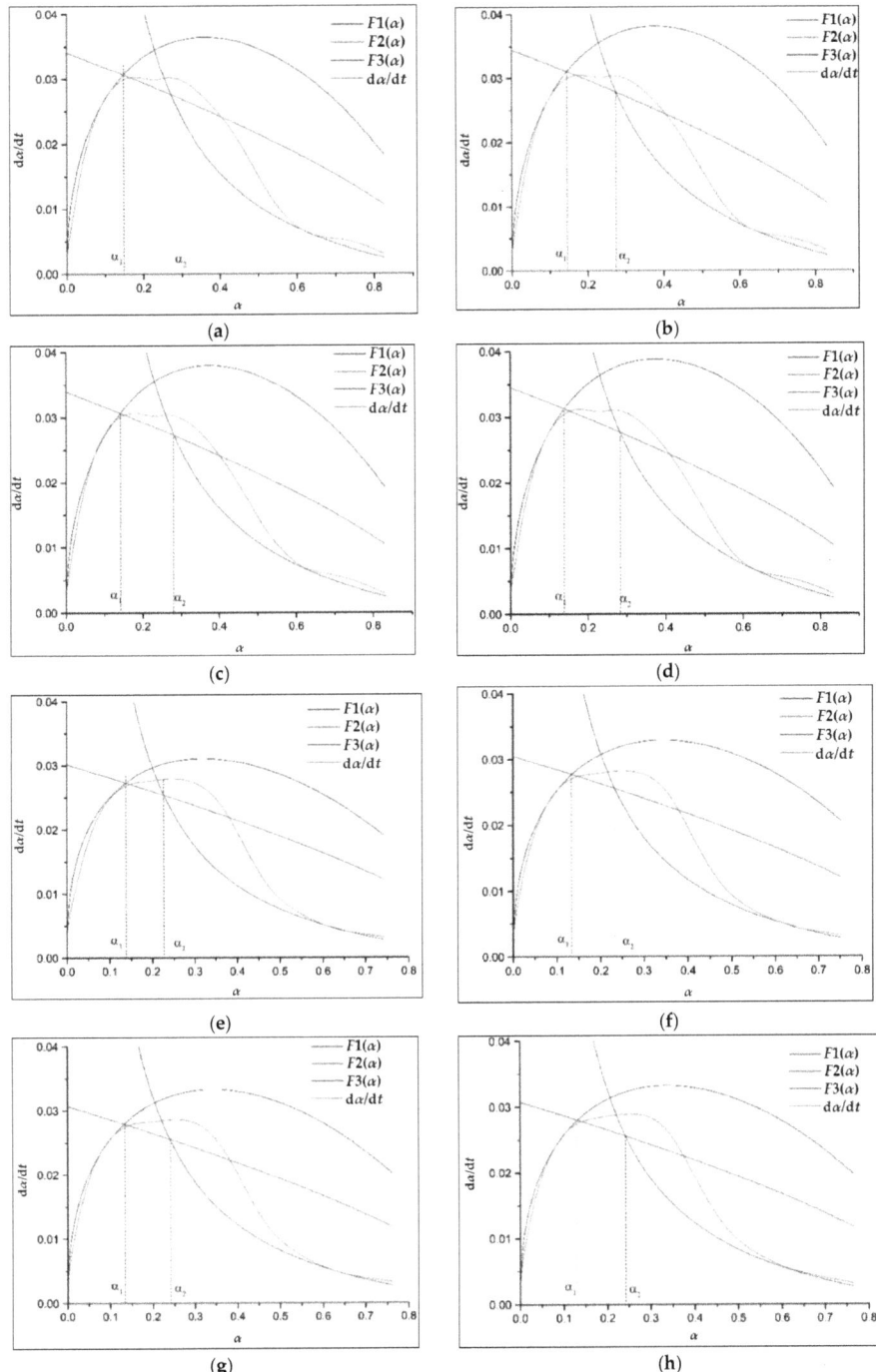

Figure 11. Hydration rate curves for samples of CGs and GCTs: (**a**) C-0; (**b**) C-1; (**c**) C-2; (**d**) C-3; (**e**) CT*-0; (**f**) CT*-1; (**g**) CT*-2; (**h**) CT*-3.

4. Conclusions

(1) Copper tailing powder, as a replacement for cement, reduces the early heat release rate and heat discharge of the cementitious system. The hydration rate and the total heat release improve with the decease of the content of copper tailing powder and the increase of the fineness of copper tailing powder. At the same time, graphene oxide can further improve the hydration rate and hydration heat of the cementitious system.

(2) The Krstulovic–Dabic kinetic model can be used to characterize the controlling process during the hydration. The hydration process of the composite cementitious materials containing copper tailing powder and graphene oxide is controlled by a multiple reaction mechanism, namely nucleation and crystal growth (NG), phase boundary reaction (I), and diffusion (D), in that order.

(3) The hydration rates, k_1', k_2', and k_3' of the three controlling processes reduce with the increase of the content of copper tailing powder, and improve with the increase of the fineness of copper tailing powder and the dosage of graphene oxide. Graphene oxide enhances the controlling effect on the nucleation and crystal growth process of the cementitious systems with or without copper tailing powder, thus promoting the early hydration.

Author Contributions: Conceptualization, S.L.; methodology, S.L.; validation, S.L.; formal analysis, Q.L. and X.Z.; investigation, Q.L.; resources, S.L.; data curation, Q.L.; writing—original draft preparation, Q.L. and X.Z.; writing—review and editing, S.L.; supervision, S.L.; project administration, S.L.; funding acquisition, S.L.

Funding: This research was funded by the National Key R&D Program of China (2016YFC0401907), and Hubei Province Technical Innovation Project (2018AAA028).

Conflicts of Interest: The authors declare no conflict of interest.

References

1. Gordon, R.B. Production residues in copper technological cycles. *Resour. Conserv. Recycl.* **2002**, *36*, 87–106. [CrossRef]
2. Onuaguluchi, O.; Özgur, E. Recycling of copper tailings as an additive in cement mortars. *Constr. Build. Mater.* **2012**, *37*, 723–727. [CrossRef]
3. Yuan, S.L. Ways and tasks of comprehensive utilization and disposal of metal mine solid wastes. *Express Inf. Mine Ind.* **2004**, *423*, 1–11.
4. Chen, J.B.; Li, R.J.; Yu, L.H. The way of the survey and assessment of copper tailings resources and their application. *J. Nat. Resour.* **2012**, *27*, 1373–1381.
5. Yu, L.H.; Jia, W.L.; Xue, Y.Z. Survey and analysis of the cooper tailing resources in China. *Met. Miner.* **2009**, *398*, 179–181.
6. Wang, Y.; Tian, J.; Zhu, Y.C. Research on development and exploitation of copper mine tailing. *Environ. Eng.* **2015**, *33*, 623–627.
7. Onuaguluchi, O.; Eren, Ö. Cement mixtures containing copper tailings as an additive: Durability properties. *Mater. Res.* **2012**, *15*, 1029–1036. [CrossRef]
8. Onuaguluchi, O.; Özgur, E. Reusing copper tailings in concrete: Corrosion performance and socioeconomic implications for the Lefke-Xeros area of Cyprus. *J. Clean. Prod.* **2016**, *112*, 420–429. [CrossRef]
9. Lv, S.H.; Ma, Y.J.; Qiu, C.C. Effect of graphene oxide nanosheets of microstructure and mechanical properties of cement composites. *Constr. Build. Mater.* **2013**, *49*, 121–127. [CrossRef]
10. Lv, S.H.; Ma, Y.J.; Qiu, C.C.; Ju, H.B. Effects of graphene oxide on microstructure of hardened cement paste and its properties. *Concrete* **2013**, *286*, 41–54.
11. Lv, S.H.; Ma, Y.J.; Qiu, C.C. Study on reinforcing and toughening of graphene oxide cement-based composites. *Funct. Mater.* **2013**, *15*, 2227–2231.
12. Lv, S.H.; Liu, J.J.; Qiu, C.C.; Ma, Y.J.; Zhou, Q.F. Microstructure and mechanism of reinforced and toughened cement composites by nanographene oxide. *Funct. Mater.* **2014**, *4*, 4084–4089.
13. Liu, Q.; Xu, Q.F.; Li, X.M. Research progress of graphene and carbon nanotubes cement-based composites. *China Concr. Cem. Prod.* **2016**, *239*, 25–30.

14. American Society for Testing and Materials, C618-15. *Standard Specification for Coal Fly Ash and Raw or Calcined Natural Pozzolan for Use in Concrete*; American Society for Testing and Materials: West Conshohocken, PA, USA, 2015.
15. Liu, S.H.; Wang, L.; Gao, Y.X. Influence of fineness on hydration kinetics of supersulfated cement. *Thermochim. Acta* **2015**, *605*, 37–42. [CrossRef]
16. Han, F.H.; Zhang, Z.; Wang, D.M. Hydration kinetics of composite binder containing slag at different temperatures. *J. Therm. Anal. Calorim.* **2015**, *121*, 815–827. [CrossRef]
17. Han, F.; He, X.; Zhang, Z.; Liu, J. Hydration heat of slag or fly ash in the composite binder at different temperatures. *Thermochim. Acta* **2017**, *655*, 202–210. [CrossRef]
18. Han, F.; Li, L.; Song, S.; Liu, J. Early-age hydration characteristics of composite binder containing iron tailing powder. *Powder Technol.* **2017**, *315*, 322–331. [CrossRef]
19. Liu, S.H.; Xie, G.S.; Wang, S. Effect of curing temperature on hydration properties of waste glass powder in cement-based materials. *J. Therm. Anal. Calorim.* **2015**, *119*, 47–55. [CrossRef]
20. Liu, S.H.; Xie, G.S.; Wang, S. Effect of glass powder on microstructure of cement pastes. *Adv. Cem. Res.* **2015**, *27*, 259–267. [CrossRef]
21. Liu, S.H.; Li, L.H. Influence of fineness on the cementitious properties of steel slag. *J. Therm. Anal. Calorim.* **2014**, *117*, 629–634. [CrossRef]
22. Bezjak, A. Nuclei growth model in kinetic analysis of cement hydration. *Cem. Concr. Res.* **1986**, *16*, 605–609. [CrossRef]
23. Krstulovic, R.; Dabic, P. A conceptual model of the cement hydration process. *Cem. Concr. Res.* **2000**, *30*, 693–698. [CrossRef]
24. Wang, X.Y.; Lee, H.S.; Park, K.B.; Golden, J.S. A multi-phase kinetic model to simulate hydration of slag–cement blends. *Cem. Concr. Compos.* **2010**, *32*, 468–477. [CrossRef]
25. Escalante-Garcia, J.I.; Fuente, A.F.; Alexer, G.; Fraire-Luna, P.E.; Guillermo, M. Hydration products and reactivity of blast-furnace slag activated by various alkalis. *J. Am. Ceram. Soc.* **2003**, *86*, 2148–2153. [CrossRef]
26. Schutter, G.D.; Taerwe, L. General hydration model for Portland cement and blast furnace slag cement. *Cem. Concr. Res.* **1995**, *25*, 593–604. [CrossRef]
27. Fernandez, J.A.; Puertas, F. Alkali-activated slag cements: Kinetic studies. *Cem. Concr. Res.* **1997**, *27*, 359–368. [CrossRef]
28. Fernandez, J.A.; Puertas, F.; Arteaga, A. Determination of kinetic equations of alkaline activation of blast furnace slag by means of calorimetric data. *J. Therm. Anal. Calorim.* **1998**, *52*, 945–955. [CrossRef]
29. Bullard, J.W. A determination of hydration mechanisms for tricalcium silicate using a kinetic cellular automaton model. *J. Am. Ceram. Soc.* **2008**, *91*, 2088–2097. [CrossRef]
30. Bullard, J.W.; Flatt, R.J. New insights into the effect of calcium hydroxide precipitation on the kinetics of tricalcium silicate hydration. *J. Am. Ceram. Soc.* **2010**, *93*, 1894–1903. [CrossRef]
31. Lothenbach, B.; Scrivener, K.; Hooton, R.D. Supplementary cementitious materials. *Cem. Concr. Res.* **2011**, *41*, 1244–1256. [CrossRef]
32. Thomas, J.J.; Biernacki, J.J.; Bullard, J.W.; Bishnoi, S.; Dolado, J.S.; Scherer, G.W.; Luttge, A. Modeling and simulation of cement hydration kinetics and microstructure development. *Cem. Concr. Res.* **2011**, *41*, 1257–1278. [CrossRef]

© 2018 by the authors. Licensee MDPI, Basel, Switzerland. This article is an open access article distributed under the terms and conditions of the Creative Commons Attribution (CC BY) license (http://creativecommons.org/licenses/by/4.0/).

Article

Scanning Rate Extension of Conventional DSCs through Indirect Measurements

Hannes Fröck [1], Michael Reich [1,*], Benjamin Milkereit [1,2] and Olaf Kessler [1,2]

[1] Chair of Materials Science, Faculty of Mechanical Engineering and Marine Technology, University of Rostock, Justus-von-Liebig-Weg 2, 18059 Rostock, Germany; hannes.froeck@uni-rostock.de (H.F.); benjamin.milkereit@uni-rostock.de (B.M.); olaf.kessler@uni-rostock.de (O.K.)
[2] Competence Centre CALOR, Department Life, Light & Matter, Faculty of Interdisciplinary Research, University of Rostock, Albert Einstein-Str. 25, 18059 Rostock, Germany
* Correspondence: michael.reich@uni-rostock.de; Tel.: +49-381-498-9490

Received: 8 March 2019; Accepted: 28 March 2019; Published: 2 April 2019

Abstract: In this work, a method is presented which allows the determination of calorimetric information, and thus, information about the precipitation and dissolution behavior of aluminum alloys during heating rates that could not be previously measured. Differential scanning calorimetry (DSC) is an established method for in-situ recording of dissolution and precipitation reactions in various aluminum alloys. Diverse types of DSC devices are suitable for different ranges of scanning rates. A combination of the various available commercial devices enables heating and cooling rates from 10^{-4} to 5 Ks^{-1} to be covered. However, in some manufacturing steps of aluminum alloys, heating rates up to several 100 Ks^{-1} are important. Currently, conventional DSC cannot achieve these high heating rates and they are still too slow for the chip-sensor based fast scanning calorimetry. In order to fill the gap, an indirect measurement method has been developed, which allows the determination of qualitative information, regarding the precipitation state, at various points of any heat treatment. Different rapid heat treatments were carried out on samples of an alloy EN AW-6082 in a quenching dilatometer and terminated at defined temperatures. Subsequent reheating of the samples in the DSC enables analysis of the precipitation state of the heat-treated samples. This method allows for previously un-measurable heat treatments to get information about the occurring precipitation and dissolution reactions during short-term heat treatments.

Keywords: EN AW-6082; AlMgSi alloy; differential scanning calorimetry; fast scanning; scanning rate extension; indirect measurements; dissolution; precipitation; time-temperature-dissolution diagram

1. Introduction

Differential scanning calorimetry is an established technique used to record precipitation and dissolution reactions of various aluminum alloys in-situ, during heating [1,2] as well as cooling [3–6]. From this data, continuous time-temperature-dissolution or precipitation diagrams can be created, showing the dissolution or precipitation behaviour of the investigated alloy. Conventional DSC devices cover a large range of both heating and cooling rates, from 10^{-4} up to 5 Ks^{-1}, as shown in Figure 1. In contrast, chip-based fast scanning calorimeters require a minimum heating rate of 10^3 Ks^{-1} [7]. Thus, it turns out that, there is a gap in measurable scanning rates between about 5 and 1000 Ks^{-1}. These scanning rates, though, are in an industrially relevant range. In different manufacturing steps, such as laser heat treatment [8] or welding [9], the material passes through a short-term heat treatment with heating rates up to several 100 Ks^{-1}. During this short-term heat treatment, the precipitation state, and thus the mechanical properties of the material, potentially changes. Furthermore, conventional calorimeters are mainly suitable for the measurement of precipitation or dissolution reactions during linear scanning steps. However, most real heat treatments show a non-linear time temperature course.

The possibilities of recording non-linear cooling processes in a DSC has been shown, but these are also limited by the device-specific maximum cooling rates [10].

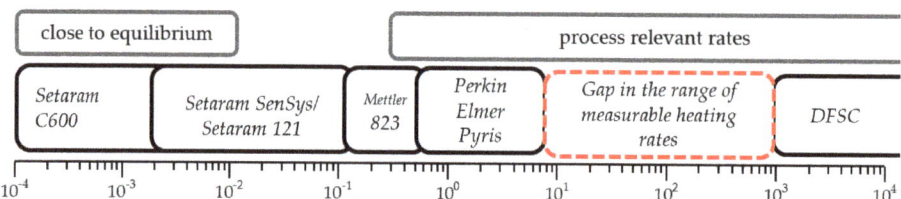

Figure 1. Scanning rate range of different conventional differential scanning calorimetry (DSC) devices and non-conventional fast scanning calorimeters.

It is known from the literature that information about the precipitation behaviour, during a heat treatment, can be obtained from a subsequent reheating in a calorimeter. Zohrabyan et al. [11] described an approach to obtain information from previously unmeasurable cooling rates. This approach has been used to determine the critical cooling rate of a high alloyed 7XXX aluminum alloy using the fast scanning chip calorimetry (FSC). Schumacher et al. [12] took up this approach and used the reheating to determine the enthalpy change following the cooling of samples, with extremely slow rates, down to 3×10^{-5} Ks^{-1}. By this method, it is only possible to measure the enthalpy change during the whole cooling step. The information at which temperatures a reaction takes place, during cooling, cannot be determined. Another development of the reheating approach was done by Yang et al. [13], who carried out the reheating after multiple interrupted quenching operations, which allowed them to obtain temperature dependent data by applying the FSC technique.

This paper aims to introduce a calorimetric method by which the dissolution and precipitation reactions, during any time-temperature curves, can be recorded in conventional DSCs. For this purpose, samples received multiple and defined previous heat treatment and subsequent reheating in a DSC. The initial heat treatments are carried out in a separate device which can realise very flexible temperature-time courses, including high heating and cooling rates, e.g., a quenching dilatometer. This produces various defined initial conditions for subsequent reheating in the DSC. The final DSC reheating curves can be used to conclude on the precipitation and dissolution reactions during the preceding heat treatments.

2. Materials and Methods

2.1. DSC Curve Reconstruction for (Fast) Linear Heating

For the direct heating experiments in a conventional DSC, the heating rates were varied from 0.01 to 5 Ks^{-1}. In order to investigate higher heating rates, a new and indirect reheating measuring methodology was developed. For this purpose, samples with the heating rate to be examined are heated to various temperatures and immediately quenched. These samples were subsequently reheated using a heating rate ideal for DSC and taking up the reheating curve. The reheating raw DSC data is treated as explained in Section 2.5. By comparing the reheating curves, which were recorded after different maximum temperatures during the previous heat treatment, conclusions about the reactions occurring during the initial heat treatment can be drawn. From the large number of investigated maximum temperatures, a virtual DSC heating curve can be reconstructed qualitatively.

To ensure the suitability of this method, some constraints must be met, as follows:

- The precipitation state of the material may not change during the cooling step. Overcritical quenching is necessary to suppress precipitation reactions during cooling.
- The precipitation state of the material may not change during the intermediate time between initial heat treatment and reheating. Store the samples in a freezer at a low temperature.

To fulfill the above requirements, samples were heated in a quenching dilatometer BÄHR 805 A/D (BÄHR Thermoanalyse GmbH, Hüllhorst, Germany), at a constant heating rate, to different maximum temperatures. After reaching the maximum temperature, without soaking, the samples were quenched using the maximum possible gas flow. The achieved cooling rates are higher than 100 Ks^{-1}. Considering the actual composition of the batch of 6082, it can be assumed that 100 Ks^{-1} is above the upper critical cooling rate [6,14]. The maximum temperatures during heating were increased in discrete steps of 25 K, within a range of 150 to 575 °C. For each maximum temperature, four fresh samples in initial state T4 were used. After the initial heat treatment, the samples were placed in a freezer at −80 °C, to prevent natural ageing. These samples were finally reheated in a DSC at a heating rate of 1 Ks^{-1}. The schematic time-temperature course applied for the indirect DSC measurements is shown in Figure 2.

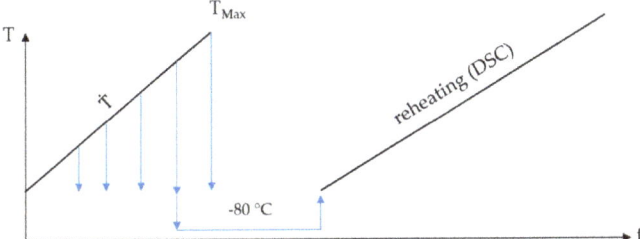

Figure 2. Schematic time-temperature profile of the heat treatments for the indirect measurements.

The reheating method is used to reconstruct DSC curves. It was validated by comparing a reconstructed DSC curve with one measured by in-situ heating DSC. For this purpose, samples were heated in the dilatometer at 1 Ks^{-1} to the different maximum temperatures and quenched. The examined heating rate of 1 Ks^{-1} can be investigated by direct measurements in the DSC. This comparison can be used to establish, and justify, how a reheating curve changes when reactions take place during a previous heat treatment.

Subsequently, higher heating rates of 20 and 100 Ks^{-1} were examined by the method of indirect measurements. As a result, the heating curves of 20 and 100 Ks^{-1} were reconstructed. Those rates were not directly assessable. Table 1 shows the heat treatment parameters used for the indirect DSC measurements. In this work the results of more than 700 single DSC measurements are reported.

Table 1. Heat treatment parameters for indirect measurements.

Previous Heating in Quenching Dilatometer				Reheating in DSC	
Heating-Rate	Investigated Temperature Range	Temperature Step Size	Quenching Rate	Reheating-Rate	Max. Reheating Temperature
1 Ks^{-1} 20 Ks^{-1} 100 Ks^{-1}	150–575 °C	25 K	>100 Ks^{-1}	1 Ks^{-1}	600 °C

2.2. Assessment of a Non-Linear Heat Treatment by DSC on the Example of a Laser Heating and a Welding Process

In addition to the investigation of linear heating processes, the presented method is also suitable for the investigation of non-linear heat treatments. This is shown by the example of a laser short-term heat treatment, as well as for the heat affected zone (HAZ) during a welding process.

Figure 3 shows the time temperature profile recorded during a laser short-term heat treatment of an aluminium extrusion profile (EN AW-6060) with a wall thickness of 2 mm. The laser heat treatment, as well as the recording of the time-temperature-profile, was carried out at the Institute of Manufacturing Technology of the University of Erlangen-Nürnberg [15]. The laser heat treatment is characterized by a high heating rate up to several 100 Ks^{-1}, with no soaking at the maximum temperature and a relatively slow, non-linear, cooling with a few 10 Ks^{-1}.

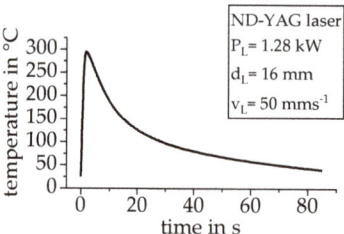

Figure 3. Time-temperature profile of a laser heat treatment at an aluminum extrusion profile with a wall thickness of 2 mm.

Another example of a heat treatment that cannot be measured directly by calorimetry is welding. In the weld, as well as in the heat affected zone, high heating rates occur. Additionally, there is no soaking at the maximum temperature and rapid cooling follows directly. In these areas, the precipitation state, and therewith, the mechanical properties, of aluminum alloy potentially change drastically during welding. It is therefore of interest to be able to characterize the dissolution and precipitation behaviour during welding. The presented indirect method is also suitable for this problem.

SARMAST et al. [16] show some characteristic simulated time-temperature curves in the heat-affected zone during arc welding of thick aluminum sheets. These non-linear temperature profiles were imitated again in the quenching dilatometer, interrupted at selected points by an overcritical quenching, and then frozen at −80 °C. Subsequently, the samples obtained were reheated in the DSC.

2.3. Investigated Aluminium Alloy

The investigated material was a hollow and quadratic aluminum extrusion profile (40 × 40 × 3) mm^3 made from an alloy EN AW-6082. As the initial state, the naturally aged T4 state was examined. The aluminum alloy EN AW-6082 was chosen as a typical extrusion alloy, which finds application in various technical fields. The chemical composition was analysed by optical emission spectroscopy (OES) and is shown in Table 2. Samples were machined from the profile under coolant supply. Preliminary tests have shown that the temperature during machining, thereby, is kept below 30 °C, such that the naturally aged initial-state remains. The method described below is later transferred to other alloys. The compositions of the other investigated alloys are also given in Table 2. The starting material of alloy EN AW-6060 was also a hollow quadratic aluminum extrusion profile (20 × 20 × 2) mm^3 in the natural-aged state T4, whereas, for welding, the alloy EN AW-6082 was a plate with a thickness of 10 mm in the artificial aged state T651.

Table 2. Mass fraction of the alloying elements in the investigated alloys in %.

Alloy	Mass Fraction in %							
	Si	Fe	Cu	Mn	Mg	Cr	Zn	Al
OES EN AW-6082 T4	0.94	0.19	0.05	0.58	0.76	0.08	0.20	balance
OES EN AW-6082 T651	0.83	0.38	0.06	0.48	0.92	0.03	0.01	balance
DIN EN 573-3 (6082)	0.7–1.3	≤0.5	≤0.1	0.4–1.0	0.6–1.2	≤0.25	≤0.2	balance
OES EN AW-6060 T4	0.40	0.22	0.07	0.14	0.56	0.02	0.02	Balance
DIN EN 573-3 (6060)	0.5–0.9	≤0.35	≤0.3	≤0.5	0.4–0.7	≤0.3	≤0.2	balance

2.4. Differential Scanning Calorimetry (DSC)

Differential scanning calorimetry has been used for two purposes, direct in-situ heating experiments for comparison with the new reheating method as well as reheating of previously heat-treated samples. Two different DSC device types were used to cover a wide range of heating rates. The slow measurements from 0.01 to 0.1 Ks^{-1} were performed in the Calvet-Type heat flow DSC Setaram S 121 (Setaram, Caluire-et-Cuire, France). For this device, typically, cylindrical samples with a diameter of 6.0 mm and a length of 21.65 mm are used. Due to the small wall thickness of the hollow profiles, such samples could not be machined. For this reason, seven samples with a diameter of 6.0 mm and a height of 3 mm were stacked into two pure aluminum crucibles. This method of stacked samples has been proven to give the same results as bulk samples [1]. The samples have a total mass of around 1580 mg. The references were also made of stacked samples of similar dimensions from high purity Al5N5 (99.9995%) aluminum.

The faster heating tests of 0.3 to 5 Ks^{-1} were performed in two power-compensated PerkinElmer DSCs, a Pyris Diamond DSC and a PerkinElmer DSC 8500 (PerkinElmer, Waltham, MA, USA). The samples for these instruments have a cylindrical shape with a diameter of 6.4 mm and a height of 1 mm. This results in a sample mass of about 80 mg. In order to keep the radiation properties of the sample as constant as possible over the whole measurement, the samples were packed in a pure aluminum crucible and covered with a pure aluminum lid. A detailed description of the different types of calorimeters used can be found in Reference [17]. All reheating experiments, for the indirect DSC measurements, were performed in the power-compensated DSC at a constant heating rate of 1 Ks^{-1}.

It should be mentioned that, for a proper evaluation, the used DSC devices must be precisely calibrated regarding heat flow [18] and temperature [19]. The latter is especially true for the direct heating measurements with device-specific high heating rates, e.g., 3 and 5 Ks^{-1} in the power compensated PerkinElmer DSC, as thermal lag correction is required here.

2.5. Data Processing of Raw Measured Heat Flow Curves

Despite all care to maintain the symmetry between sample and the reference during the measurement, the results show a device-specific basic curvature. This base curvature can be removed by subtracting the heat flow of a baseline measurement (reference sample versus reference sample, $\dot{Q}_{Bl}$) from the sample measurement of heat flow (alloyed sample versus reference sample, $\dot{Q}_S$). The basic curvature of the different DSC-devices changes slightly with time, especially in the power-compensated DSCs. In order to have a close in time baseline for each sample measurement, the measurement scheme sample-baseline-sample was used.

In order to be able to compare the results of different masses and scan rates, it is necessary to normalize the recorded curves. For this purpose, the differential heat flow is normalised by the sample mass (m_S) and the scan rate (β) according to Equation (1) [17]. Further evaluations are carried out on the resulting excess specific heat capacity curves ($C_{p_{excess}}$).

$$C_{p_{excess}} = \frac{\dot{Q}_S - \dot{Q}_{Bl}}{m_S \cdot \beta} \text{ in } (J\,g^{-1}K^{-1}) \tag{1}$$

The raw data treatment is illustrated in Figure 4. The basis is the high-quality raw DSC data, as shown in Figure 4A. The baseline is subtracted from the raw heat flow data of the sample measurement, eliminating the device specific curvature, as seen in Figure 4B. When switching between isothermal soaking to heating and again to isothermal soaking at high temperatures, overshoot artefacts appear in the curves. These overshoot artefacts must be removed from the measurement data to avoid misinterpretations. In the next step, the data is normalised, according to Equation (1), to the unit of specific excess heat capacity. Despite due diligence when conducting the measurements, there may be small deviations between the individual measurements. This is, for example, caused by slight variations of the sample or furnace lid positions. These deviations cannot be avoided and lead to a further slight curvature of the measured data, particularly its zero level [1]. By fitting and subtracting a 3rd order polynomial, the measured data can be corrected for this zero level curvature, as shown in Figure 4C [1]. Figure 4D shows the average curve from the four processed single measurements and the scattering of the measurements. The data scattering is illustrated by plotting the DSC curves with minimum, as well as maximum, values and shading the area in-between. In the further results, only the mean value curves are shown, because the deviations between the individual measurements are small.

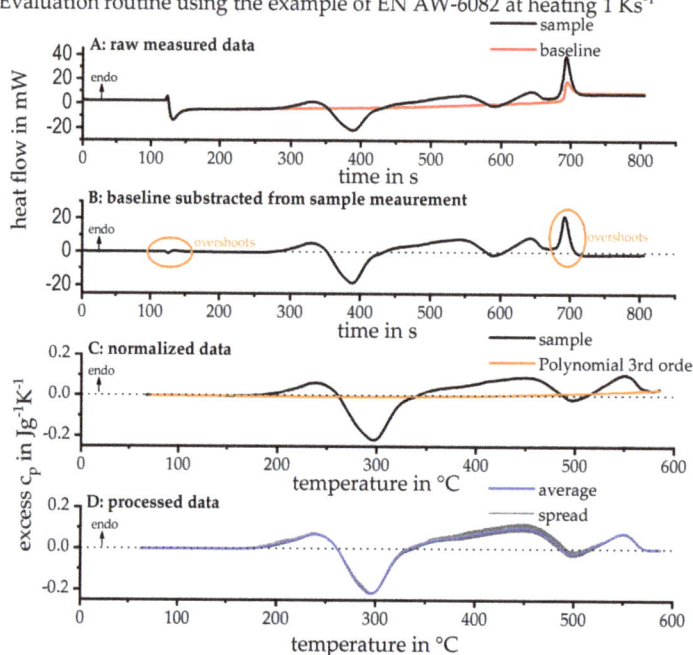

Figure 4. Raw data treatment of DSC heating experiments. (**A**) Raw heat flow measurement data; (**B**) subtraction of baseline measurement from the sample measurement; (**C**) normalization of the data; (**D**) processed data ready for discussion.

It should be noted that the polynomial fit can only be performed if there is a reaction-free zone at the beginning, as well as at the end, of the graph. In order to achieve this, it is necessary to carry out the heating experiments up to the highest possible temperatures, particularly above the heating rate specific solvus temperature. This allows the completion of all the precipitation and dissolution reactions. However, melting of the sample could, potentially, damage the calorimeter. Therefore, DSC melt tests were carried out before, using ceramic crucibles. These preliminary tests show a solidus temperature of 607 °C at a heating rate of 1 Ks^{-1} for the batch 6082 T4. Thus, for the heating experiments, a maximum temperature of 600 °C has been used.

In the illustrations, the associated zero-level for each individual DSC heating curve is indicated by a dashed line. A deviation of the measurement curve above this zero-level represents the predominance of endothermic reactions, i.e. the dissolution of precipitates. A deviation below the zero-level represents the predominance of exothermic reactions, and thus the formation of precipitates.

During heating of an age hardening Al alloy in a certain metastable initial condition, typically, a sequence of alternating precipitation and dissolution reactions is seen [1,20–23]. This report intends to create continuous time-temperature dissolution diagrams. For this purpose, the single precipitation and dissolution reactions need to be separated in a certain way. This evaluation was carried out as illustrated in Figure 5 and initially described by Osten et al. [1]. For this, zero crossings of the DSC curves are evaluated. However, due to the severe superposition of the individual microstructural reactions, the applied peak separation is physically meaningless. A more meaningful separation of the distinct reactions would require kinetic modelling of the superimposed reactions, which is beyond the scope of the present paper. Therefore, at present, the applied peak separation is the method of best practice.

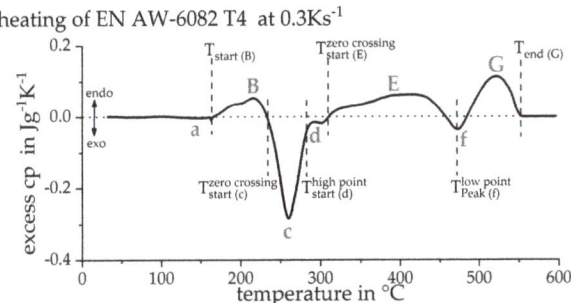

Figure 5. Separation and designation of the individual peaks using the example of a heating rate of 0.3 Ks^{-1}.

As a further complication, due to the severe superposition of exothermic precipitation and endothermic dissolution, some peaks do not reach the zero line. However, over the whole range of heating rates studied, it becomes clear that these reactions are occurring. In such cases, the peak temperature of the reaction is evaluated. From the characteristic temperatures determined, the associated time is calculated by means of the heating rate. These values are then plotted into the continuous time-temperature dissolution diagram for each heating rate [1].

2.6. Precipitation and Dissolution Reactions

In order to interpret the reheating curves, it is necessary to understand the occurring precipitation and dissolution reactions. It cannot be deduced from the DSC data which phases are precipitated or dissolved in a certain temperature range. For this purpose, further investigations, such as electron microscopy, are necessary. However, the AlMgSi alloy system has been investigated to a wide extent and the occurring peaks can be assigned to occurring reactions, based on literature data.

In the initial naturally aged state, there might be solved alloying elements in solid solution. With a slight increase in temperature, these can form clusters. The weak exothermic peak (a) is therefore generally interpreted as the precipitation of clusters resulting from the residual potential of solved alloying elements [24]. This is followed by the endothermic peak (B), which probably consists of two separate reactions. It represents the dissolution of clusters and Guinier Preston (GP)-zones that have formed during previous natural aging [25]. The dissolution reactions underlying this peak have an enormous influence on the mechanical properties. In this temperature range, the strength-increasing nano-particles of natural ageing are dissolved, causing a significant softening of the material [2]. At higher temperatures, the exothermic peaks (c) and (d) appear. These two peaks are commonly interpreted as the precipitation of the metastable phases β″ and β′ [26]. Precipitates of the β″ phase can provide the highest strength, while the precipitation of the phase β′ corresponds to the overaged state. The endothermic peak (E) is considered to be the dissolution of the previously formed metastable phases β″ and β′ [27]. Following, the exothermic peak (f) represents the precipitation of the equilibrium phase β-Mg$_2$Si [28]. At high temperatures, the endothermic peak (G) appears. This is considered to be the dissolution of the equilibrium phase β-Mg$_2$Si and other remained phases [29]. With the finish of the peak (G), the DSC signal drops to zero, which indicates the scanning rate specific solvus temperature. Thus, the evaluated finish temperature of peak (G) at least has a physical meaning.

2.7. Previous Heat Treatment in a Quenching Dilatometer

In order to realize very flexible initial heat treatments, a highly dynamic measuring setup is necessary. Conventional DSCs cannot meet these requirements. For this reason, the initial heat treatment, before reheating, is carried out in a Bähr 805A quenching dilatometer. In this device, the samples are placed in the centre of an induction coil. Defined heating rates of several 100 Ks^{-1} can be achieved. The sample temperature is continuously controlled via a spot-welded thermocouple on the surface of the sample. The samples are clamped between two quartz glass rods and held in position. The induction coil is double-walled. Gas can be passed through the inner perforated wall. Thus, cooling can take place via gas quenching and cooling rates of a few 100 Ks^{-1} can be achieved. The schematic heat treatment setup in the dilatometer is shown in Figure 6.

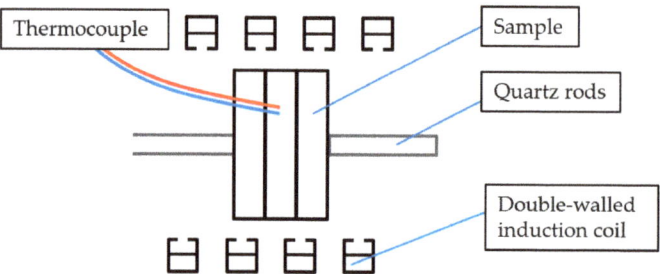

Figure 6. Schematic of the Bähr 805A dilatometer.

During heat treatment in the dilatometer, three individual DSC samples are stacked, with the thermocouple being connected to the centre sample. During the resistance spot-welding process, the sample already undergoes a certain heat treatment, with locally high temperatures changing its initial state. Thus, further investigation of this sample is not reasonable. For this reason, the centre sample, equipped with a thermocouple, is used as a temperature control dummy and is not considered for further evaluation.

In order to evaluate the temperature distribution between the three individual samples, all three samples were provided with thermocouples in preliminary tests. Figure 7 shows the maximum deviation from the target temperature (ΔT), during a heating experiment, towards distinct maximum temperatures, using a heating rate of 20 Ks^{-1}. It can be seen that the maximum temperature deviates from the target temperature by, at most, 5 K. However, most samples deviate by less than 2 K. This error is considered to be low.

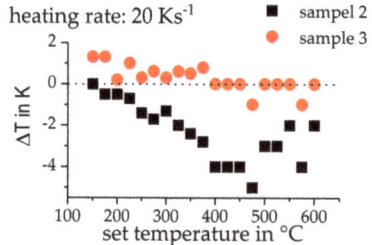

Figure 7. Deviation of the maximum temperature of the unregulated samples from the target temperature.

3. Results and Discussion

3.1. Direct DSC-Measurements

Figure 8 shows the directly measured DSC heating curves of the alloy EN AW-6082, initially in the naturally aged T4 state. A heating rate range of 0.01 to 5 Ks^{-1} is considered. It can be seen that the peak (B) is shifted to higher temperatures with an increasing heating rate. It also appears as if the peak (B) becomes stronger with an increasing heating rate. Peak (B) represents the dissolution of clusters and GP-zones from the initial state. The content of these phases is constant before each measurement. Assuming that all these phases are dissolved during heating, this dissolution peak should also be constant in intensity over all the heating rates. Though, it is to be expected that any diffusion-controlled dissolution or precipitation reaction is suppressed kinetically. The behaviour of peak (B) is therefore explained by the overlap with the opposite exothermic peak (c). Peak (c) is also shifted to higher temperatures with an increasing heating rate. The intensity of peak (c) is also increasing, up to the heating rate of 0.5 Ks^{-1}. At low heating rates, the peak temperatures of both reactions are very close to each other. For this reason, it can be considered that peaks (B) and (c) overlap very strongly at low heating rates and partly compensate each other in the sum of their heat flows. By increasing the heating rate, the peak temperatures drift apart. As a result, the overlap of the two peaks decreases and both seem to gain in intensity. With a further increase of the heating rate above 0.5 Ks^{-1}, it becomes clear that peak (c) becomes blurred with peak (d) and becomes weaker overall. This is due to the increasing suppression of precipitation reactions with increasing heating rate, associated with less time for diffusion. It also becomes clear that the precipitation reaction (c + d) is suppressed earlier than the dissolution reaction (B), as already described in the literature [1].

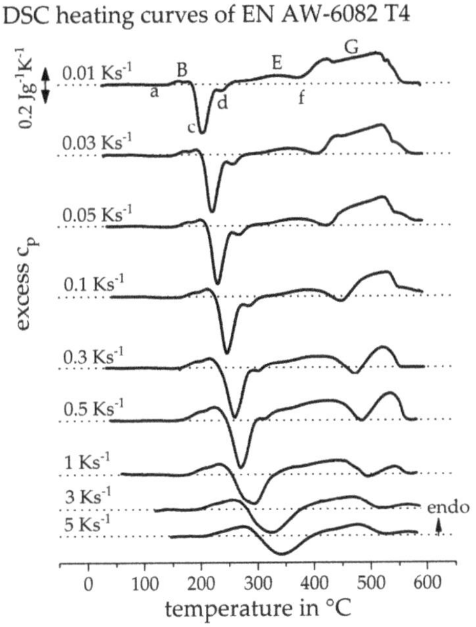

Figure 8. Direct DSC heating curves of EN AW-6082 T4 with the conventional DSC. Peak assignments are stated for the heating rate of 0.01 Ks^{-1}.

The endothermic peak (E) is also strongly influenced by the superposition with the exothermic peaks (c), (d), and (f). The precipitation peak (f) of the equilibrium phase Mg$_2$Si, at low heating rates, remains completely in the endothermic area. At this point, the strong overlap with other peaks is evident. Peak (E) appears to have a low intensity, even at low cooling rates. During this peak, the precipitates, formed in peaks (c + d), are dissolved. Peak (E) should, therefore, have a similar intensity as peaks (c + d). At low heating rates, the intensity of peak (E) is much lower than peak (c + d). It can be considered, therefore, that the dissolution of phases β″and β′ is not yet complete when the precipitation of β already begins. The endothermic dissolution reactions, as well as the exothermic precipitation reaction, overlap. Nevertheless, peak (f) shifts to higher temperatures with an increasing heating rate and is only very weak at high heating rates. It can be seen that the precipitation reaction is increasingly suppressed at higher heating rates.

The final dissolution reaction (G) also appears to lose intensity with an increasing heating rate. During this dissolution reaction, it is mainly the precipitates that have been formed during previous heating which are dissolved. This peak should become weaker if the previous precipitation reactions are suppressed with an increasing heating rate. During slow heating, coarse precipitates are formed, which must be dissolved afterwards. Increasing the heating rate should result in finer precipitates, which dissolve faster. For this reason, final dissolution of precipitates seems to be completed even at higher heating rates.

3.2. Validation of Indirect Rehaeting Method Versus the Known Heating Curve of 1 Ks^{-1}

The results of the newly developed indirect measurement methodology are presented in Figure 9 for previous heating with 1 Ks^{-1}. On the left side, the reheating DSC curves after certain previous heat treatments up to different maximum temperatures, are shown. The reheating curves, after the lowest maximum temperature, during the initial heat treatment are shown at the bottom and, with increasing maximum temperature, the curves are arranged in ascending order. The maximum temperature during the initial heat treatment is given beneath each reheating curve, In order to analyze the reheating curves different sections are defined, which are separated by grey lines. The sections are labelled with Roman numerals I–V. On the right side, the directly measured DSC heating curve for 1 Ks^{-1} is shown. This curve is rotated by 90° compared to the reheating curves so that its temperature scale coincides with the maximum temperatures of the initial heat treatment and a direct comparison is possible.

In section I, up to 175 °C, no change in the DSC reheating curves can be seen compared to the initial state. It can be concluded that no reactions took place during the initial heat treatment. This is confirmed by the direct DSC heating curve.

In section II, up to 250 °C, it can be seen that the peak (B) becomes continuously weaker as the maximum temperature rises during the initial heat treatment. It can be concluded that during the initial heat treatment, some clusters and GP-zones were already dissolved. The proportion of dissolved phases rises, in section II, with increasing maximum temperature. During the subsequent reheating, less or none of these phases must be dissolved and peak (B) weakens steadily. This can also be proved by the direct measured DSC.

In the reheating curves in section III, up to 325 °C, peak (B) does not appear. All precipitates of the initial state were dissolved in the temperature range of section II. It becomes clear that peak (c + d) in the reheating curves becomes smaller in section III as the maximum temperature, during the initial heat treatment, rises. The precursor precipitates β″and β′ are formed during this peak and with a rising maximum temperature, their fraction increases. If these precipitates are already formed during the initial heat treatment, the potential for their precipitation during reheating decreases. It can be concluded that in the temperature range of section III, during the initial heat treatment, the phases β″ and β′ are formed. This is again confirmed by the directly measured DSC heating curve.

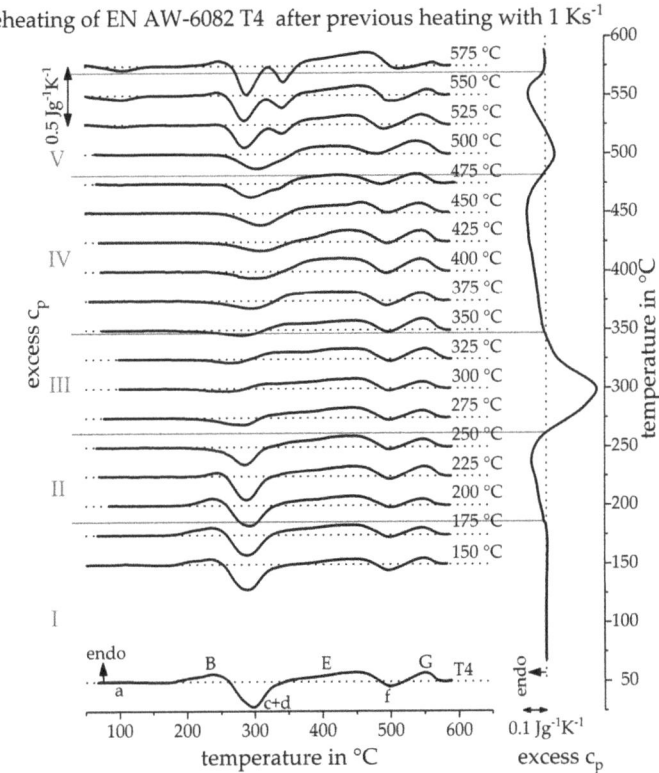

Figure 9. DSC reheating curves, after an initial heat treatment with a heating rate of 1 Ks^{-1}, up to different maximum temperatures and the direct measured DSC heating curve at 1 Ks^{-1} in comparison.

In section IV, up to 475 °C, the intensity of peak (c + d) increases, again with increasing maximum temperature during the initial heat treatment. This behavior can be explained by the dissolution of the phases β″ and β′. When these are dissolved during the initial heat treatment, the concentration of the aluminum solid solution, before the reheating step, increases. Thus, during reheating, the precipitates of peak (c + d) may form again. The more phases are dissolved during the initial heat treatment, the more alloying elements are solved, and the more precipitates can form during reheating. Therefore, it can be concluded, from the reheating curves, that mainly the phases β″ and β′ dissolve in section IV during the initial heat treatment. This is indicated by the directly measured reheating curve.

In the reheating curves of section V, a new peak (a) occurs at about 100–150 °C, peak (B) appears weak, peak (c + d) divides into two single peaks, and peak (G) becomes progressively smaller with an increasing maximum temperature of the initial heat treatment. Due to high maximum temperatures, of at least 500 °C during the initial heat treatment in section V, a large part of the phases has already been dissolved and a high content of solved alloying elements is present in the solid solution before reheating. Already, at about 100 °C during reheating, some clusters can form, which is reflected in peak (a). It can be concluded that, during the initial heat treatment, phases with high solvus temperatures dissolve. This is supported by the directly measured heating curve in which the final dissolution peak (G) occurs.

This series of measurements has shown the feasibility of the proposed indirect DSC method. From reheating, the temperature ranges can be identified at which precipitation and dissolution reactions take place during the initial heat treatment. Therefore, by applying this procedure, unknown or unavailable DSC heating curves can be reconstructed qualitatively.

3.3. Reconstruction of DSC Heating Curve for 20 Ks^{-1} and 100 Ks^{-1}

Figure 10 shows the DSC reheating curves after an initial heat treatment, with a heating rate of 20 Ks^{-1}, up to different maximum temperatures (left) and the resulting, reconstructed, direct DSC heating curve at 20 Ks^{-1} (right). In section I, up to a maximum temperature of 200 °C, the reheating curves do not change in their course. It can be concluded that during heating of the alloy, with a rate of 20 Ks^{-1} up to the temperature of 200 °C, no reactions occur. This finding was adopted in the reconstructed heating curve, as seen on the right side of the diagram.

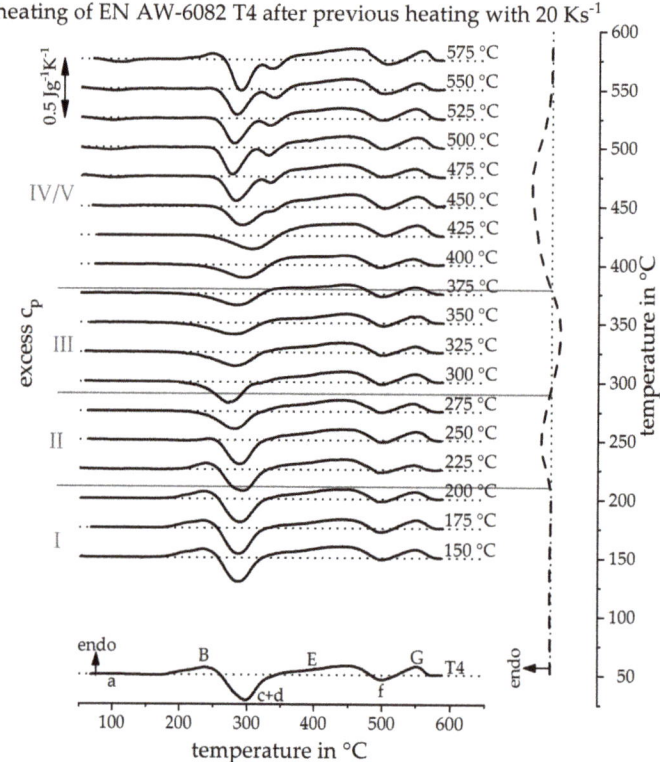

Figure 10. Reheating curves after a heat treatment with a heating rate of 20 Ks^{-1} up to different maximum temperatures and the resulting reconstructed heating curve at 20 Ks^{-1}.

In section II, from 225 °C to 275 °C, the heating curves show that the dissolution peak (B) becomes continuously smaller with increasing maximum temperature of the initial heat treatment. It can be concluded that, during the initial heat treatment with 20 Ks^{-1}, in this temperature range, the clusters and GP zones of the initial state are dissolved. For this reason, the endothermic dissolution peak (B) is plotted in the reconstructed DSC heating curve in this temperature range.

In section III, with reheating curves after an initial heat treatment with the maximum temperatures of 300 °C up to 375 °C, peak (c + d) continuously gets smaller with an increasing maximum temperature during the initial heat treatment. It can be concluded that in this temperature range, during the initial heat treatment, β″ and β′ precipitates were formed. For this reason, in the temperature range of 300 °C up to 375 °C in the reconstructed DSC heating curve, an exothermic precipitation reaction of phases β″ and β′ is assumed.

Sections IV and V are considered together. In the reheating curves after a maximum temperature above 375 °C during the initial heat treatment, peak (c + d), on the one hand, regains its intensity with increasing maximum temperature and, on the other hand, splits into two individual peaks (c) and (d). Furthermore, a weak precipitation reaction, peak (a), occurs during reheating at low temperatures. Peak (G) also loses some of its intensity in the reheating curves after the high maximum temperatures of the initial heat treatment. From these observations, it can be concluded that during the initial heat treatment, a dissolution of phases β″ and β′ as well as phases with a higher solvus temperature has taken place. An area in which the precipitation of equilibrium phase β-Mg_2Si dominates could not be identified. Probably, this precipitation reaction is greatly suppressed at heating with 20 Ks^{-1}. Thus, from the findings, a broad dissolution reaction in the reconstructed heating curve was assumed.

The analysis of the DSC reheating curves allowed a direct DSC heating curve to be reconstructed at the previously immeasurable heating rate of 20 Ks^{-1}, as seen in Figure 10 on the right. It is important to mention, that the temperature sections for different reactions can be reconstructed quantitatively, whereas reaction intensities can only be estimated qualitatively. Therefore, the reconstructed curve is plotted as a dashed line. Due to the discrete reconstruction of the heating curve, the start or end temperatures of a reconstructed peak are determined with uncertainties. The temperature uncertainty is at least as great as the examined temperature step size. Figure 11 shows the DSC curve reconstruction for a heating rate of 100 Ks^{-1}. The procedure to reconstruct the heating curve at 100 Ks^{-1} is equivalent to the reconstruction of the heating curve at 20 Ks^{-1}. The resulting dissolution and precipitation sections I–V are just shifted to slightly higher temperatures.

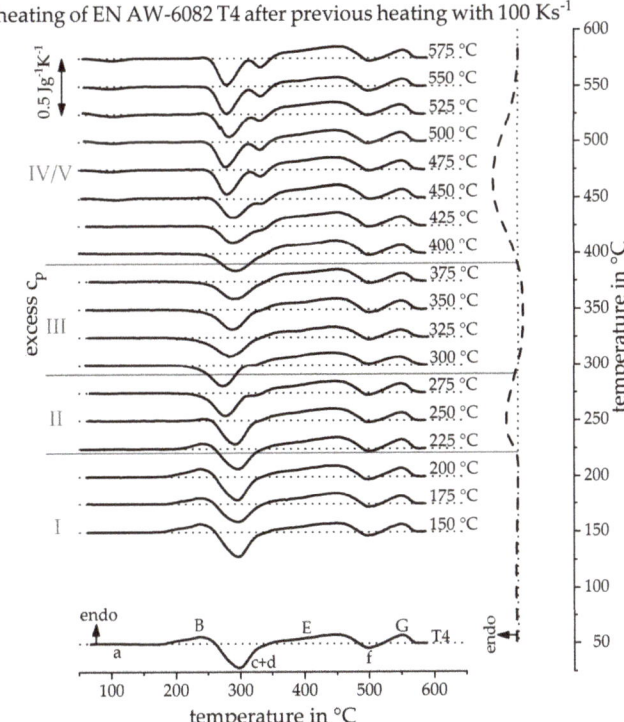

Figure 11. Reheating curves after a heat treatment with a heating rate of 100 Ks^{-1} up to different maximum temperatures and the resulting reconstructed heating curve at 100 Ks^{-1}.

3.4. The Continuous Heating Dissolution Diagram of EN AW-6082 T4

Figure 12 shows the direct DSC heating curves of the investigated alloy EN AW-6082, in the naturally aged initial state T4, up to 5 Ks^{-1}, as well as the reconstructed DSC heating curves at 20 and 100 Ks^{-1}. It turns out that the reconstructed heating curves are a good fit with the directly measured heating curves. Remember that the intensities of the reconstructed DSC heating curves are only estimated qualitatively. However, it is seen that the derived temperature ranges of the individual peaks agree with those of the directly measured DSC heating curves of higher heating rates.

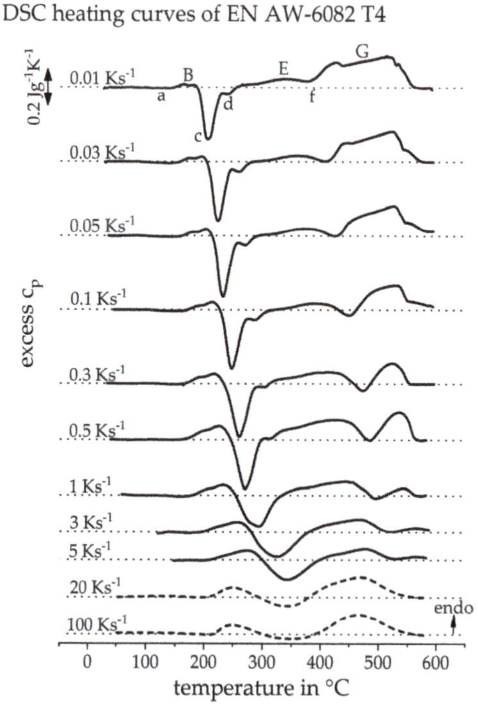

Figure 12. Heating curves of EN AW-6082 T4 with the conventional DSC and the reconstructed heating curves at 20 and 100 Ks^{-1}.

Figure 13 shows the continuous heating dissolution diagram of the investigated alloy 6082 T4 obtained from the results given in Figure 12. This diagram is suitable for the selection of reasonable heat treatment parameters during production, as well as a necessary input for heat treatment simulations. The continuous heating dissolution diagrams are valid only for continuous heating steps and, therefore, are to be read at a selected heating rate (grey line) from low to high temperatures only. During heating, different reaction areas are run through. Finally, the heating rate specific solvus temperature is reached. Above this temperature for the specific heating rate, it is considered that all major alloying elements are in solid solution. Due to the proposed indirect DSC method, the heating rate range could be enlarged significantly towards higher rates.

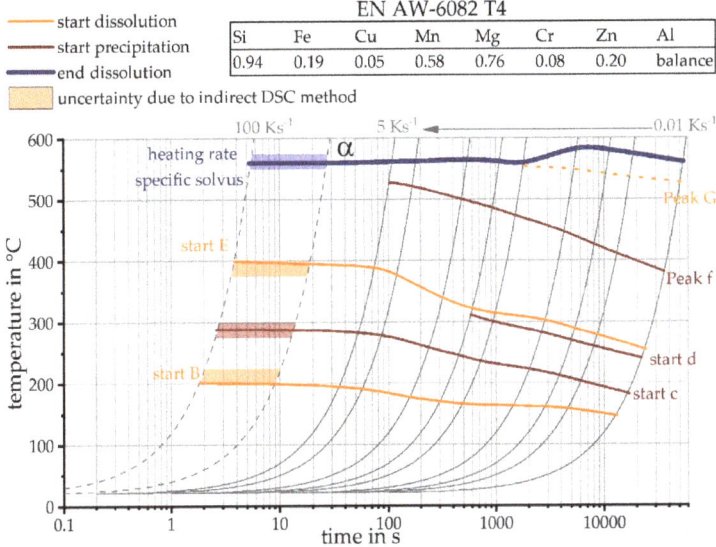

Figure 13. Continuous heating dissolution diagram of EN AW-6082 T4, including the findings of the indirect measurements.

4. Transfer of This Measurement Methodology

4.1. Transfer to Other Alloys

The method described was transferred to other alloys as well as non-linear heat treatments. Figure 14 shows heating DSC curve reconstruction for the alloy EN AW-6060 T4 with a heating rate of 20 Ks^{-1}. In comparison to the alloy EN AW-6082 T4, this alloy has a lower content of alloying elements. For this reason, the individual reactions are weaker, compared to EN AW-6082 T4. However, the main alloying elements of both alloys are the same and the initial heat treatment state is comparable. For this reason, the recorded reheating curves have a very similar course. The sequence of exothermic and endothermic peaks during heating from the initial state is identical and the individual peaks can also be assumed to refer to the same reactions. Unlike the alloy EN AW-6082, the peak (a) does not show up in this alloy during reheating, even at high peak temperatures, due to the low content of alloying elements. The procedure to reconstruct the heating curve, at 20 Ks^{-1}, of this alloy is equivalent to the reconstruction of the heating curve described before.

Figure 15 shows the heating DSC curve reconstruction for a heating rate of 100 Ks^{-1}. It is noticeable that, above a maximum temperature of 300 °C, the reheating curves show little change. From these results it can be concluded that the precipitation state no longer changes above 300 °C during this very rapid heating. It can be seen, from section II, that the T4 strength increasing precipitates are dissolved between 225 °C and 300 °C. However, the heating rate of 100 Ks^{-1} is obviously high enough to suppress further precipitation reactions. Figure 16 shows the continuous heating dissolution diagram of EN AW-6060 T4, including the findings of the indirect measurements. The heating curves of direct DSC measurements have already been published and can be seen from Fröck et al. [2]. It also becomes clear, that with this alloy, the results of the indirect measurements fit very well into the results of the direct DSC. The results have shown that the described method is also suitable for extending the bandwidth of scanning rates of other alloys.

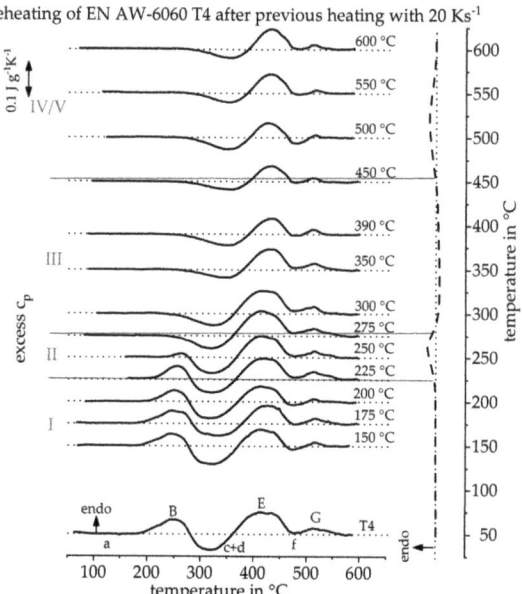

Figure 14. Reheating curves of the alloy EN AW-6060 T4 after a heat treatment with a heating rate of 20 Ks^{-1} up to different maximum temperatures and the resulting reconstructed DSC heating curve at 20 Ks^{-1}.

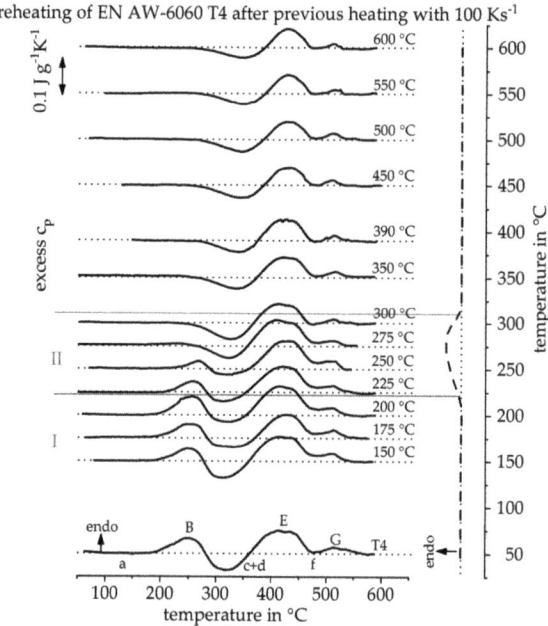

Figure 15. Reheating curves of the alloy EN AW-6060 T4 after a heat treatment with a heating rate of 100 Ks^{-1} up to different maximum temperatures and the resulting reconstructed DSC heating curve at 100 Ks^{-1}.

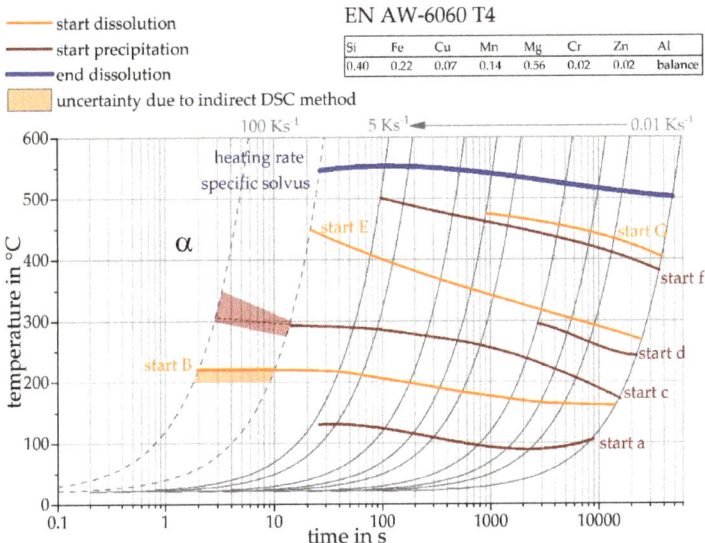

Figure 16. Continuous heating dissolution diagram of EN AW-6060 T4, including the findings of the indirect measurements.

4.2. Transfer to Non-Linear Heating and Cooling

The non-linear temperature profile of the laser heat treatment was analyzed using the presented indirect measurement methodology. For this purpose, DSC samples were heat treated in the quenching dilatometer, with the temperature profile given in Figure 3. At defined points, this heat treatment was interrupted by an overcritical gas quenching. Reheating curves were subsequently recorded in the DSC.

Figure 17 shows the investigated time temperature profile on the left-hand, the recorded reheating curves of the alloy EN AW-6060 T4 in the middle, and the concluded precipitation and dissolution reactions on the right-hand side. The reconstruction of a DSC curve for the investigated heat treatment would be purposeless, because of the non-linearity and as well as the fact that heating and cooling are examined. The reheating curves shown in this chapter are to be read chronologically from top to bottom.

It becomes clear that, during heating of the initial state, a pronounced dissolution peak (B) takes place, as can be seen in section I. After a short-term heat treatment up to 255 °C within ≈2 s, this peak became much weaker and almost disappeared, as can be seen in section II. From this we can conclude that during the described laser heat treatment, during heating up to 255 °C, a large part of the strength increasing T4 state precipitates are dissolved. By heating to 297 °C, this peak completely disappears in the reheating curves, from which we can conclude that nearly all clusters and GP zones of the initial state have been dissolved during this heat treatment. Subsequently, the cooling starts. The reheating curves of the cooling step do not change much, as can be seen in section III. Only the peak (c + d) is slightly weaker in the reheat curves, with increasing time of the initial heat treatment. It can be concluded that a weak precipitation reaction takes place during the cooling of the laser heat treatment. Thus, alloying elements are bound in particles and the potential for precipitation during reheating decreases.

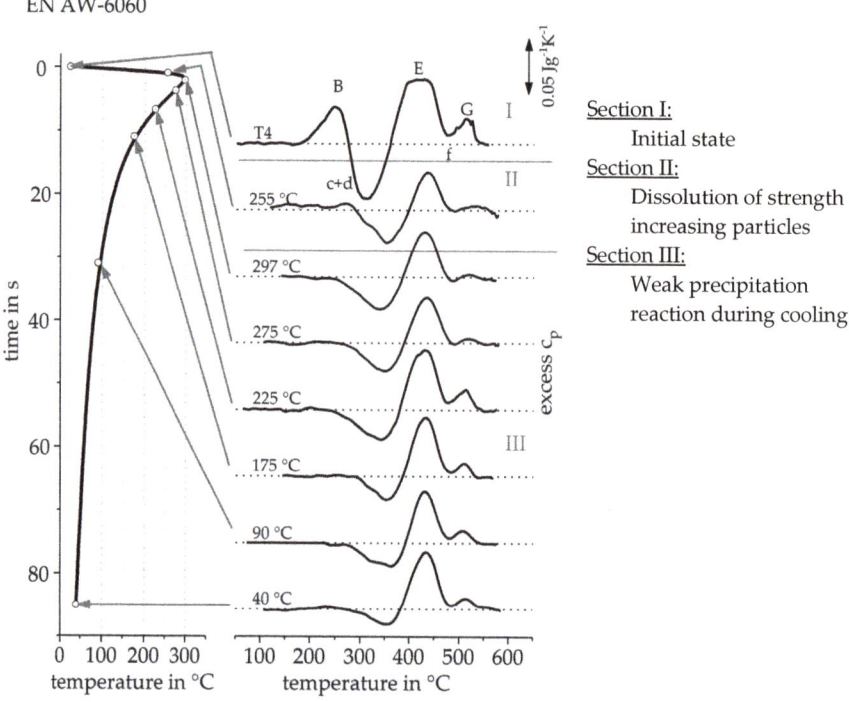

Figure 17. Reheating curves of the alloy EN AW-6060 T4 after an imitated laser short-term heat treatment.

The cooling is much slower compared to the heating, however, with respect to the upper critical cooling rate of EN AW-6060, it still has to be considered as fast for this alloy. The cooling from 297 °C to 40 °C, in total, takes 83 s, which equals an average cooling rate of 3 Ks^{-1}. Compared to the results in Reference [14], the upper critical cooling rate of this batch of EN AW-6060 can be estimated to be in the order of 1 to 6 Ks^{-1}. However, the reheating DSC for the cooling path indicates the occurrence of a very weak precipitation reaction.

Figure 18 shows the investigated time temperature profile for the HAZ of an arc welding process on the left side [16], the recorded reheating curves of the alloy EN AW-6082 T6 in the middle, and the concluded precipitation and dissolution reactions on the right side. It is clearly visible that the reheating curves of the initial artificial aging T651 have a different course than those of the initial natural aging T4. From the reheating curves seen in Figure 18 Section I, up to a temperature of 250 °C, no obvious change is seen. From this it can be concluded that the precipitation state does not change up to this temperature. Section II, between 250 °C and 490 °C, shows that peak (B) disappears and peak (d) occurs more pronounced with increasing temperature. It can be concluded that during the HAZ heat treatment the T6 precipitates are increasingly dissolved. The dissolution of precipitates results in solved alloying elements which can form precipitates during reheating, which explains the increase of the peak (d). In Section III, the cooling, only weak reactions take place. It becomes clear that the two reactions (d) are steadily weakening. It can be concluded that a slight precipitation reaction occurs during cooling. This nicely correlates with findings on the upper critical cooling rate for this batch of EN AW-6082 of about 30 Ks^{-1} [6]. Considering the non-linear cooling within the simulated HAZ, an average cooling rate of about 12 Ks^{-1} is achieved in the relevant temperature range of 490–200 °C. As this cooling rate is below the upper critical cooling rate, a certain volume fraction of quench induced precipitation is to be expected.

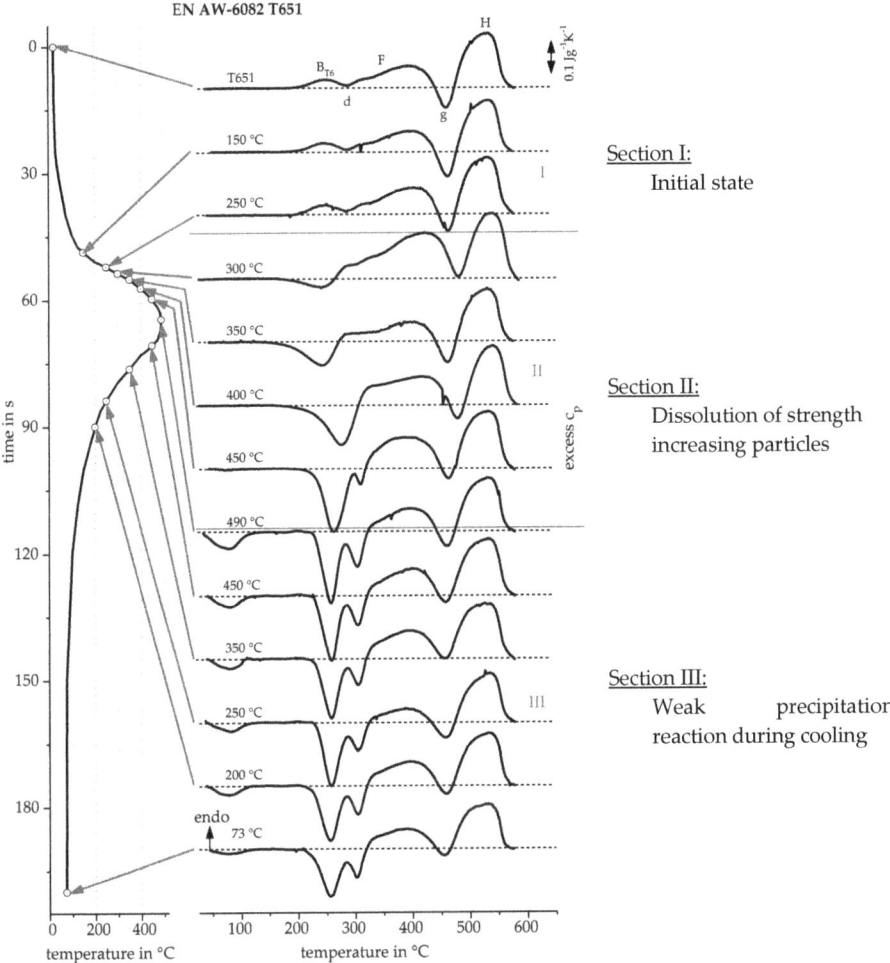

Figure 18. Reheating curves of the alloy EN AW-6082 T651 after heating like in the heat affected zone caused by an arc welding.

5. Conclusions

In this work, a new indirect DSC measuring method is presented, which allows the reconstruction of DSC heating curves of fast and/or non-linear heat treatments, which were not previously assessable with DSC. Due to the combination of direct and indirect DSC, a very wide range of heating rates from 0.01 to a few 100 Ks^{-1} has been investigated, using the example of aluminum alloys, including short-term laser heat treatment and welding.

For this purpose, a large number of samples, with unknown transformational behavior, were subjected to the heat treatment of interest. This initial heat treatment must not be done in a DSC, but can be performed in any suitable controlled device, e.g., a quenching dilatometer. The heat treatment of interest was interrupted at certain points and the samples were quenched as quickly as possible in order to freeze the existing material state. These samples were then reheated at a measurable heating rate in a conventional DSC. The reconstruction of the pertinent heating curves is possible at discrete temperature steps. In this work, a temperature step of 25 K was used, which resulted in about 100 individual DSC measurements per reconstructed DSC curve.

Advantages of indirect DSC are as follows:

- Reconstruction of DSC curves for fast or non-linear heat treatments, which previously were not assessable, is now possible.
- This method is applicable for linear and non-linear heat treatments, as well as for heating and cooling.
- Temperature ranges of the main reactions can be reconstructed quantitatively.

Disadvantages of indirect DSC are as follows:

- Very time-consuming method due to many individual measurements.
- Additional device is necessary for a defined heat treatment, including the necessity of overcritical quenching to allow for process interruption.
- Validation based on a known DSC heating curve is required. This will be necessary for new alloys and new heat treatment processes. As long as alloys and processes are similar, as in in the above examples, one typical validation will be sufficient.

Author Contributions: H.F., M.R., B.M., and O.K. conceived and designed the experiments; H.F. performed the experiments and analysed the data; H.F., M.R., B.M., and O.K. discussed and interpreted the results together; H.F. wrote the paper.

Funding: This research was funded by the German Research Foundation (DFG), within the scope of the research project Improvement of formability of extruded aluminium profiles by a local short-term heat treatment (DFG KE616/22-2).

Conflicts of Interest: The authors declare no conflict of interest.

References

1. Osten, J.; Milkereit, B.; Schick, C.; Kessler, O.; Je, J.H. Dissolution and Precipitation Behaviour during Continuous Heating of Al-Mg-Si Alloys in a Wide Range of Heating Rates. *Materials* **2015**, *8*, 2830–2848. [CrossRef]
2. Fröck, H.; Graser, M.; Reich, M.; Lechner, M.; Merklein, M.; Kessler, O. Influence of short-term heat treatment on the microstructure and mechanical properties of EN AW-6060 T4 extrusion profiles: Part A. *Prod. Eng.* **2016**, *10*, 383–389. [CrossRef]
3. Milkereit, B.; Kessler, O.; Schick, C. Continuous Cooling Precipitation Diagrams of Aluminium-Magnesium-Silicon Alloys. In Proceedings of the 11th International Conference on Aluminium Alloys, Aachen, Germany, 22–26 September 2008; pp. 1232–1237.
4. Milkereit, B.; Fröck, H.; Schick, C.; Kessler, O. Continuous cooling precipitation diagram of cast aluminium alloy Al-7Si-0.3Mg. *Trans. Nonferrous Met. Soc.* **2014**, *24*, 2025–2033. [CrossRef]
5. Zohrabyan, D.; Milkereit, B.; Schick, C.; Kessler, O. Continuous cooling precipitation diagram of high alloyed Al-Zn-Mg-Cu 7049A alloy. *Trans. Nonferrous Met. Soc.* **2014**, *24*, 2018–2024. [CrossRef]
6. Fröck, H.; Milkereit, B.; Wiechmann, P.; Springer, A.; Sander, M.; Kessler, O.; Reich, M. Influence of Solution-Annealing Parameters on the Continuous Cooling Precipitation of Aluminum Alloy 6082. *Metals* **2018**, *8*, 265. [CrossRef]
7. Schick, C.; Mathot, V. *Fast Scanning Calorimetry*; Springer: Basel, Switzerland, 2016.
8. Merklein, M.; Böhm, W.; Lechner, M. Tailoring Material Properties of Aluminum by Local Laser Heat Treatment. *Phys. Procedia* **2012**, *39*, 232–239. [CrossRef]
9. Wiechmann, P.; Panwitt, H.; Heyer, H.; Reich, M.; Sander, M.; Kessler, O. Combined Calorimetry, Thermo-Mechanical Analysis and Tensile Test on Welded EN AW-6082 Joints. *Materials* **2018**, *11*, 1396. [CrossRef]
10. Milkereit, B.; Beck, M.; Reich, M.; Kessler, O.; Schick, C. Precipitation kinetics of an aluminium alloy during Newtonian cooling simulated in a differential scanning calorimeter. *Thermochim. Acta* **2011**, *522*, 86–95. [CrossRef]
11. Zohrabyan, D.; Milkereit, B.; Kessler, O.; Schick, C. Precipitation enthalpy during cooling of aluminum alloys obtained from calorimetric reheating experiments. *Thermochim. Acta* **2012**, *529*, 51–58. [CrossRef]

12. Schumacher, P.; Pogatscher, S.; Starink, M.J.; Schick, C.; Mohles, V.; Milkereit, B. Quench-induced precipitates in Al-Si alloys: Calorimetric determination of solute content and characterisation of microstructure. *Thermochim. Acta* **2015**, *602*, 63–73. [CrossRef]
13. Yang, B.; Milkereit, B.; Zhang, Y.; Rometsch, P.A.; Kessler, O.; Schick, C. Continuous cooling precipitation diagram of aluminium alloy AA7150 based on a new fast scanning calorimetry and interrupted quenching method. *Mater. Charact.* **2016**, *120*, 30–37. [CrossRef]
14. Milkereit, B.; Wanderka, N.; Schick, C.; Kessler, O. Continuous cooling precipitation diagrams of Al-Mg-Si alloys. *Mater. Sci. Eng. A* **2012**, *550*, 87–96. [CrossRef]
15. Graser, M.; Fröck, H.; Lechner, M.; Reich, M.; Kessler, O.; Merklein, M. Influence of short-term heat treatment on the microstructure and mechanical properties of EN AW-6060 T4 extrusion profiles—Part B. *Prod. Eng.* **2016**, *10*, 391–398. [CrossRef]
16. Sarmast, A.; Kokabi, A.H.; Serajzadeh, S. A study on thermal responses, microstructural issues, and natural aging in gas tungsten arc welding of AA2024-T4. *Proc. Inst. Mech. Eng. B J. Eng. Manuf.* **2013**, *228*, 413–421. [CrossRef]
17. Höhne, G.W.H.; Hemminger, W.F.; Flammersheim, H.J. *Differential Scanning Calorimetry*, 2nd ed.; Springer: Berlin, Germany, 2003.
18. Sarge, S.M.; Hemminger, W.; Gmelin, E.; Höhne, G.W.H.; Cammenga, H.K.; Eysel, W. Metrologically based procedures for the temperature, heat and heat flow rate calibration of DSC. *J. Therm. Anal.* **1997**, *49*, 1125–1134. [CrossRef]
19. Höhne, G.; Cammenga, H.; Eysel, W.; Gmelin, E.; Hemminger, W. The temperature calibration of scanning calorimeters. *Thermochim. Acta* **1990**, *160*, 1–12. [CrossRef]
20. Birol, Y. The effect of sample preparation on the DSC analysis of 6061 alloy. *J. Mater. Sci.* **2005**, *40*, 6357–6361. [CrossRef]
21. Birol, Y. DSC Analysis of the precipitation reactions in the alloy AA6082. *J. Therm. Anal.* **2006**, *83*, 219–222. [CrossRef]
22. Birol, Y. DSC analysis of the precipitation reaction in AA6005 alloy. *J. Therm. Anal.* **2008**, *93*, 977–981. [CrossRef]
23. Birol, Y. A calorimetric analysis of the precipitation reactions in AlSi1MgMn alloy with Cu additions. *Thermochim. Acta* **2017**, *650*, 39–43. [CrossRef]
24. Kim, S.N.; Kim, J.H.; Tezuka, H.; Kobayashi, E.; Sato, T. Formation Behavior of Nanoclusters in Al-Mg-Si Alloys with Different Mg and Si Concentration. *Mater. Trans.* **2013**, *54*, 297–303. [CrossRef]
25. Murayama, M.; Hono, K. Pre-precipitate clusters and precipitation processes in Al-Mg-Si alloys. *Acta Mater.* **1999**, *47*, 1537–1548. [CrossRef]
26. Tsao, C.-S.; Chen, C.-Y.; Jeng, U.-S.; Kuo, T.-Y. Precipitation kinetics and transformation of metastable phases in Al-Mg-Si alloys. *Acta Mater.* **2006**, *54*, 4621–4631. [CrossRef]
27. Edwards, G.; Stiller, K.; Dunlop, G.; Couper, M. The precipitation sequence in Al-Mg-Si alloys. *Acta Mater.* **1998**, *46*, 3893–3904. [CrossRef]
28. Doan, L.C.; Ohmori, Y.; Nakai, K. Precipitation and Dissolution Reactions in a 6061 Aluminum Alloy. *Mater. Trans. JIM* **2000**, *41*, 300–305. [CrossRef]
29. Gupta, A.K.; Lloyd, D.J. Study of precipitation kinetics in a super purity Al-0.8 pct Mg-0.9 pct Si alloy using differential scanning calorimetry. *Metall. Mater. Trans. A* **1999**, *30*, 879–884. [CrossRef]

© 2019 by the authors. Licensee MDPI, Basel, Switzerland. This article is an open access article distributed under the terms and conditions of the Creative Commons Attribution (CC BY) license (http://creativecommons.org/licenses/by/4.0/).

Article

Effect of Intermetallic Compounds on the Thermal and Mechanical Properties of Al–Cu Composite Materials Fabricated by Spark Plasma Sintering

Kyungju Kim [1,2], Dasom Kim [3,4], Kwangjae Park [3], Myunghoon Cho [2], Seungchan Cho [5] and Hansang Kwon [2,4,*]

1. The Industrial Science Technology Research Center, Pukyong National University, 365, Sinseon-ro, Nam-Gu, Busan 48547, Korea; ngm13@ngm.re.kr
2. Department of R&D, Next Generation Materials Co., Ltd., 365, Sinseon-ro, Nam-Gu, Busan 48547, Korea; ngm15@ngm.re.kr
3. Department of Hard Magnets Research, The National Institute of Advanced Industrial Science and Technology (AIST), Shimo-Shidami, Moriyama-ku, Nagoya, Aichi 463-8560, Japan; ds-kim@aist.go.jp (D.K.); k.park@aist.go.jp (K.P.)
4. Department of Materials System Engineering, Pukyong National University, 365, Sinseon-ro, Nam-Gu, Busan 48547, Korea
5. Department of Composites Research, Korea Institute of Materials Science, Changwon-daero, Seongsan-gu, Changwon-si 51508, Gyeongsangnam-do, Korea; sccho@kims.re.kr
* Correspondence: kwon13@pknu.ac.kr

Received: 9 April 2019; Accepted: 9 May 2019; Published: 10 May 2019

Abstract: Aluminium–copper composite materials were successfully fabricated using spark plasma sintering with Al and Cu powders as the raw materials. Al–Cu composite powders were fabricated through a ball milling process, and the effect of the Cu content was investigated. Composite materials composed of Al–20Cu, Al–50Cu, and Al–80Cu (vol.%) were sintered by a spark plasma sintering process, which was carried out at 520 °C and 50 MPa for 5 min. The phase analysis of the composite materials by X-ray diffraction (XRD) and energy-dispersive spectroscopy (EDS) indicated that intermetallic compounds (IC) such as $CuAl_2$ and Cu_9Al_4 were formed through reactions between Cu and Al during the spark plasma sintering process. The mechanical properties of the composites were analysed using a Vickers hardness tester. The Al–50Cu composite had a hardness of approximately 151 HV, which is higher than that of the other composites. The thermal conductivity of the composite materials was measured by laser flash analysis, and the highest value was obtained for the Al–80Cu composite material. This suggests that the Cu content affects physical properties of the Al–Cu composite material as well as the amount of intermetallic compounds formed in the composite material.

Keywords: aluminium composite; copper composite; spark plasma sintering; thermal properties; powder metallurgy; intermetallic compound

1. Introduction

Heat dissipation and the development of lightweight materials are important concerns for the automobile, aerospace, optical material panel, electronic packaging, and semiconductor component industries, among others [1–8]. Given the increasing global enforcement of carbon dioxide emission regulations, these lightweight materials should be eco-friendly and economical, reduce carbon dioxide emissions, and improve the fuel efficiencies of automobiles, shipbuilding, and aviation applications. Aluminium, which has a low density, and copper, which has a high heat dissipation capacity, have attracted attention as suitable materials to satisfy various industrial needs. Recently, the demand for high-functionality materials that are lightweight and have high heat dissipation, low thermal stress, and high strength characteristics has increased to improve the fuel efficiency of transportation and realise miniaturisation of integrated circuits [9–16]. As a result, alloy or composite-related studies have been conducted owing to the demands for various functions in a single material. Low-density aluminium has a relatively lower hardness than other metals; thus, materials such as Fe, Mg, SiC, B, Ti, V, diamond, and carbon nanotubes (CNT) are added as composites to increase the hardness [17–33]. Kwon et al. [34] fabricated Al–CNT composites through a combination of spark plasma sintering and a hot extrusion process. The tensile strength of the Al–CNT composite was 194 MPa, which is twice that of pure bulk Al. The increase in the mechanical characteristics was attributed to the effects of the particular strengthening by CNT and the regularly oriented CNTs achieved through the nanoscale dispersion method. Metal alloys and composite materials such as Al, Cu, Cu–W, Cu–Mo, and Al–SiC have been widely used as heat dissipation materials with high thermal conduction and low thermal expansion coefficient characteristics [35–37]. However, cost is an important factor in industrial applications, and the complex manufacturing process and high price of these materials limits their suitability for industrial applications. Recently, interest in Al and Cu composites is increasing, as these composites combine the lightweight properties of Al and the thermal characteristics of Cu. Studies of these composites have been performed using the accumulative roll bonding (ARB) process and the squeeze casting method. Eizadjou et al. [38] used Al 1100 and Cu strips to investigate the mechanical characteristics of the modified structure of a multi-layered Al/Cu composite made with the ARB process. It was reported that as the average thickness of the Cu layer decreased from 100 μm to 7 μm, the strength and hardness increased. However, the thermal characteristics of the Al–Cu composite for use as a heat dissipation material were not investigated. Wu et al. [39] investigated the effects of adding Cu on the thermal characteristics of an Al–Cu/diamond composite manufactured by squeeze casting. It was reported that the thermal conductivity of the Al–3.0 wt% Cu/diamond composite was 330 $W·m^{-1}·K^{-1}$ which is 57% higher than that of the Al/diamond composite. Nevertheless, studies on composites with high Al and Cu content are rarely performed because these manufacturing processes have about three times the density difference between Al and Cu and lead to compound formation between the metals.

In this study, the mechanical ball milling and spark plasma sintering (SPS) composite manufacturing processes were used to fabricate a composite material combining the advantages of Al and Cu. The microstructure and form of the composite material were analysed using X-ray diffraction (XRD), scanning electron microscopy (SEM), field-emission SEM (FE-SEM), and energy-dispersive spectroscopy (EDS). The mechanical characteristics of the Al–Cu composite material were measured using a Vickers hardness tester, and the thermal characteristics were measured using laser flash analysis.

2. Materials and Methods

The raw powders used in this study were pure Al (99.9%, Metalplayer Co., Ltd., Incheon, Korea) and Cu (99.9%, Metalplayer Co., Ltd.) powders with an average particle size of about 45μm. A mixture of zirconia (diameter of 15 mm) and stainless balls (diameter of 8 mm) in a stainless-steel jar were used for the ball milling process and with the ball mill (SMBL-6, SciLabMix™, Programmable Ball Mill). In all experiments, a specific amount of starting materials were used, giving balls to powder weight ratios of 3:1. Pure Al powders and pure Cu powders were combined as Al–20Cu, Al–50Cu, and Al–80Cu (vol.%) and mixed with 50 ml heptane as a process control agent (PCA) in the stainless-steel jar; the ball milling process was then performed for 24 h at 420 rpm under the ambient atmosphere. The PCA was eliminated by natural evaporation. The composite powders were placed in a graphite mold (diameter of 20 mm), held for 5 min at 520 °C, and sintered using spark plasma sintering equipment (Fuji Electronic Industrial Co., Ltd., SPS-321Lx, Saitama, Japan) with a compacting pressure of 50 MPa. Al–Cu sintered bodies were fabricated with a diameter of 20 mm and a thickness of 5 mm. The density of the composite material was measured with densitometer using the Archimedes method, and the theoretical density was calculated based on the mixture of pure Al and Cu. XRD patterns for the Al–Cu composites were obtained using an X-ray diffractometer (Ultima IV, Rigaku, Tokyo, Japan) with a Cu Kα radiation source (λ = 1.5148 Å, 40 kV, and 40 mA) in the 2-theta range of 20–80° using a linear detector (D/tex ultra, Rigaku).

The microstructures and relative composition of the composite materials were analysed with SEM (VEGA II LSU, TESCAN, Czech Republic), FE-SEM (MIRA 3 LMH In-Beam, TESCAN, Czech Republic), and EDS (HORIBA, EX-400, Kyoto, Japan). Measurements of the area fraction of each component in the Al–Cu composites were performed through digital image analysis using the ImageJ software, which is a version of NIH Image (US National Institutes of Health, http://rsb.info.nih.gov/nih-image). The mechanical properties of the composite materials were determined according to JIS B 7725 and ISO 6507-2 standard using a load of 0.3 kg for 5 s (HM-101 Vickers hardness tester, Mitutoyo Corporation, Kawasaki, Japan); at least five measurements were performed for each sample. The thermal diffusivities and heat capacity of the composites were measured at room temperature with a laser flash apparatus (LFA467, Netzsch, Selb, Germany) according to ISO 22007-4, ISO 18755 and ASTM E 1461 standard. The accuracy of the measuring device is according to the manufacturer ±3% for thermal diffusivity measurements and ±5% for heat capacity measurements. The laser flash method is used to measure thermal diffusivity in a variety of different materials. An energy pulse heats one side of a plane-parallel sample, and the resulting time-dependent temperature increase on the backside due to the energy input is detected. The thermal conductivity, λ, is defined as the ability of a material to transmit heat, and it is measured in watts per square metre of surface area for a temperature gradient of 1 K per unit thickness of 1 m. The thermal diffusivity (α) and heat capacity (Cp) measured by the laser flash method has the following relationship to the thermal conductivity (κ):

$$\lambda(T) = \alpha(T) \times Cp(T) \times \varrho(T) \tag{1}$$

where Cp is the heat capacity, and ϱ is the density.

3. Results and Discussions

Figure 1 shows the morphologies of the pure Al and Cu powders. The pure Al powder exhibits mostly spherical particles, with some ellipse forms, as shown in Figure 1a. As shown in Figure 1b, the pure Cu powder is composed of a dispersion of various sizes of spherical particles. At this point, it was assumed that the pure Al had a natural oxide on the particle surfaces. The morphologies of the Al–20Cu, Al–50Cu, and Al–80Cu composite powders created through the ball milling of pure Al and Cu powders are shown in Figure 1c–e, respectively. In the ball milling process, the powders were milled with plastic deformation as stresses accumulate in the powders through continuous impact with the balls. The Al–Cu composite powders exhibit morphologies containing a mixture of flake and plate-like particles and particles with surface deformation. Ductile particles are easily deformed by the impact energy of the balls, and thus the Al particles with higher ductility than the Cu particles exhibit flake and plate forms. This shows that the impact energy employed during the ball milling under this process condition has enough energy to achieve plastic deformation of the Al particles and deform only the surfaces of the Cu particles. In addition, ductile materials are generally easily aggregated by ball milling, but the Al–Cu composite powders exhibited no aggregate formation and a suitable distribution of Al and Cu. It is desirable for the dispersion in the matrix of the Al–Cu composite material to yield a relatively equal distribution. Figure 1f shows the XRD patterns for the Al–Cu composite powders prior to the SPS process. The XRD pattern results suggest that there were no reactions between the powders during the ball milling process, as the spectra for the Al–20Cu, Al–50Cu, and Al–80Cu composite powders all contained only peaks corresponding to Al and Cu.

Figure 2a shows a photograph of the Al–20Cu, Al–50Cu, and Al–80Cu composites after SPS. The Al–Cu composite material in this study was composed of a circular plate having an identical inner structure of a circular carbon mould. As the Cu content in the Al–Cu composite material increases from 20 to 50 and 80 vol.%, the colour changes from light silver to dark copper. Figure 2b shows the relative density of Al–20Cu, Al–50Cu, and Al–80Cu composites after SPS. The theoretical density of the composite materials was calculated based on the Al and Cu mixture composition and the density of Al and Cu. The relative density was calculated by dividing the theoretical density by the measured density (Rd = Ed/Td). Al–20Cu composite exhibits relative densities of about 100% or more and achieved full density during the SPS process in a short period of time. Composites with high relative densities at relatively low sintering temperatures could be successfully obtained through the SPS process. At this point, the Al–Cu composites with a relative density of greater than 100% were considered to no longer have the form of only pure Al and Cu, but rather to contain a different phase. The Al–Cu composites with Cu contents of greater than 50 vol.% exhibited a decrease in the relative density to approximately 90% with increasing Cu content. This indicates that the densification of the composite at relatively low temperature was difficult to achieve for high density owing to the higher melting point of pure Cu than pure Al.

The results indicate that in the Al–20Cu and Al–50Cu composites, Al performs the matrix role because it does have a continuous connection, and Cu functions as a dispersed phase, showing that it is an Al-matrix composite. In contrast, Cu performs the matrix role in the Al–80Cu composite. Furthermore, during the SPS process, it is considered that the Al and Cu reacted to form a different phase than pure Al and Cu at the interface between the Al and Cu. Both the Al–50Cu and Al–80Cu composites formed an identical new phase, which has a pore area corresponding to the relative density. It is suggested that Al–Cu composites can be fabricated by the SPS process without Cu agglomeration and with a uniformly dispersed structure.

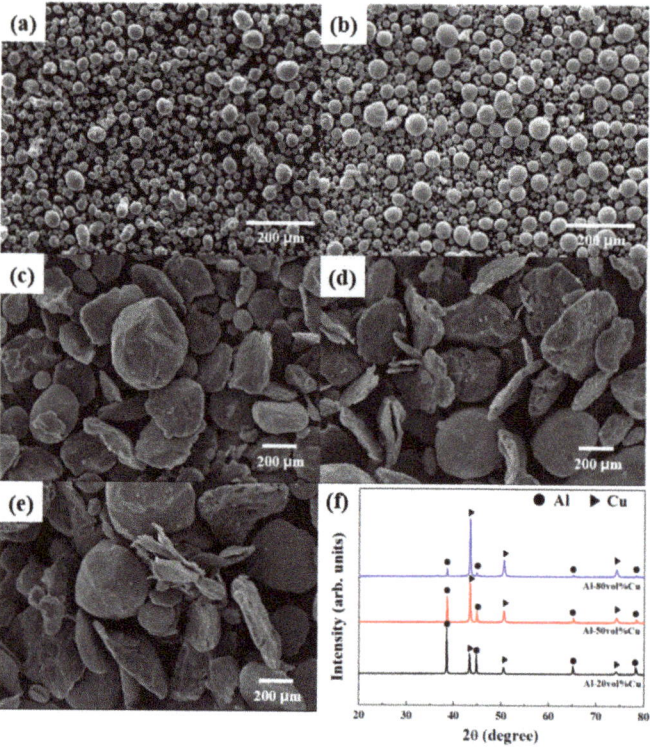

Figure 1. Scanning electron microscopy (SEM) images of (**a**) pure Al, (**b**) pure Cu, (**c**) Al–20Cu, (**d**) Al–50Cu, and (**e**) Al–80Cu powders, and (**f**) X-ray diffraction (XRD) patterns of the composite powders.

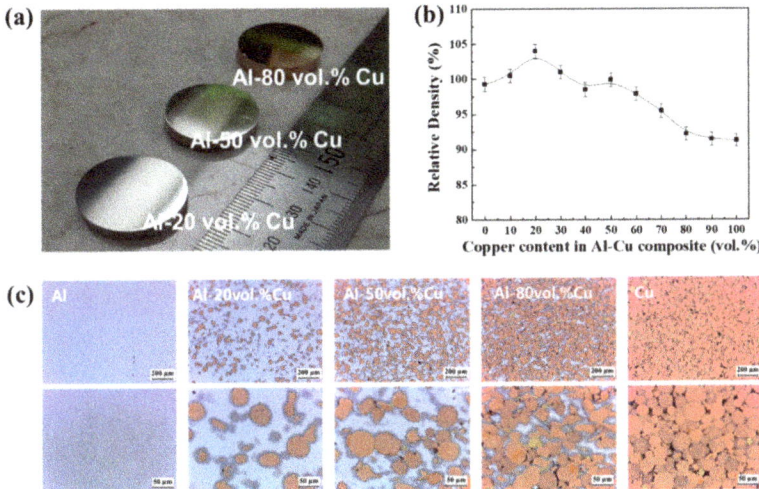

Figure 2. (**a**) Photograph of the Al–Cu composites after spark plasma sintering (SPS), (**b**) evolution of the relative density of composites, and (**c**) cross-sectional light microscopy images.

The displacement profile recorded during the SPS of Al–Cu powders is shown in Figure 3a,b. During the SPS process, the pressure was set to 50 MPa. The temperature was maintained constant at 520 °C for 5 min. The sintering temperature was raised to be similar to the set temperature. As it can be seen during the sintering process, the displacement increases to approximately 3 mm. The displacement was attributed to the particle rearrangement facilitated by the temperature increase. The displacement of Al–Cu composites remains almost constant as the material reaches its maximum density.

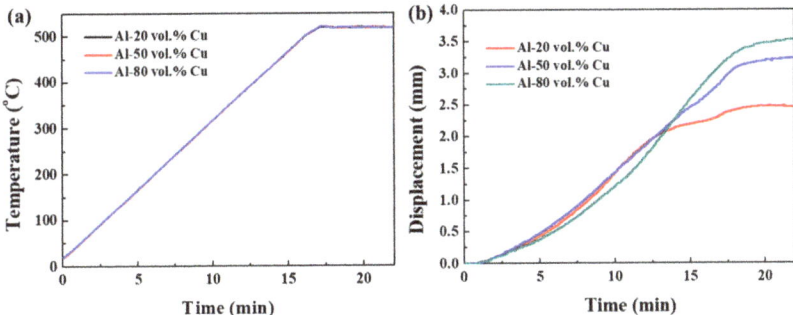

Figure 3. Variation of (**a**) the temperature and (**b**) displacement, as function of holding time SPS.

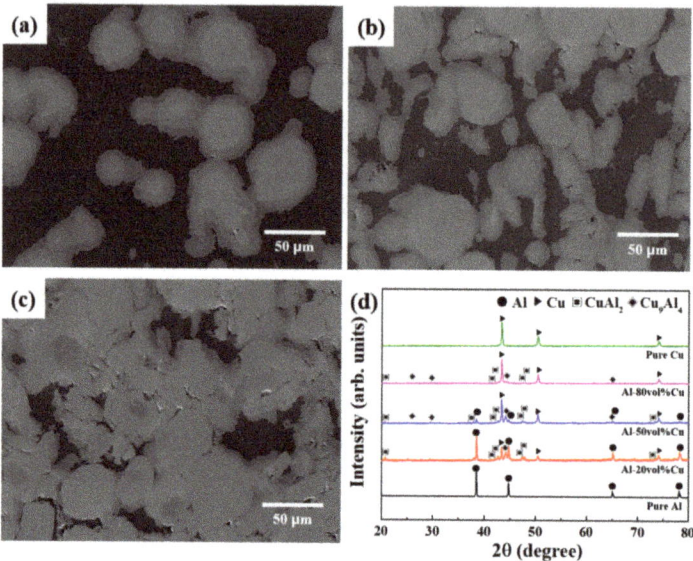

Figure 4. SEM micrographs of (**a**) Al–20Cu, (**b**) Al–50Cu, and (**c**) Al–80Cu composites, and (**d**) XRD patterns of Al–Cu composites.

Figure 4a–c shows SEM micrographs of the Al–20Cu, Al–50Cu, and Al–80Cu composite materials. In Figure 4, the Al–20Cu and Al–50Cu composites fabricated by SPS exhibit fully densified behaviour. SPS is capable of sintering in a short period of time and can provide microstructural control that cannot be expected in conventional sintering processes. In addition, SPS surface treatment can reduce the impurities and oxides present on particle surfaces. Consequently, high-quality and high-density sintered materials can be obtained in a short amount of time. The Al–Cu composites produced by SPS exhibit Al as the darkest phase, Cu as the brightest phase, and two layers between the Al and Cu interface. Figure 4d shows the XRD pattern for the Al–Cu composite powder subjected to SPS. Intermetallic compounds composed of the Al phase and Cu phase as well as $CuAl_2$ and Cu_9Al_4 phases were detected in the XRD patterns of the Al–20Cu, Al–50Cu, and Al–80Cu composites. In the Al–Cu composite material, the natural air-formed oxide layer on the Al surface was removed by the micro-plasma generated between particles during the SPS process. In addition, the formation of intermetallic compounds is induced by the activation of intermetallic reactions by local high temperatures. The possible chemical reactions in the Cu–Al binary system are as follows:

$$4Al + 9Cu = Cu_9Al_4 \quad (2)$$

$$2Al + Cu = CuAl_2 \quad (3)$$

$$Al + Cu = CuAl \quad (4)$$

According to the standard Gibbs free energy of formation values for chemical reactions (2) to (4), they are all forward reactions, indicating the possibility for formation of intermetallic compounds. The heats of formation for the phases in these chemical reactions are $CuAl_2$: -6.1 kJ·mol^{-1}, CuAl: -5.1 kJ·mol^{-1}, and Cu_9Al_4: -4.1 kJ·mol^{-1} [40,41]. The heats of formation for these intermetallic compounds can thus be arranged in order from smallest to largest as $CuAl_2$, CuAl, and Cu_9Al_4. The $CuAl_2$ phase will be formed first, followed by CuAl and Cu_9Al_4. However, in the XRD diffraction patterns for the Al–Cu composites subjected to SPS, only the $CuAl_2$ and Cu_9Al_4 phases were detected; the CuAl phase was not observed. This result occurred because the Cu and Al atoms more readily diffuse into

Cu$_9$Al$_4$ than CuAl, and thus the CuAl phase is not formed. This suggests that kinetic factors are more dominant than thermodynamic factors for the formation of intermetallic compounds. In addition, these results support previous studies that have suggested that kinetic factors are more dominant than thermodynamic factors for the formation of intermetallic compounds.

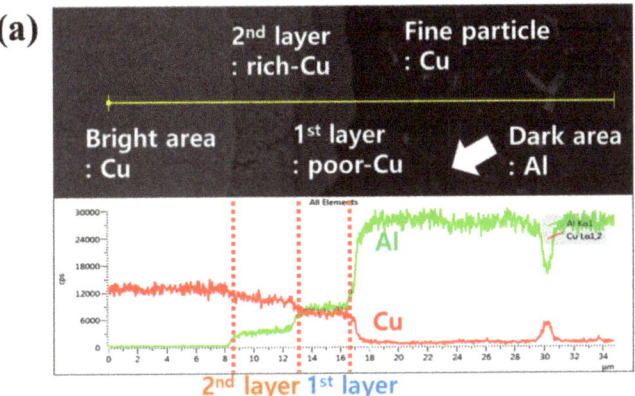

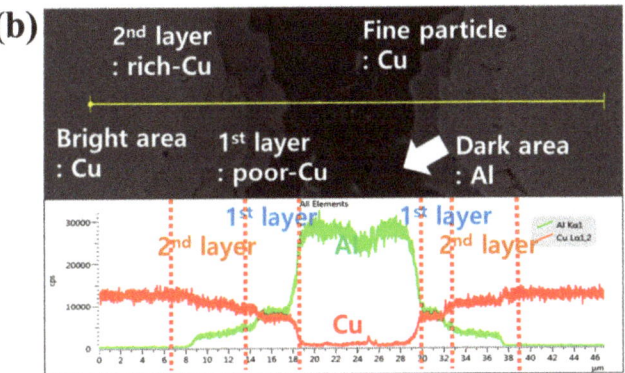

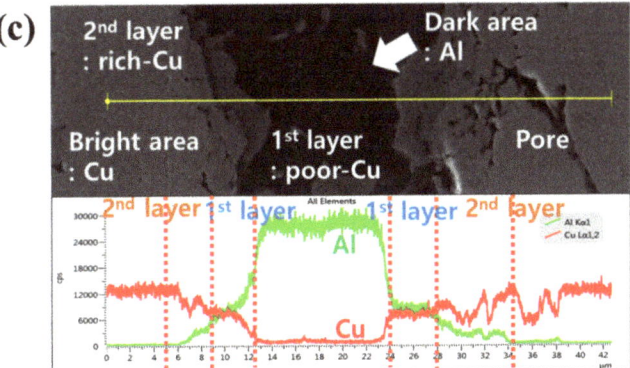

Figure 5. SEM micrograph and energy-dispersive spectroscopy (EDS) line scanning results along the yellow line for (**a**) Al–20Cu, (**b**) Al–50Cu, and (**c**) Al–80Cu.

Figure 5 shows the line mapping images for the Al–20Cu, Al–50Cu, and Al–80Cu composites after SPS. The brightest grey particles confirm the formation of pure Cu phase in the Al–Cu composite material, while the dark area shows the pure Al phase. Of the two IC layers between the Cu and Al phases, the IC layer closer to the Cu is observed to have a higher Cu content than the IC layer closer to the Al. The IC layer that formed closer to the Cu is the Cu-rich Cu_9Al_4 phase detected in the XRD pattern in Figure 4d, while the layer closer to the Al is the Al-rich $CuAl_2$ phase. The Al–20Cu, Al–50Cu, and Al–80Cu composites exhibited a thicker Cu_9Al_4 phase than $CuAl_2$ phase. According to the heat of formation for the IC phase, the $CuAl_2$ phase is formed first on the interface. However, the Al–20Cu, Al–50Cu, and Al–80Cu composites showed greater growth of the Cu_9Al_4 phase than the $CuAl_2$ phase. This should be considered a kinetic rather than thermodynamic effect. Xu et al. [39] reported activation energies based on the growth of $CuAl_2$ and Cu_9Al_4 at Cu–Al interfaces with Cu–Al wire bonds. The activation energy for the growth of the Cu_9Al_4 phase is 75.61 kJ·mol^{-1} and that for the $CuAl_2$ phase is 60.66 kJ·mol^{-1}, which means that the $CuAl_2$ phase with a lower activation energy readily participates in the IC phase growth. This suggests that the $CuAl_2$ phase is formed first at the Cu–Al interface, and then Cu_9Al_4 is formed through the diffusion of Cu atoms into the $CuAl_2$ at the $CuAl_2$–Cu interface, after which Cu_9Al_4 and $CuAl_2$ grow simultaneously. At high temperatures, the Cu_9Al_4 phase grows more rapidly due to the difference in the diffusion rates of Al atoms and Cu atoms into the formed $CuAl_2$ phase, which facilitates the growth of the Cu_9Al_4 phase until all of the $CuAl_2$ phase is consumed. However, it is considered that two phases coexist because the Al–Cu composites were subjected to the SPS process with a relatively high temperature and short heating period.

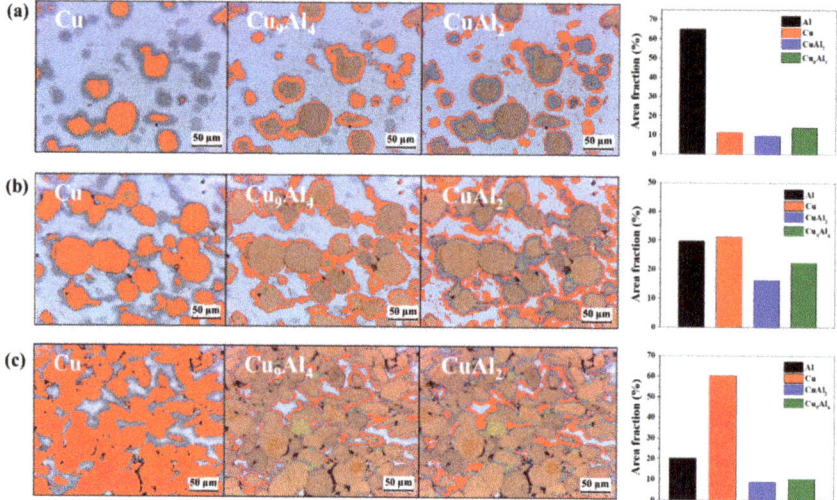

Figure 6. Area fractions of Al, Cu, Cu_9Al_4, and $CuAl_2$ obtained using image analysis software ImageJ: (**a**) Al–20Cu, (**b**) Al–50Cu, and (**c**) Al–80Cu.

As shown in the XRD diffraction pattern in Figure 4 and the EDS analysis in Figure 5, it was confirmed that not only Cu and Al phases, but also intermetallic compounds of $CuAl_2$ and Cu_9Al_4 phases, were present in the Al–Cu composites fabricated by SPS. The area analysis for each formed phase was performed using the image analyser tool in the ImageJ software, which measures the area fraction of the phases. The area fractions of the Al–20Cu, Al–50Cu, and Al–80Cu composites were measured using the ImageJ programme. For the light micrographs of Al–Cu composites fabricated by SPS, analysis using the ImageJ analysis toolbox was used to obtain the following area fractions: (i) Al area, (ii) Cu area, (iii) $CuAl_2$ area in the intermetallic compounds, and (iv) Cu_9Al_4 area in the intermetallic compounds. Figure 6 and Table 1 present the analysis results from the ImageJ programme for a section of the Al–Cu composites after SPS. The area analysis results confirm that the Al and Cu contents of the Al–20Cu, Al–50Cu, and Al–80Cu composites were approximately similar to the desired compositions. The area of the Cu_9Al_4 phase in the Al–Cu composites is approximately 1.2 times larger than area of the $CuAl_2$ phase. This confirms that the growth of the Cu_9Al_4 phase in the SPS process at high temperature is faster. The total IC phase areas in the Al–20Cu, Al–50Cu, and Al–80Cu composites are approximately 23.3%, 38.7%, and 19.3%, respectively; the area of the Al–50Cu composite is the largest. This indicates that the growth of Cu_9Al_4 is kinetically faster as the effective contact area between $CuAl_2$ and Al or Cu increases.

Table 1. Area fraction analysis results for $CuAl_2$ and Cu_9Al_4 in the Al–Cu composites.

Phase	Partition Fraction (%)		
	Al–20Cu	Al–50Cu	Al–80Cu
Aluminium	64.9 (±3.24)	29.8 (±1.49)	20.2 (±1.01)
Copper	11.5 (±0.57)	31.3 (±1.56)	60.3 (±3.01)
$CuAl_2$	9.7 (±0.62)	16.3 (±0.97)	8.9 (±0.45)
Cu_9Al_4	13.9 (±0.54)	22.6 (±0.96)	10.6 (±0.51)

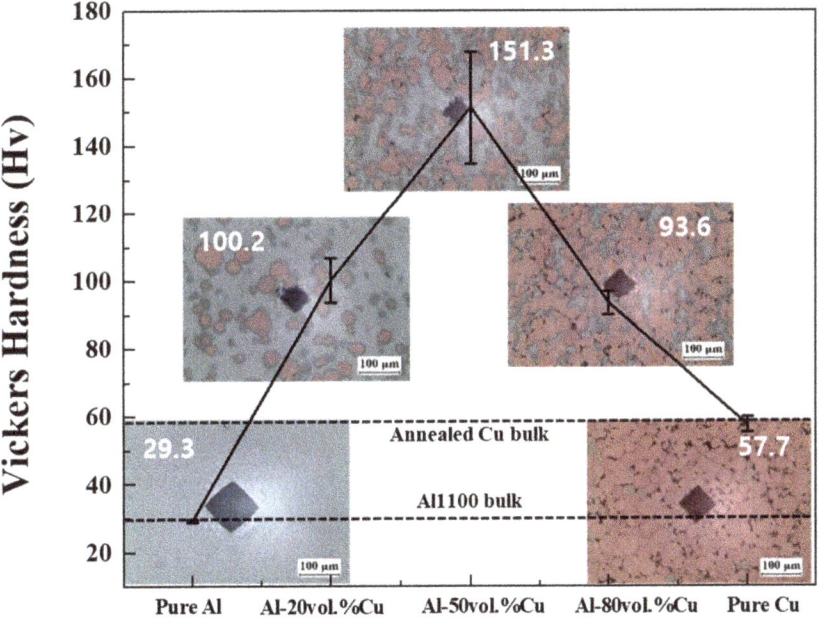

Figure 7. Vickers hardness of the fabricated pure Al, pure Cu, and Al–Cu composites.

Figure 7 shows the Vickers hardness of the Al–Cu composites and pure Al and Cu. The hardness value of a pure Al body fabricated by SPS was similar to the hardness value of Al1100 bulk, while the harness of the Cu sintered body was lower than that of annealed Cu bulk owing to existence of pores [42,43]. However, the Vickers hardness values of the Al–Cu composites were higher than that of pure Al and Cu regardless of the composition. Interestingly, the Vickers hardness of the Al–50Cu composite was about 151.3 HV, which is approximately five times greater than that of the pure Al sintered body and about 2.6 times greater than that of the pure Cu sintered body. It is suggested that this strengthening of the Al–Cu composites was affected by the presence of the ICs. This is mainly because at a higher temperature, Cu and Al will react to form intermetallic compounds, and these intermetallic compounds have a much higher hardness, thus resulting in a significant increase in the hardness at the Cu–Al interface. Under the same sintering condition, Cu and Al form a substitutional solid solution, thus resulting in lattice distortions, which can increase the resistance to dislocation movement. Therefore, plastic deformation becomes difficult, eventually leading to a significant increase in the hardness at the bonding interface. Moreover, the distribution and amount of intermetallic compounds in the Al–Cu composite materials may also be an important factor. As presented in Figure 6 and Table 1, the amount of intermetallic compounds in the Al–50Cu composite was greater than that in the other composites. Thus, the Vickers hardness of the Al–50Cu composite material was the highest because the intermetallic compounds surrounding Cu particles were uniformly dispersed. Therefore, SPS can be considered an effective process for the fabrication of composites of dissimilar materials.

Table 2 lists the thermal conductivities calculated from the thermal diffusivity and heat capacity values for the Al–Cu composites measured with the laser flash method. The pure Al and Cu sintered bodies have values that are similar to the theoretical thermal conductivities based on the mixture composition. However, the thermal conductivity of the Al–Cu composites were lower than the theoretical values and lower than that of the pure Al sintered body. In particular, the Al–50Cu composite with a higher content of intermetallic compounds exhibited a lower thermal conductivity. The interfaces of the Al, Cu, and intermetallic compounds play a critical role in determining both the microstructure and thermal conduction behaviour of the composites. The thermal conductivity of the composites decreased with increasing intermetallic compound content, because the movement of electrons and phonons is disturbed by phase interfaces, and scattering occurs with the formation and the amount of intermetallic compounds. The results show that the intermetallic compounds formed during SPS have a positive effect on the hardness of the Al–Cu composites, but a negative effect on their thermal conductivity. However, investigations using X-ray photoelectron spectroscopy, transmission electron spectroscopy, and on controlling the formation of intermetallic compounds are needed to investigate the enhancement of the material properties, and will be the focus of future studies.

Table 2. Thermal properties and densities of Al, Cu, and Al–Cu composites at room temperature.

Sample	Density			Heat Capacity $(J \cdot g^{-1} \cdot K^{-1})$	Diffusivity		Thermal Conductivity	
	Theoretical Density $(g \cdot cm^{-3})$	Experimental Density $(g \cdot cm^{-3})$	Relative Density (%)		Theoretical Diffusivity (mm^2/s)	Experimental Diffusivity (mm^2/s)	Theoretical Thermal Conductivity $(W \cdot m^{-1} \cdot K^{-1})$	Experimental Thermal Conductivity $(W \cdot m^{-1} \cdot K^{-1})$
Pure Al	2.70	2.680	99.3	0.99	97	81.96	230	219
Al–20Cu	3.95	4.111	104.0	0.84	100	46.09	264	158
Al–50Cu	5.83	5.826	99.9	0.67	105	33.55	316	130
Al–80Cu	7.71	7.110	92.2	0.59	110	45.62	367	191
Pure Cu	8.96	8.178	91.3	0.50	113	83.9	401	341

4. Conclusions

Al–Cu composite materials were successfully prepared using mechanical ball milling and SPS. The Cu powder was dispersed in Al powder through the ball milling process, and the obtained composite powders were analysed using SEM. It was determined that the ball milling process was suitable for preparing composite powders. ICs were created from reactions between the Al and Cu during SPS, and their presence in the composites was confirmed by the XRD and EDS analysis results. Regardless of the composition, the Al–Cu composite materials exhibited higher Vickers hardness values than pure Al and Cu. The Al–50Cu composite exhibited the highest Vickers hardness of approximately 151 HV. The observed strengthening effects are considered to be related to the formation of intermetallic compounds, which were formed from the reaction between Al and Cu via micro-plasma sparks during the SPS process and detected in the X-ray diffraction and EDS analyses. The Al–50Cu composite with a higher content of intermetallic compounds exhibited lower thermal conductivity. It is suggested that the properties of the Al–Cu composites were affected by the presence of the ICs. Nevertheless, SPS can be considered an effective process for fabricating Al–Cu composite materials.

Author Contributions: Conceptualization, K.K. and H.K.; Methodology, K.K.; Validation, K.K. and D.K.; Formal analysis, K.K.; Investigation, K.K., K.P., D.K., M.C. and S.C.; Writing—original draft preparation, K.K.; Writing—review and editing, K.K. and H.K.; Visualization, K.K., D.K. and M.C.; Supervision, H.K.; Project administration, H.K.; Funding acquisition, H.K.

Funding: This research was funded by the Commercializations Promotion Agency for R&D Outcomes (COMPA-2019K000076) funded by the Ministry of Science and ICT (MSIT).

Conflicts of Interest: The authors declare no conflict of interest.

References

1. Ambrogio, G.; Filice, L.; Gagliardi, F. Formability of lightweight alloys by hot incremental sheet forming. *Mater. Des.* **2012**, *34*, 501–508. [CrossRef]
2. Chen, B.; Liu, J. Contribution of hybrid fibers on the properties of the high-strength lightweight concrete having good workability. *Cem. Concr. Res.* **2005**, *35*, 913–917. [CrossRef]
3. Cole, G.; Sherman, A. Light weight materials for automotive applications. *Mater. Charact.* **1995**, *35*, 3–9. [CrossRef]
4. Froes, F.H.; Friedrich, H.; Kiese, J.; Bergoint, D. Titanium in the family automobile: The cost challenge. *JOM* **2004**, *56*, 40–44. [CrossRef]
5. Immarigeon, J.-P.; Holt, R.; Koul, A.; Zhao, L.; Wallace, W.; Beddoes, J. Lightweight materials for aircraft applications. *Mater. Charact.* **1995**, *35*, 41–67. [CrossRef]
6. Joost, W.J.; Krajewski, P.E. Towards magnesium alloys for high-volume automotive applications. *Scr. Mater.* **2017**, *128*, 107–112. [CrossRef]
7. Kulekci, M.K. Magnesium and its alloys applications in automotive industry. *Int. J. Adv. Des. Manuf. Technol.* **2008**, *39*, 851–865. [CrossRef]
8. Witik, R.A.; Payet, J.; Michaud, V.; Ludwig, C.; Månson, J.-A.E. Assessing the life cycle costs and environmental performance of lightweight materials in automobile applications. *Compos. Part A: Appl. Sci. Manuf.* **2011**, *42*, 1694–1709. [CrossRef]
9. Macke, A.; Schultz, B.F.; Rohatgi, P. Metal matrix composites. *Adv. Mater. Processes* **2012**, *170*, 19–23.
10. Huang, R.; Riddle, M.; Graziano, D.; Warren, J.; Das, S.; Nimbalkar, S.; Cresko, J.; Masanet, E. Energy and emissions saving potential of additive manufacturing: the case of lightweight aircraft components. *J. Clean. Prod.* **2016**, *135*, 1559–1570. [CrossRef]
11. Jambor, A.; Beyer, M. New cars—new materials. *Mater. Des.* **1997**, *18*, 203–209. [CrossRef]
12. Kim, T.; Zhao, C.; Lu, T.; Hodson, H. Convective heat dissipation with lattice-frame materials. *Mech. Mater.* **2004**, *36*, 767–780. [CrossRef]
13. Rawal, S.P. Metal-matrix composites for space applications. *JOM* **2001**, *53*, 14–17. [CrossRef]
14. Tian, J.; Kim, T.; Lu, T.; Hodson, H.; Queheillalt, D.; Sypeck, D.; Wadley, H. The effects of topology upon fluid-flow and heat-transfer within cellular copper structures. *Int. J. Heat Mass Transf.* **2004**, *47*, 3171–3186. [CrossRef]
15. Yang, X.; Yan, Y.; Mullen, D.; Yan, Y. Recent developments of lightweight, high performance heat pipes. *Appl. Eng.* **2012**, *33*, 1–14. [CrossRef]
16. Zinn, W.; Scholtes, B. Mechanical Surface Treatments of Lightweight Materials—Effects on Fatigue Strength and Near-Surface Microstructures. *J. Mater. Eng. Perform.* **1999**, *8*, 145–151. [CrossRef]
17. Park, K.; Park, J.; Kwon, H. Effect of intermetallic compound on the Al-Mg composite materials fabricated by mechanical ball milling and spark plasma sintering. *J. Alloy. Compd.* **2018**, *739*, 311–318. [CrossRef]
18. Moghadam, A.D.; Schultz, B.F.; Ferguson, J.B.; Omrani, E.; Rohatgi, P.K.; Gupta, N. Functional Metal Matrix Composites: Self-lubricating, Self-healing, and Nanocomposites-An Outlook. *JOM* **2014**, *66*, 872–881. [CrossRef]
19. Moghadam, A.D.; Omrani, E.; Menezes, P.L.; Rohatgi, P.K. Mechanical and tribological properties of self-lubricating metal matrix nanocomposites reinforced by carbon nanotubes (CNTs) and graphene—A review. *Compos. Part B: Eng.* **2015**, *77*, 402–420. [CrossRef]

20. Fathy, A.; El-Kady, O.; Mohammed, M.M. Effect of iron addition on microstructure, mechanical and magnetic properties of Al-matrix composite produced by powder metallurgy route. *Trans. Nonferrous Met. Soc. China* **2015**, *25*, 46–53. [CrossRef]
21. Hong, S.Y.; Markus, I.; Jeong, W.-C. New cooling approach and tool life improvement in cryogenic machining of titanium alloy Ti-6Al-4V. *Int. J. Mach. Tools Manuf.* **2001**, *41*, 2245–2260. [CrossRef]
22. Jiang, L.; Yang, H.; Yee, J.K.; Mo, X.; Topping, T.; Lavernia, E.J.; Schoenung, J.M. Toughening of aluminum matrix nanocomposites via spatial arrays of boron carbide spherical nanoparticles. *Acta Mater.* **2016**, *103*, 128–140. [CrossRef]
23. Liu, Q.; Mo, Z.; Wu, Y.; Ma, J.; Tsui, G.C.P.; Hui, D. Crush response of CFRP square tube filled with aluminum honeycomb. *Compos. Part B: Eng.* **2016**, *98*, 406–414. [CrossRef]
24. Kim, D.; Park, K.; Kim, K.; Miyazaki, T.; Joo, S.; Hong, S.; Kwon, H. Carbon nanotubes-reinforced aluminum alloy functionally graded materials fabricated by powder extrusion process. *Mater. Sci. Eng. A* **2019**, *745*, 379–389. [CrossRef]
25. Park, K.; Kim, D.; Kim, K.; Cho, S.; Kwon, H. Behavior of Intermetallic Compounds of Al-Ti Composite Manufactured by Spark Plasma Sintering. *Materials* **2019**, *12*, 331. [CrossRef]
26. Kim, D.; Park, K.; Chang, M.; Joo, S.; Hong, S.; Cho, S.; Kwon, H. Fabrication of Functionally Graded Materials Using Aluminum Alloys via Hot Extrusion. *Metals* **2019**, *9*, 210. [CrossRef]
27. Sahin, Y. Preparation and some properties of SiC particle reinforced aluminium alloy composites. *Mater. Des.* **2003**, *24*, 671–679. [CrossRef]
28. Selvakumar, S.; Dinaharan, I.; Palanivel, R.; Babu, B.G. Characterization of molybdenum particles reinforced Al6082 aluminum matrix composites with improved ductility produced using friction stir processing. *Mater. Charact.* **2017**, *125*, 13–22. [CrossRef]
29. Shirvanimoghaddam, K.; Khayyam, H.; Abdizadeh, H.; Akbari, M.K.; Pakseresht, A.; Ghasali, E.; Naebe, M. Boron carbide reinforced aluminium matrix composite: Physical, mechanical characterization and mathematical modelling. *Mater. Sci. Eng. A* **2016**, *658*, 135–149. [CrossRef]
30. Szlancsik, A.; Katona, B.; Bobor, K.; Májlinger, K.; Orbulov, I.N. Compressive behaviour of aluminium matrix syntactic foams reinforced by iron hollow spheres. *Mater. Des.* **2015**, *83*, 230–237. [CrossRef]
31. Ueno, T.; Yoshioka, T.; Ogawa, J.-I.; Ozoe, N.; Sato, K.; Yoshino, K. Highly thermal conductive metal/carbon composites by pulsed electric current sintering. *Synth. Met.* **2009**, *159*, 2170–2172. [CrossRef]
32. Wu, J.; Zhang, H.; Zhang, Y.; Li, J.; Wang, X. Effect of copper content on the thermal conductivity and thermal expansion of Al–Cu/diamond composites. *Mater. Des.* **2012**, *39*, 87–92. [CrossRef]
33. Yang, Y.; Li, X. Ultrasonic Cavitation Based Nanomanufacturing of Bulk Aluminum Matrix Nanocomposites. *J. Manuf. Sci. Eng.* **2007**, *129*, 497–501. [CrossRef]
34. Kwon, H.; Leparoux, M.; Kawasaki, A. Functionally Graded Dual-nanoparticulate-reinforced Aluminium Matrix Bulk Materials Fabricated by Spark Plasma Sintering. *J. Mater. Sci. Technol.* **2014**, *30*, 736–742. [CrossRef]
35. Selvakumar, N.; Vettivel, S. Thermal, electrical and wear behavior of sintered Cu–W nanocomposite. *Mater. Des.* **2013**, *46*, 16–25. [CrossRef]
36. Chwa, S.O.; Klein, D.; Liao, H.; Dembinski, L.; Coddet, C. Temperature dependence of microstructure and hardness of vacuum plasma sprayed Cu–Mo composite coatings. *Surf. Coatings Technol.* **2006**, *200*, 5682–5686. [CrossRef]
37. Rao, B.; Hemambar, C.; Pathak, A.; Patel, K.; Rödel, J.; Jayaram, V. Al/SiC Carriers for Microwave Integrated Circuits by a New Technique of Pressureless Infiltration. *IEEE Trans. Electron. Packag. Manuf.* **2006**, *29*, 58–63. [CrossRef]
38. Eizadjou, M.; Kazemitalachi, A.; Daneshmanesh, H.; Shahabi, H.S.; Janghorban, K. Investigation of structure and mechanical properties of multi-layered Al/Cu composite produced by accumulative roll bonding (ARB) process. *Compos. Sci. Technol.* **2008**, *68*, 2003–2009. [CrossRef]
39. Xu, H.; Liu, C.; Silberschmidt, V.V.; Pramana, S.; White, T.; Chen, Z.; Acoff, V.; Pramana, S. Behavior of aluminum oxide, intermetallics and voids in Cu–Al wire bonds. *Acta Mater.* **2011**, *59*, 5661–5673. [CrossRef]
40. Pretorius, R.; Vredenberg, A.M.; Saris, F.W.; de Reus, R. Prediction of phase formation sequence and phase stability in binary metal-aluminum thin-film systems using the effective heat of formation rule. *J. Appl. Phys.* **1991**, *70*, 3636–3646. [CrossRef]

41. Pretorius, R.; Theron, C.C.; Vantomme, A.; Mayer, J.W. Compound Phase Formation in Thin Film Structures. *Crit. Rev. Solid State Mater. Sci.* **1999**, *24*, 1–62. [CrossRef]
42. Hatch, J.E. *Aluminum: Properties and Physical Metallurgy*; American Society for Metals: Materials Park, OH, USA, 1984; p. 424.
43. Goodfellow. Available online: http://www.goodfellow.com/pdf/1207_1111010.pdf (accessed on 22 January 2019).

 © 2019 by the authors. Licensee MDPI, Basel, Switzerland. This article is an open access article distributed under the terms and conditions of the Creative Commons Attribution (CC BY) license (http://creativecommons.org/licenses/by/4.0/).

Article

A Different Approach to Estimate Temperature-Dependent Thermal Properties of Metallic Materials

Luís Felipe dos Santos Carollo, Ana Lúcia Fernandes de Lima e Silva and Sandro Metrevelle Marcondes de Lima e Silva *

Heat Transfer Laboratory—LabTC, Institute of Mechanical Engineering—IEM, Federal University of Itajubá—UNIFEI, Campus Prof. José Rodrigues Seabra, Av. BPS, 1303, Itajubá 37500-903, MG, Brazil
* Correspondence: metrevel@unifei.edu.br; Tel.: +55-35-3629-1069

Received: 19 July 2019; Accepted: 10 August 2019; Published: 13 August 2019

Abstract: Thermal conductivity, λ, and volumetric heat capacity, ρc_p, variables that depend on temperature were simultaneously estimated in a diverse technique applied to AISI 1045 and AISI 304 samples. Two distinctive intensities of heat flux were imposed to provide a more accurate simultaneous estimation in the same experiment. A constant heat flux was imposed on the upper surface of the sample while the temperature was measured on the opposite insulated surface. The sensitivity coefficients were analyzed to provide the thermal property estimation. The Broydon-Fletcher-Goldfarb-Shanno (BFGS) optimization technique was applied to minimize an objective function. The squared difference objective function of the numerical and experimental temperatures was defined considering the error generated by the contact resistance. The temperature was numerically calculated by using the finite difference method. In addition, the reliability of the results was assured by an uncertainty analysis. Results showing a difference lower than 7% were obtained for λ and ρc_p, and the uncertainty values were above 5%.

Keywords: temperature-dependent thermal properties; simultaneous estimation; optimization; sensitivity coefficients; uncertainty analysis

1. Introduction

At present, globalization provides newer, faster, more reliable, and more accurate techniques to estimate thermal properties of materials depending on temperature. The cost of obtaining the parameters is another important issue, since it determines the reliability to compete in the internal and external markets. This paper proposes a technique that may be applied, for example, to accurately select, from the point of view of thermal properties, which materials will be employed in the manufacturing of heat exchangers. The methodology that leads to correct values of the thermal properties allows for saving of energy and other consequent environmental benefits; matters which, recently, have been largely considered. The machining process can be cited as an example for the aforementioned saving. A large amount of heat produced during the cutting process is dissipated to the tool holder. Knowing the values of the thermal conductivity of the tool and the tool holder leads to the correct choice of their material. Therefore, researchers have developed several procedures in this field [1,2].

A number of methods are available to estimate the thermal properties considering precision, speed, and cost, among other characteristics. In this context, Jannot et al. [3] presented a study in which the thermal conductivity of insulated materials was determined based on a pulsed method with a good precision, Xamán et al. [4] applied a guarded hot plate apparatus for the same purpose, and Thomas et al. [5] determined thermal conductivity and specific heat of insulated materials by applying a new experimental design. These thermal properties may be determined individually or together, and a great part of the estimations happen safely, precisely, and rapidly; however, few were used for temperature-dependent estimation or metallic materials. Other researchers present techniques that allow the estimation of only one temperature-dependent thermal property, for example, Aksöz et al. [6] estimated the thermal conductivity of Al-Cu alloys by using a radial heat flow apparatus and changed the initial temperature and Karimi et al. [7] determined the thermal conductivity of silver alloys by varying the temperature based on a linear heat flux apparatus. Recently, many researchers have presented techniques to simultaneously estimate temperature-dependent thermal properties, such as Sadeghi et al. [8] who determined thermal conductivity and diffusivity of SiC samples by applying the microwave heating process, Zamel et al. [9] who presented an improved firework algorithm to solve inverse problems allowing simultaneous estimation of properties of molten salt, and Öztürk et al. [10] who presented a method to estimate thermal conductivity and specific heat temperature-dependent of thermal protective fabric with good results. Other studies were performed [11–16], but none of them included the possibility of estimating thermal properties of metals depending on temperature simultaneously. Moreover, the experimental apparatus for most of these techniques is usually expensive.

Thus, this work presents a technique to simultaneously determine volumetric heat capacity, ρc_p, and thermal conductivity, λ, of AISI 1045 and AISI 304 samples; variables that depend on temperature. Some advantages of this method are the low cost, the precision, and the speed when compared with the techniques cited. Additionally, the uncertainty analysis presented in this work considers the influence of the numerical and experimental temperature errors and contact resistance. This work presents the betterments carried out concerning Carollo et al. [17].

2. Materials and Methods

2.1. Thermal Design Model

The representation of the one-dimensional (1D) heat diffusion model is presented in Figure 1. This thermal model is obtained by using a resistive heater between two samples, and the sample-heater set is insulated. The thickness of the sample is much smaller than the other dimensions to ensure the one dimension.

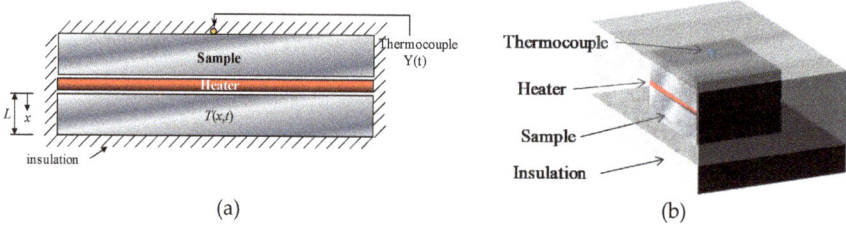

Figure 1. A one-dimensional (1D) representation of the model. (**a**) thermal model; (**b**) three-dimensional view of the thermal model.

The conduction equation for the problem in Figure 1 is:

$$\frac{\partial}{\partial x}\lambda(T)\frac{\partial T(x,t)}{\partial x} = \rho c_p(T)\frac{\partial T(x,t)}{\partial t}, \qquad (1)$$

In accordance with the literature, there are two methods to estimate thermal properties dependent on temperature. In the first method, presented in Özisik [18], the thermal model is based on nonlinear heat conduction. The second method adopts constant thermal properties within a temperature range to solve the thermal model [19]. Thus, the initial temperature was defined (T_0) and the estimation of the properties occurred considering 5 °C as the maximum range of temperature. This condition was performed for all the desired temperatures.

Therefore, the heat diffusion equation for the thermal problem with constant thermal properties is expressed as below:

$$\frac{\partial T(x,t)}{\partial x^2} = \frac{\rho c_p}{\lambda} \frac{\partial T(x,t)}{\partial t}, \quad (2)$$

subjected to the following conditions of boundary:

$$-\lambda \frac{\partial T(x,t)}{\partial x} = \varphi(t) \text{ at } x = 0, \quad (3)$$

$$\frac{\partial T(x,t)}{\partial x} = 0 \text{ at } x = L, \quad (4)$$

and the initial condition:

$$T(x,t) = T_0 \text{ at } t = 0, \quad (5)$$

where x is the heat direction, t the temporal interval, φ the applied heat flow, T_0 the temperature in the beginning of the process, and L the thickness.

The finite difference method was used to calculate the numerical temperature of the conduction (Equation (2)).

2.2. Objective Function

Equation (6) presents the objective function applied to estimate ρc_p and λ:

$$F = (R_c'' \varphi_m)^2 + \sum_{j=1}^{m} (Y_j - T_j)^2, \quad (6)$$

where m is the number of points where temperature was measured, Y is the measured temperature, R_c'' is the heater thermal contact resistance, and φ_m is the weighted average heat flow.

The optimum values of ρc_p and λ are required to minimize Equation (6). To perform this procedure, the BFGS sequential optimization technique [20] is used in this work.

2.3. Experimental Procedure

The experimental apparatus sketch used is shown in Figure 2. As can be seen, all the components are numbered following this configuration: 1: Micro-computer; 2: Data acquisition; 3: Oven; 4: Multimeter; 5: Power supply; and 6: Multimeter.

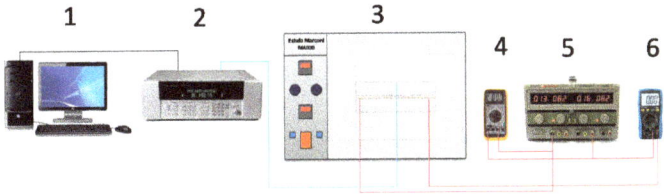

Figure 2. Experimental apparatus sketch used to estimate ρc_p and λ.

In this work, two materials were analyzed: AISI 1045 Steel (99.9 × 99.9 × 11.9 mm^3) and AISI 304 stainless steel (49.9 × 49.9 × 10.5 mm^3). Due to the different dimensions of the samples, a 99.5 × 99.5 × 0.2 mm^3 resistive kapton heater with 23.2 Ω was necessary and another of 48.5 × 48.5 × 0.2 mm^3 with 24.4 Ω. The resistive kapton heater was chosen due to its thinness, which allows a faster and more uniform warming. An Instrutemp ST-305 D-II power supply (Instrutemp Instrumentos de Medição Ltda, São Paulo, Brazil) was used to provide the adequate heat flow to perform the experiments. To ensure the correct values of current and resistance, calibrated multimeters were used (Minipa ET-2042-C (Minipa do Brasil, Joinvile, Brazil) for the current and Instrutherm MD-380 (Instrutherm Instrumentos de Medição, São Paulo, Brazil) for the resistance. Additionally, a symmetrical assembly was set up to reduce the errors caused by the measurement of heat flux on the top surface. A data acquisition system Agilent 34980A (Agilent, Santa Clara, CA, USA) was used to connect Type T thermocouples (30 AWG), welded by a capacitive discharge. To provide different initial conditions, the heater-sample set was placed inside a MA030 Marconi oven (Marconi Equipamentos para Laboratórios, Piracicaba, Brazil). The whole set was insulated by ceramic fiber plates with two purposes (Figure 1b): To guarantee a 1D heat flow and to reduce the convection effects. Lastly, all the experiments were performed in a temperature-controlled room.

3. Results and Discussion

All the experiments were performed following the procedure defined by Carollo et al. [17].

3.1. AISI 1045 Steel

In order to achieve significant results to simultaneously estimate λ and ρc_p, 15 experiments were performed for each initial condition (25 °C, 50 °C, 75 °C, 100 °C, 125 °C, and 150 °C). The experiment lasted 80 s each, following these conditions: 0–10 s with heat flux of 7709 Wm^{-2}; 10–70 s with 1854 Wm^{-2}, and 70–80 s without heat flux. These conditions provide the best estimation of the thermal properties [17] and respect the hypotheses of constant thermal properties for each initial condition, since the difference between the initial and final temperature must be lower than 5 K. Lastly, the temperature was monitored in a time interval of 0.1 s so as to have more data.

To ensure the best region and the ideal condition to determine the thermal properties, two analyses were done: The first analysis corresponded to the sensitivity and objective function and the second concerned the best condition and design for the experiments [17].

Figure 3 presents the sensitivity coefficient (sens. coef.) of ρc_p and λ at $x = L$. These coefficients are important to indicate the best conditions to estimate the properties, such as, experimental time, time interval, number of points analyzed, and others. The sensitivity coefficient of λ increased only in the beginning of the experiment and remained constant after the heat flux was changed, until the power supply was turned off. By analyzing the sensitivity coefficient of ρc_p it was possible to confirm that the value increased while there was heat flux. Because of this behavior, the best condition to simultaneously estimate the properties was applying two diverse heat flux intensities, the higher of which was applied in the beginning to maximize the sensitivity for thermal conductivity estimation and the lower was applied to guarantee enough sensitivity for the volumetric heat capacity estimation. It is important to claim that simultaneous estimation was possible because there was no dependence between the sensitivity curves.

The evaluation of Equation (6) for each property is shown in Figure 4. λ and ρc_p were estimated simultaneously due to a minimum value for each property. It is important to inform that the contact resistance of the heater on the sample was considered in Equation (6) to find its influence on the temperature measurements. In this study, a temperature difference of 0.23 °C corresponded to the influence of this contact resistance.

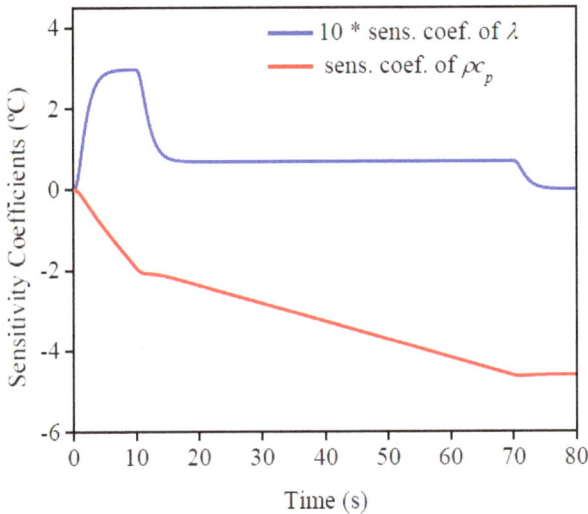

Figure 3. AISI 1045 steel with its corresponding sensitivity coefficients.

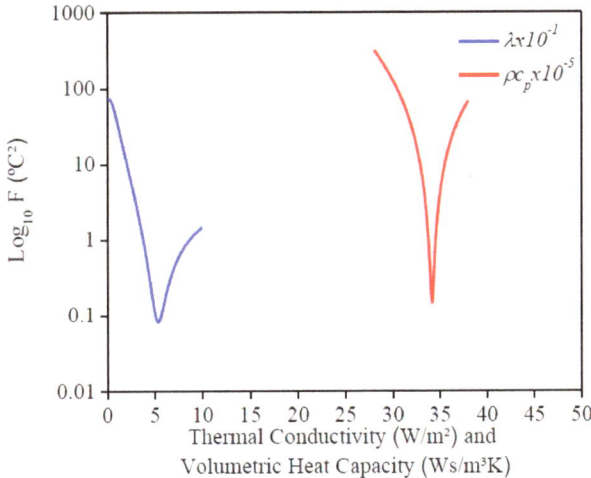

Figure 4. Values of F (Equation (6)) for AISI 1045 steel.

The best condition and design for the experiment is presented in Figure 5. This analysis indicates that the best quality for the experiment was when the sensitivity coefficient of ρc_p and λ plus the temperature difference was close to 0 ($X_1 + X_2 + Y - Y_0 \cong 0$). This analysis is relevant because it is a complement to the sensitivity analysis, in other words, it is a confirmation that all the established conditions allow a precise property estimation. A good condition of the experiment may be seen here, since the highest difference was around 0.12 °C. This affirmation can be checked when the obtained difference is compared with the difference between the final and initial temperature of each experiment, which is around 4 °C.

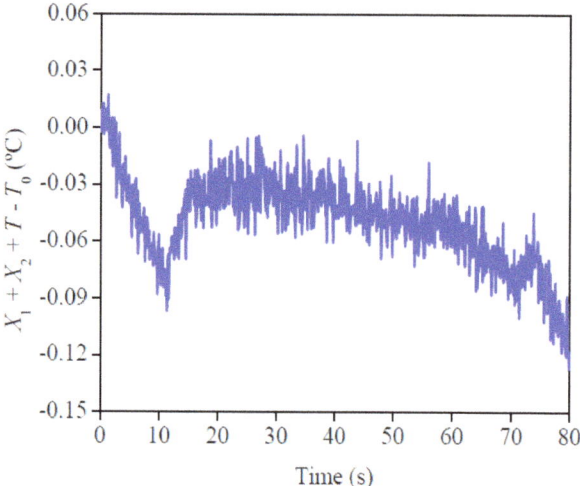

Figure 5. Results of the ideal condition and design for the experiment on AISI 1045.

Figure 6 shows the applied heat flux at $x = 0$ and the temperatures values at $x = L$. It is possible to see the good agreement between the experimental temperature and numerical temperature, which was calculated by using the obtained thermal properties. To confirm this affirmation, Figure 7 presents the residuals between these temperatures. Once the maximum difference found was around 0.10 °C, it was possible to confirm the good quality of the methodology. This affirmation can be validated by comparing the obtained difference with the thermocouple uncertainty, that it is around 0.10 °C. Lastly, this small difference can be attributed to the isolation condition.

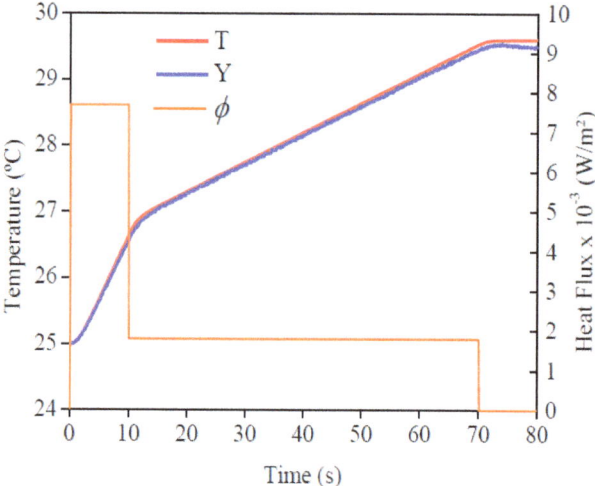

Figure 6. Experimental heat flow (ϕ) for AISI 1045. Comparison of temperatures obtained numerically (T) and experimentally (Y).

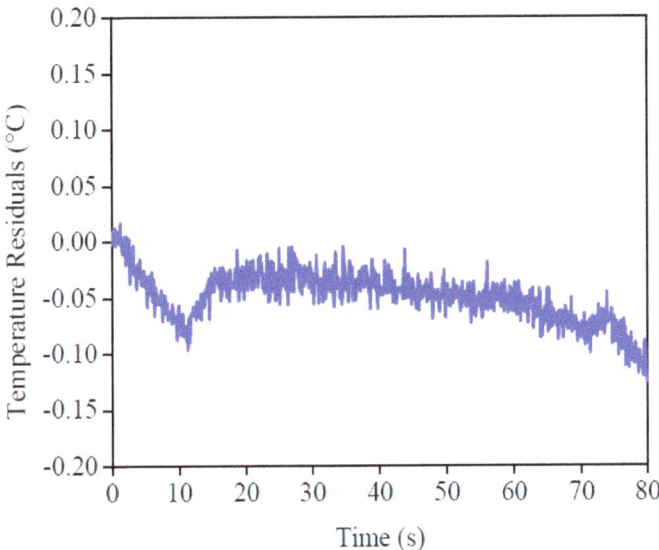

Figure 7. Residuals of temperatures of AISI 1045.

The results of ρc_p and λ on the AISI 1045 steel sample for each initial temperature are presented in Table 1. The percentage difference between the average and the literature value was considered to calculate the error. Based on the lower standard deviation and the error found, the estimated values of ρc_p and λ show a good conformity when compared to the literature values. One can also see that the results of ρc_p are more precise due to its higher sensitivity (Figure 3).

Table 1. Average results, standard deviation, and error of ρc_p and λ for AISI 1045.

Average T_0 (°C)	Thermal Properties	Mean	Obtained value from Grzesik et al. [21]	Standard Deviation	Error (%)
25.4	$\rho c_p \times 10^{-6}$ (J/m³K)	3.48	3.43	±0.02	1.44
	λ (W/mK)	52.04	51.80	±0.49	0.46
50.1	$\rho c_p \times 10^{-6}$ (J/m³K)	3.54	3.50	±0.05	1.18
	λ (W/mK)	51.39	51.4	±0.25	0.02
75.4	$\rho c_p \times 10^{-6}$ (J/m³K)	3.59	3.54	±0.04	1.36
	λ (W/mK)	50.92	51.00	±0.37	0.16
100.4	$\rho c_p \times 10^{-6}$ (J/m³K)	3.61	3.66	±0.03	0.99
	λ (W/mK)	50.47	50.40	±0.28	0.14
125.3	$\rho c_p \times 10^{-6}$ (J/m³K)	3.66	3.61	±0.05	1.42
	λ (W/mK)	49.75	49.80	±0.38	0.10
150.2	$\rho c_p \times 10^{-6}$ (J/m³K)	3.74	3.77	±0.02	0.86
	λ (W/mK)	49.55	49.40	±0.36	0.30

Figures 8 and 9 present the literature and experimental result values. One can see the good agreement of the curves, which present a correlation factor of 0.98 for λ and 0.99 for ρc_p. In accordance with Montgomery and Runger [22], the correlation factor indicates a quantitative measurement between two factors. Moreover, when the correlation factor presents value from +0.9 up to +1.0, it is possible to say that the correlation is direct and reliable.

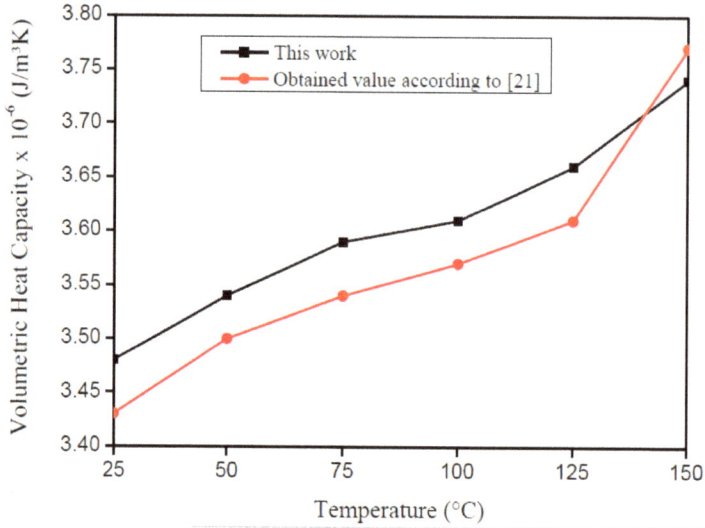

Figure 8. Comparison between the literature values with the estimated results of ρc_p on the AISI 1045 steel plate.

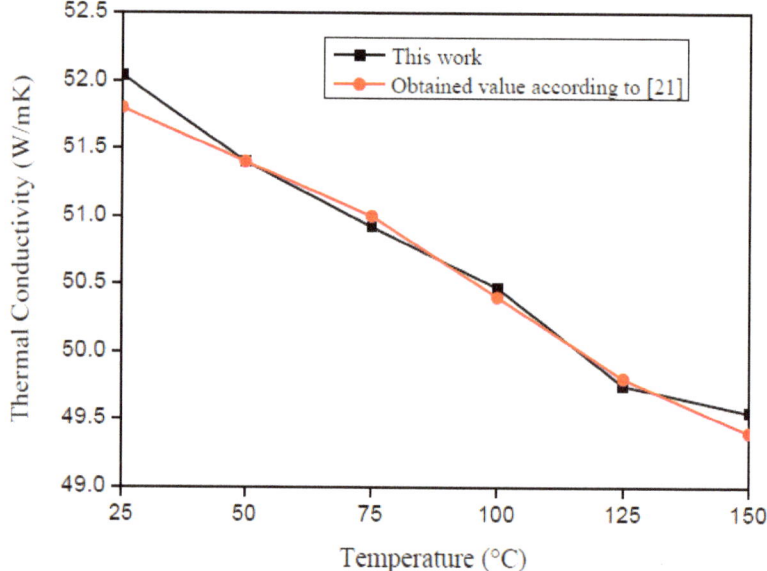

Figure 9. Comparison between the literature values with the estimated results of λ on the AISI 1045 steel sample.

From these results, Equations (7) and (8) can be written as follows for ρc_p and λ, respectively:

$$\lambda(T) = -0.02228 \times T + 52.500 \; [\text{W/mK}], \qquad (7)$$

$$\rho c_p(T) = (0.00210 \times T + 3.43253) \times 10^6 \; [\text{J/m}^3\text{K}], \qquad (8)$$

These equations can be used in the range of 25 °C up to 150 °C.

3.2. AISI 304 Stainless Steel

Following the same procedure that was applied for the AISI 1045 steel sample, 15 experiments were performed for each initial condition (25 °C, 50 °C, 75 °C, 100 °C, 125 °C, and 150 °C) to estimate ρc_p and λ simultaneously. Each experiment lasted 150 s following this condition: 0–20 s with a heat flux of 2672 Wm^{-2}; 20–140 s with 668 Wm^{-2}, and 140–150 s without a heat flux.

Figures 10 and 11 present the sensitivity coefficients and the objective function, respectively, for each property. It can be seen that the behavior found was the same as the AISI 1045 steel, so it is possible to estimate the properties simultaneously. By analyzing the sensitivity coefficient of both materials, it is possible to affirm that the thermal conductivity estimation could be more precisely for AISI 304 stainless steel than 1045 steel. This is because there is more time for information, in other words, more points to analyze, and the difference between the values of the sensitivity coefficients for both properties are lower. This behavior is a consequence of the lower thermal conductivity of stainless steel when it is compared to 1045 steel.

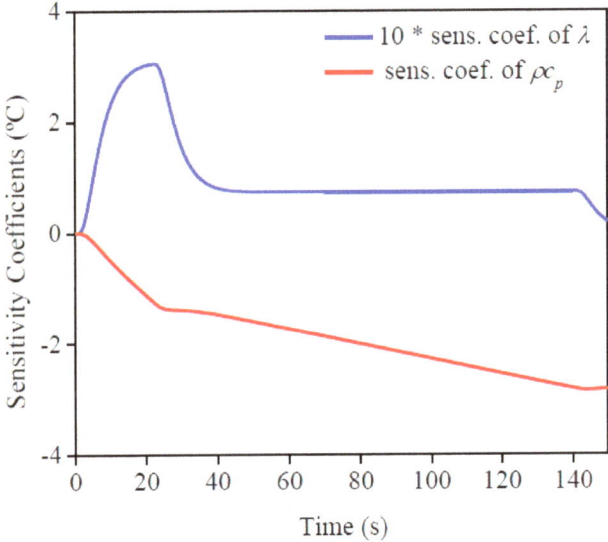

Figure 10. Sensitivity coefficient of the AISI 304 stainless steel sample.

Figure 12 shows the results for the analysis of the best experimental configuration. The maximum deviation found, around 0.05 °C, was lower than the uncertainty of the thermocouple, which confirms the reliability of the results and the good experimental configuration defined based on Figures 10 and 11.

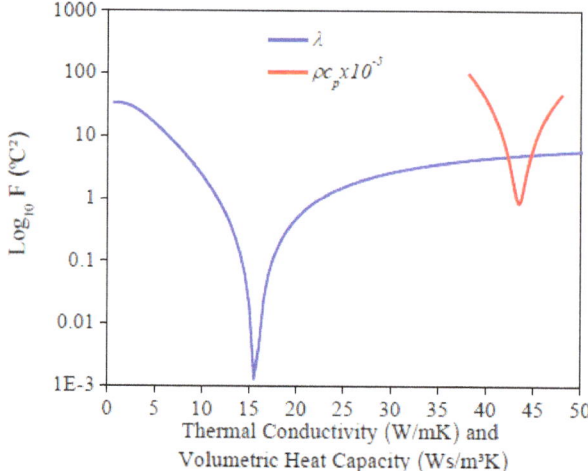

Figure 11. Values of the objective function for the AISI 304 sample.

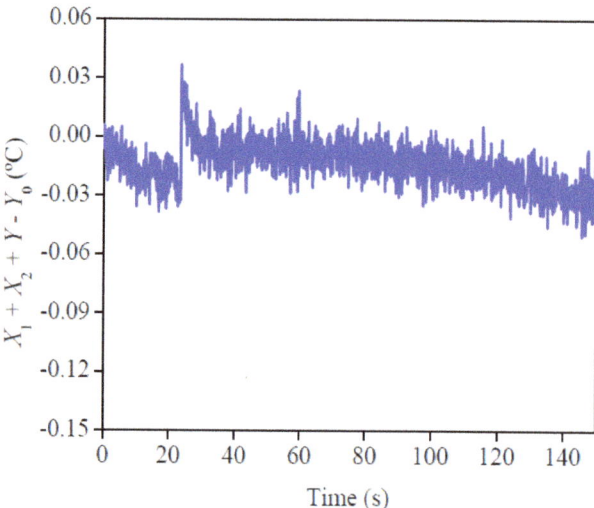

Figure 12. Results of the ideal condition and design for the experiment with AISI 304.

The imposed heat flux and the temperatures are presented in Figure 13. By analyzing this figure, one can see the good concordance between the temperatures. To validate this affirmation, the temperature residuals, where the maximum deviation was 0.05 °C, are presented in Figure 14. If this value was compared to the temperature difference, around 3 °C, one could see the good quality of the obtained results.

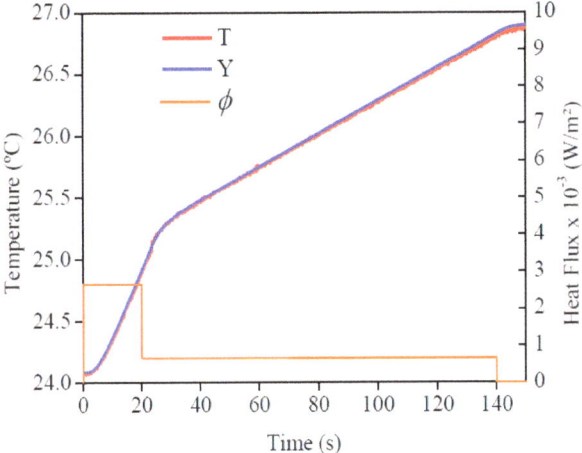

Figure 13. Experimental heat flow (ϕ) for AISI 304. Comparison of temperatures obtained numerically (T) and experimentally (Y).

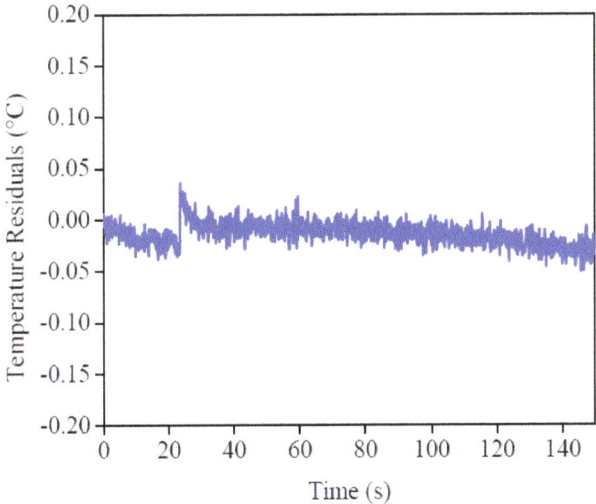

Figure 14. Residuals of temperature of AISI 304.

The results of ρc_p and λ on the AISI 304 stainless steel sample for each initial temperature are presented in Table 2. Based on the lower standard deviation and the error found, the estimated values of ρc_p and λ show a good conformity when compared to the literature values. This affirmation is based on the low standard deviation and error found. Similar to AISI 1045 steel, one may see that the results of ρc_p are more precise due to its higher sensitivity.

Table 2. Results obtained for the AISI 304 stainless steel sample.

Average T_0 (°C)	Thermal Properties	Mean	Obtained value from Abas et al. [23]	Standard Deviation	Error (%)
24.8	$\rho c_p \times 10^{-6}$ (J/m^{3}K)	4.35	4.35	±0.12	0.03
	λ (W/mK)	15.49	14.8	±0.39	4.64
49.7	$\rho c_p \times 10^{-6}$ (J/m^{3}K)	4.55	4.40	±0.09	3.31
	λ (W/mK)	16.04	15.0	±0.58	6.95
74.9	$\rho c_p \times 10^{-6}$ (J/m^{3}K)	4.61	4.45	±0.06	3.64
	λ (W/mK)	16.33	15.3	±0.47	6.76
99.2	$\rho c_p \times 10^{-6}$ (J/m^{3}K)	4.63	4.46	±0.07	3.77
	λ (W/mK)	16.63	15.6	±0.53	6.62
124.9	$\rho c_p \times 10^{-6}$ (J/m^{3}K)	4.65	4.50	±0.06	3.22
	λ (W/mK)	16.97	15.9	±0.57	6.74
149.5	$\rho c_p \times 10^{-6}$ (J/m^{3}K)	4.80	4.60	±0.09	4.38
	λ (W/mK)	16.73	16.4	±0.67	2.01

Figures 15 and 16 present the literature and experimental result values. On analyzing Figures 15 and 16, it is possible to verify the good agreement between the obtained results and those from the literature. To validate this affirmation, a correlation study was performed, and the correlation factor was 0.86 for λ and 0.95 for ρc_p.

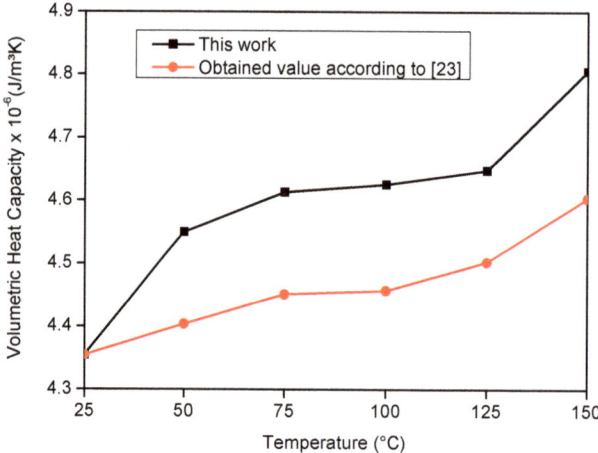

Figure 15. Comparison between the literature values with the estimated results of ρc_p on the AISI 304 stainless steel sample.

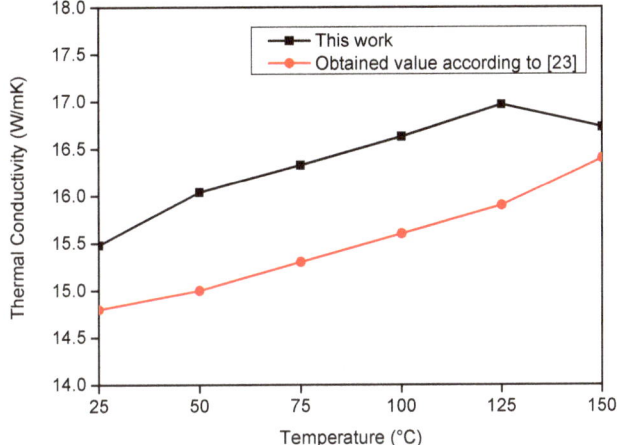

Figure 16. Comparison between the literature values with the estimated results of λ on the AISI 304 stainless steel sample.

From these results, Equations (9) and (10) can be written as follows for ρc_p and λ respectively:

$$\lambda(T) = 0.01067 \times T + 15.4293 \; [\text{W}/\text{mK}], \tag{9}$$

$$\rho c_p(T) = (0.0029266 \times T + 4.34313) \times 10^6 \; [\text{J}/\text{m}^3\text{K}], \tag{10}$$

These equations can be used in the range from 25 °C to 150 °C.

4. Uncertainty Analysis

The uncertainty propagation was considered to perform this analysis, as described in Carollo et al. [17] and Taylor [24], and it is important to assure the reliability of the estimated results.

Equations (11) and (12) show the uncertainty estimation based on the objective function (Equation (6)):

$$U_{final}^2 = U_Y^2 + U_T^2 + U_{BFGS}^2, \tag{11}$$

$$U_{final}^2 = U_{aquis.}^2 + U_{therm.}^2 + U_{contact\;resist.}^2 + U_{insul.}^2 + U_{current}^2 + U_{resistance}^2 + U_{MDF}^2 + U_{BFGS}^2 \tag{12}$$

Individual uncertainty, which was divided by the mean value of the parameter, was used to calculate the partial uncertainty. Therefore, Table 3 presents the final uncertainty. One can see that these values are acceptable once they are around 5%.

Table 3. Uncertainty values for each analyzed material.

Material	Uncertainty (%)
AISI 1045 steel	5.45
AISI 304 stainless steel	4.79

5. Conclusions

This paper presents a different approach for the estimation of λ and ρc_p in metallic materials simultaneously depending on temperature. The materials analyzed were the AISI 304 stainless steel and AISI 1045 steel. The good results found can be confirmed since the difference between the estimated values and literature is small, that is, lower than 7%, the standard deviation is the low, and the good uncertainty values are lower than 6%.

This work is validated to estimate λ and ρc_p simultaneously in metals. However, this technique may be applied to reliably estimate λ and ρc_p of metals that present thermal conductivity from 10 W/mK to 60 W/mK in a range of 25 °C up to 150 °C.

For future work, the use of a thermal model designed in three dimensions should be used to analyze the locations of temperature sensors in different positions to determine the areas that display better sensitivity to estimate λ and ρc_p.

Author Contributions: L.F.d.S.C. developed the methodology, conceived and wrote the paper; A.L.F.d.L.e.S. was the co-advisor of this work; and S.M.M.d.L.e.S. is a specialist in heat transfer and was the advisor of this work.

Funding: The authors would like to thank CNPq, FAPEMIG, and CAPES for their financial support.

Conflicts of Interest: The authors declare no conflict of interest. The founding sponsors had no role in the design of the study; in the collection, analyses, or interpretation of data; in the writing of the manuscript, or in the decision to publish the results.

References

1. Santos, M.R.; Lima e Silva, S.M.M.; Machado, A.R.; Silva, M.B.; Guimarães, G.; Carvalho, S.R. Analysis of Effects of Cutting Parameters on Cutting Edge Temperature Using Inverse Heat Conduction Technique. *Math. Prob. Eng.* **2014**, *2014*, 1–11. [CrossRef]
2. Brito, R.F.; Carvalho, S.R.; Lima e Silva, S.M.M. Investigation of Thermal Aspects in a Cutting Tool Using Comsol and Inverse Problem. *Appl. Therm. Eng.* **2015**, *86*, 60–68. [CrossRef]
3. Jannot, Y.; Degiovanni, A.; Payet, G. Thermal conductivity measurement of insulating materials with a three layers device. *Int. J. Heat Mass Transf.* **2009**, *52*, 1105–1111. [CrossRef]
4. Xamán, J.; Lira, L.; Arce, J. Analysis of the temperature distribution in a guarded hot plate apparatus for measuring thermal conductivity. *Appl. Therm. Eng.* **2009**, *29*, 617–623. [CrossRef]
5. Thomas, M.; Boyard, N.; Lefèvre, N.; Jarny, Y.; Delaunay, D. An experimental device for the simultaneous estimation of the thermal conductivity 3-D tensor and the specific heat of orthotropic composite materials. *Int. J. Heat Mass Transf.* **2010**, *53*, 5487–5498. [CrossRef]
6. Aksöz, S.; Ocak, Y.; Marasli, N.; Cadirli, E.; Kaya, H.; Böyük, U. Dependency of the Thermal and Electrical Conductivity on the Temperature and Composition of Cu in the Al based Al–Cu Alloys. *Exp. Therm. Fluid Sci.* **2010**, *34*, 1507–1516. [CrossRef]
7. Karimi, G.; Li, X.; Teerstra, P. Measurement of Through-plane Effective Thermal Conductivity and Contact Resistance in PEM Fuel Cell Diffusion. *Media Electrochim. Acta* **2010**, *55*, 1619–1625. [CrossRef]
8. Sadeghi, E.; Djilali, N.; Bahrami, M. A Novel Approach to Determine the In-plane Thermal Conductivity of Gas Diffusion Layers in Proton Exchange Membrane Fuel Cells. *J. Power Sources* **2011**, *196*, 3565–3571. [CrossRef]
9. Zamel, N.; Becker, J.; Wiegmann, A. Estimating the Thermal Conductivity and Diffusion Coefficient of the Microporous Layer of Polymer Electrolyte Membrane Fuel Cells. *J. Power Sources* **2012**, *207*, 70–80. [CrossRef]
10. Öztürk, E.; Aksöz, S.; Keslioglu, K.; Marasli, N. The Measurement of Thermal Conductivity variation with Temperature for Sn-20 wt.% in based Lead-free Ternary Solders. *Thermochim. Acta* **2013**, *554*, 63–70. [CrossRef]
11. Zgraja, J.; Cieslak, A. Induction heating in estimation of thermal properties of conductive materials. *Int. J. Comput. Math. Electr. Electron. Eng.* **2017**, *36*, 458–468. [CrossRef]
12. Somasundharam, S.; Reddy, K.S. Simultaneously estimation of thermal properties of orthotropic material with non-intrusive measurement. *Int. J. Heat Mass Transf.* **2018**, *126*, 1162–1177. [CrossRef]
13. García, E.; Amaya, I.; Correa, R. Estimation of thermal properties of a solid sample during a microwave heating process. *Appl. Therm. Eng.* **2018**, *129*, 587–595. [CrossRef]
14. Lembcke, L.G.M.; Roubinet, D.; Gidel, F.; Irving, J.; Pehme, P.; Parker, B.L. Analytical analysis of borehole experiments for the estimation of subsurface thermal properties. *Adv. Water Resour.* 2016 *91*, 88–103.
15. Ren, Y.; Qi, H.; He, M.; Ruan, S.; Ruan, L.; Tan, H. Application of an improved firework algorithm for simultaneous estimation of temperature-dependent thermal and optical properties of molten salt. *Int. Commun. Heat Mass Transf.* **2016**, *77*, 33–42. [CrossRef]

16. Udayraj; Talukdar, P.; Das, A.; Alagirusamy, R. Simultaneous estimation of thermal conductivity and specific heat of thermal protective fabrics using experimental data of high heat flux exposure. *Appl. Therm. Eng.* **2016**, *107*, 785–796. [CrossRef]
17. Carollo, L.F.S.; Lima e Silva, A.L.F.; Lima e Silva, S.M.M. Applying Different Heat Flux Intensities to Simultaneously Estimate the Thermal Properties of Metallic Materials. *Meas. Sci. Technol.* **2012**, *23*, 1–10. [CrossRef]
18. Özisik, M.N. *Heat Conduction*; John Wiley & Sons: Etobicoke, ON, Canada, 1993; p. 692.
19. Tillmann, A.R.; Borges, V.L.; Guimarães, G.; Lima e Silva, A.L.F.; Lima e Silva, S.M.M. Identification of Temperature-Dependent Thermal Properties of Solid Materials. *J. Braz. Soc. Mech. Sci. Eng.* **2008**, *30*, 269–278. [CrossRef]
20. Vanderplaats, G.N. *Numerical Optimization Techniques for Engineering Design*; McGraw: New York, NY, USA, 2005; p. 465.
21. Grzesik, W.; Nieslony, P.; Bartoszuk, M. Modelling of the Cutting Process Analytical and Simulation Methods. *Adv. Manuf. Sci. Technol.* **2009**, *33*, 6–29.
22. Montgomery, D.C.; Runger, G.C. *Applied Statistic and Probability for Engineers*; John Wiley & Sons: New York, NY, USA, 2013; p. 811.
23. Abas, R.A.; Hayashi, M.; Seetharaman, S. Thermal Diffusivity Measurements of Some Industrially Important Alloys by a Laser Flash Method. *Int. J. Mater. Res.* **2009**, *98*, 1–6. [CrossRef]
24. Taylor, J.R. *An Introduction to Error Analysis: The Study of Uncertainties in Physical Measurements*; University Science Books: Sausalito, CA, USA, 1997; p. 488.

© 2019 by the authors. Licensee MDPI, Basel, Switzerland. This article is an open access article distributed under the terms and conditions of the Creative Commons Attribution (CC BY) license (http://creativecommons.org/licenses/by/4.0/).

Article

Thermophysical Measurements in Liquid Alloys and Phase Diagram Studies

Yuri Kirshon [1,†], Shir Ben Shalom [1,†], Moran Emuna [2], Yaron Greenberg [2], Joonho Lee [3], Guy Makov [1,*] and Eyal Yahel [2]

1. Department of Materials Engineering, Ben-Gurion University of the Negev, Beer Sheva 84105, Israel; iurik@post.bgu.ac.il (Y.K.); shirbe@bgu.ac.il (S.B.S.)
2. Physics Department, Nuclear Research Centre Negev, Beer Sheva 84190, Israel; morankm131@gmail.com (M.E.); yaron300@gmail.com (Y.G.); eyalyahel@gmail.com (E.Y.)
3. Department of Materials Science and Engineering, Korea University, Seoul 02841, Korea; joonholee@korea.ac.kr
* Correspondence: makovg@bgu.ac.il
† These authors contributed equally to this work.

Received: 12 November 2019; Accepted: 28 November 2019; Published: 2 December 2019

Abstract: Towards the construction of pressure-dependent phase diagrams of binary alloy systems, both thermophysical measurements and thermodynamic modeling are employed. High-accuracy measurements of sound velocity, density, and electrical resistivity were performed for selected metallic elements from columns III to V and their alloys in the liquid phase. Sound velocity measurements were made using ultrasonic techniques, density measurements using the gamma radiation attenuation method, and electrical resistivity measurements were performed using the four probe method. Sound velocity and density data, measured at ambient pressure, were incorporated into a thermodynamic model to calculate the pressure dependence of binary phase diagrams. Electrical resistivity measurements were performed on binary systems to study phase separation and identify phase transitions in the liquid state.

Keywords: thermal analysis; sound velocity; electrical resistivity; density; liquid metals; calculation of phase diagrams (CALPHAD)

1. Introduction

Phase diagrams are employed extensively in material science and associated technological applications in industry. For decades, phase diagrams at ambient pressure have been explored by measuring thermophysical quantities and by theoretical modeling. Phase boundaries can be determined directly by experimental techniques such as thermal analysis, including using differential thermal analysis (DTA) or differential scanning calorimetry (DSC), volume changes (e.g., dilatometry), electrical resistivity measurements, or X-ray diffraction (XRD) to identify structural changes [1].

Experimental methods are costly when mapping a phase diagram of binary and ternary systems over a wide range of compositions and temperatures, due to the large number of experiments required. Hence, direct measurements of phase transitions are often complemented by thermodynamic modeling of phase diagrams, typically determined by the Gibbs free energy, which describes the lowest energy, most stable phases, as a function of the thermodynamic parameters, temperature, composition and occasionally pressure. Calculation of phase diagrams (CALPHAD) methodology is commonly used to model phase diagrams. It is based on thermodynamic considerations and empirical databases, extrapolating the thermodynamic parameters to calculate these complex systems [2,3]. Several software packages have been developed (e.g., ChemSage, WinPhad, PANDAT, and ThermoCalc) that employ the CALPHAD methodology to calculate the phase diagrams of complex systems at ambient pressure [4].

A major challenge of this approach is the need for reliable thermodynamic property databases to calculate the phase diagram. Thus, extensive thermodynamic measurements of enthalpy, heat capacity, and activity (through electromotive force (EMF) measurements and vapor pressure studies) have been conducted and analyzed [5] to support this venture.

Pressure can affect physical properties and, in many cases, stabilize new phases. Physical properties of alloys under high pressure are at the focus of planetary research, as well as having advanced technological applications [6–8]. The emergence of high pressure technologies enables the scientific community to explore regions of phase diagrams at elevated pressures and high temperatures, which were inaccessible in the past [9,10]. However, numerous experimental challenges remain. The versatile and popular high-pressure apparatus of the diamond anvil cell (DAC) is widely used to measure structural changes under high pressure. Currently, measurements can be performed at a pressure range of 0–500 GPa and at temperatures ranging from cryogenic to thousands of degrees. At a lower pressure range one can find other techniques such as piston cylinder, multi-anvil press, and the Paris–Edinburgh large volume cell. To date, phase diagrams of elements under pressure are relatively well-known and established [11] but the database for binary systems is limited. For ternary systems, the database is practically nonexistent.

To establish phase diagrams in the liquid state, additional experimental challenges are presented, for example, chemical reactivity of the liquid at high temperatures with structural materials and effects associated with high vapor pressure. Hence, information on the properties of liquids at high temperatures, and even more so at high pressures, is limited. The scarce data that exists has relatively large experimental error, and as a result, there are discrepancies between reported thermophysical values. Even for elemental systems there are open questions, for example, the shape of the melting curve [12] and the possible existence of liquid–liquid transitions [13]. Direct measurements of the changes in the liquid structure and their interpretation can be challenging [14]. Indirect studies of phase transformations in the liquid using classical thermal analysis approaches are rare.

The study of liquid binary systems requires even greater experimental effort, due to the need to carefully measure a range of compositions and include phase separation phenomena. Therefore, this area of research remains relatively unexplored. Recently, we have developed a methodology to determine the pressure and temperature dependence of the excess interaction parameters in liquid solution, from ambient pressure measurements of density, heat capacity, and thermal expansion. This approach enables the construction of binary alloy phase diagrams at low pressures of up to several GPa [15,16].

The purpose of this paper is to provide an overview of thermophysical measurements and associated models that support the construction of phase diagrams in liquid binary alloys of columns III to V and their constituent elemental systems. These systems exhibit complex phase diagrams under pressure and their thermophysical properties are often anomalous. Their low melting points make them easily accessible for experimental study, and thus they are ideal candidates for developing new approaches to phase diagram studies. In particular, we have developed techniques to measure the density and sound velocity to high accuracy in liquid metals, which is required to establish reliable diagrams. The results of such measurements have been incorporated into a model, describing the free energy of the liquid alloys and predicting the pressure dependence of binary phase diagrams. Electrical measurements provide a complementary technique to explore phase transitions and the limits of liquid phase stability. Such measurements have been applied at both ambient and high pressures to identify phase stability limits. This set of techniques and accurate measurements provides new pathways to study the liquid phase diagram under pressure.

2. Density

Density is a fundamental thermodynamic property of matter, relevant to the determination of other physical parameters, such as viscosity, surface tension, heat capacity, and even to the description of the radial density function (RDF) from which the short-range order can be evaluated. In the lieterature, several experimental techniques have been described for measuring the density of liquid metals. These include the maximum bubble pressure, Archimedean, liquid drop, dilatometric, and gamma radiation attenuation methods [17]. In our research we implemented a high-accuracy measurement based on gamma radiation attenuation.

2.1. Gamma Radiation Attenuation Method

This method is based on measuring the attenuation of gamma radiation passing through a sample. A γ radiation source is located on one side of the furnace, containing the sample, and a γ radiation detector is placed on the other side. The beam, passing through the liquid sample, is attenuated, and the density can be calculated by:

$$I = I_0 e^{(-\mu\rho x)} \tag{1}$$

where I_0 and I are the intensity of the beam before the sample and the intensity measured at the detector, respectively; μ, ρ, and x are the absorption coefficient, the density, and the optical thickness of the entire path from the source to the detector. The contribution of the liquid can be deduced from premeasuring the transmission in the exact experimental configuration without the sample. To eliminate the thermal expansion of the experimental setup, we measured relative density. This method is mostly used for materials with high density, in which the attenuation due to the liquid is significant as compared with the structural materials, to obtain a large signal-to-noise ratio.

The experimental error is due to the following two main factors: (i) statistical error of the detector counts and (ii) the radiation background intensity. The first is assumed to have Lorentz distribution, and hence follows $\Delta I = \sqrt{I}$. To reduce the statistical error, we significantly increased the measurement time. To reduce background radiation we subtracted reference measurements of the radiation intensity measured without a sample as a function of temperature, with the same temperature profile. As a result, the relative error of the density is proportional to the squared sum of the relative errors of the intensities with and without the liquid sample.

2.2. Density Results

The density of pure lead and bismuth was measured by the gamma radiation attenuation method. The radiation source used was ^{137}Cs with a characteristic energy of 662 Kev. This source is monoenergetic, and has a relatively long half-life and a large cross-section for absorption in the liquid samples. The crucible was made of fused quartz, which has low thermal expansion, with a square cross-section for simple geometry. The detector chosen was a CsI scintillator due to its high sensitivity and dynamic range.

Density measurements in elemental lead (99.999% purity) were carried out in the temperature range of 330 to 950 °C every 20 °C, as shown in Figure 1. It can be seen that the density decreases monotonically as the temperature is increased, without discontinuities or changes in slope. The uncertainty (standard deviation) in these measurements was estimated to be 1% due to the sampling statistics. The thermal expansion of liquid Pb can be evaluated, and was found to be 1.22×10^{-4} K^{-1}. Comparing the present data with previous measurements [18] of the density obtained by the Archimedean method shows a small difference that is contained by the experimental error. The difference in the density at melting point was found to be less than 1%.

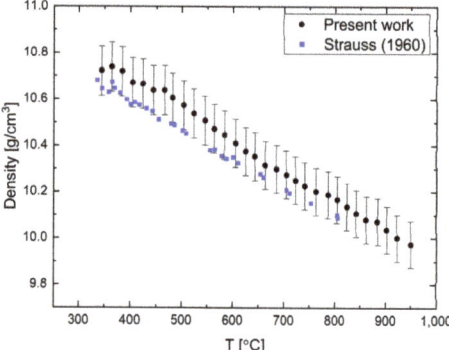

Figure 1. Lead density as a function of temperature compared with data from reference [18].

The density of elemental bismuth (99.999% purity) was measured and reported in [19] in a temperature range of 200–1000 °C every 2 °C to 5 °C (Figure 2). A good agreement between the previously published data and our measurements can be seen in Figure 2. Similar to liquid Pb, liquid Bi expands as the temperature increases. The errors in the measurements are calculated to be 0.1%. Density changes are observed at ca. 550 °C and 720 °C, which might point to structural changes in the liquid phase of Bi [19].

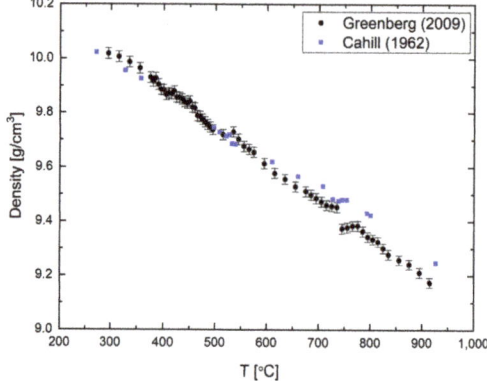

Figure 2. Bismuth density as a function of temperature reproduced with permission from Ref. [19] compared with data from Ref. [20].

3. Sound Velocity

Sound velocity is an important thermodynamic quantity, sensitive to changes in the material properties. To measure the velocity of sound in liquids, ultrasonic or laser methods can be employed [21,22]. In optical systems, stress waves are generated, which result in surface motion. This method is suitable for measuring sound velocity at very high temperatures and pressures. The two systems used in our studies are modifications of the well-known ultrasonic method designed for high-accuracy measurement. The first is based on transmission and includes two transducers to measure the transmitted wave, one to generate the acoustic wave and the other to receive it (i.e., the wave travels only once in the liquid sample). The second apparatus is based on reflection of the emitted wave from the base of the sample container back to the transducer. Both these tabletop systems are presented schematically in Figure 3 and operate at ambient pressure and have a unique high accuracy that is preserved at elevated temperatures.

3.1. Sound Velocity Measurements by Ultrasonic Techniques

The pulse transmission system is composed of two ceramic buffer rods that serve as waveguides and two electronic ultrasound transducers that are attached to them, the first transmitting an acoustic wave and the second receiving it. The buffer rods allow the transducers to operate below the maximum working temperature. The crucible containing the liquid sample is machined at the top of the lower buffer rod. An ultrasonic elastic wave travels through the upper buffer rod, which is immersed in the liquid metal, traveling through the liquid sample and continuing to the lower buffer rod. The upper rod is attached to an accurate linear motor. By moving this rod a known distance of Δx, the traveling time of the sound wave is increased in Δt, and the sound velocity is deduced from the ratio $C = \Delta x/\Delta t$ (i.e., it is a differential measurement). The uncertainty of the measurement estimated to be 0.1% to 0.35%, mainly due to the finite precision of the linear motor. More details on this measurement method are in reference [23].

The pulse-echo ultrasonic technique is, in principle, similar to the pulse transmission method. However, in this method we used a single buffer rod and a transducer that operates as both transmitter and receiver of the acoustic wave. The liquid sample is held in a ceramic crucible, the bottom of which serves as a reflector. To eliminate uncertainty as to the distance the ultrasonic wave travels through the liquid, the rod is translated by a known distance, Δx. As the wave travels a distance of $2\Delta x$ in the liquid metal with extra time Δt, the sound velocity is determined by $c = 2\Delta x/\Delta t$. This method has a similar error, as the limited accuracy of the linear motor is the same, and it is estimated to be 0.2%. More details on this measurement technique are in reference [24].

In both cases, the experimental apparatus was placed in a glovebox with a protective gaseous atmosphere of high-purity argon in a constant flow mode.

The transmission technique benefits from a higher signal-to-noise ratio, due to the short distance the sound wave travels through the liquid, and due to the attenuated shear waves in the thicker lower rod. The pulse-echo technique is easier to apply and has a simple setup, however, its main drawback is that the bottom of the crucible needs to be polished to a high quality to minimize losses upon reflection of the sound wave. In addition, the amplitude of the wave is more attenuated since the sound wave travels through the liquid twice.

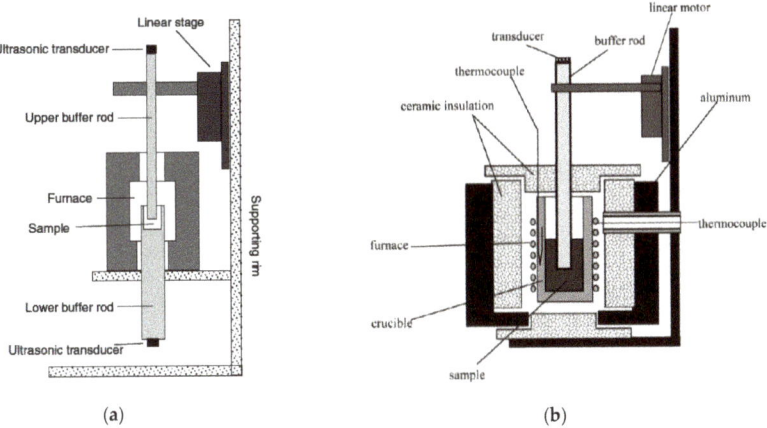

Figure 3. Schematic view of (**a**) the ultrasonic pulse transmission experimental measurement apparatus reproduced with permission from [23] and (**b**) the ultrasonic pulse-echo experimental measurement apparatus reproduced with permission from [24].

3.2. Sound Velocity Results

3.2.1. Elemental Systems

The sound velocity in pure liquid lead and bismuth has been measured by both of the ultrasonic techniques presented in the previous section as reported in [23,25]. For liquid Pb, the sound velocity decreases as the temperature is increased with a constant rate, as shown in Figure 4a. The difference between the measured and literature data increases with temperature and reaches about 1.5% at ca. 1000 °C. Our two methods agree reasonably well within the overlapping measurement range. For the liquid Bi (Figure 4b), the overall tendency is a negative temperature coefficient, but it shows a more complex behavior, namely, that the temperature coefficient changes with temperature. There is a good agreement between the two techniques.

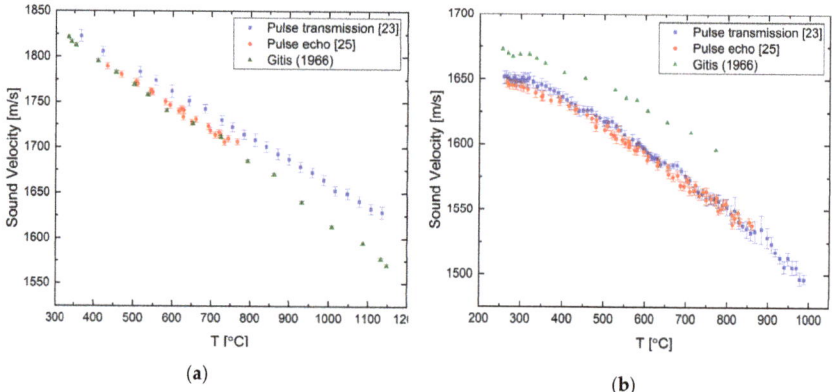

Figure 4. The temperature dependence of the sound velocity in (**a**) pure liquid lead [23,25] compared with data from [26] and in (**b**) pure liquid bismuth [23,25] compared with data from [26].

The velocity of sound was measured for elemental tin and antimony in [23] using the pulse transmission method only. For tin, we observe a normal sound velocity dependency, with an excellent agreement with previous data, as shown in Figure 5a. Antimony has an anomalous behavior as presented in Figure 5b. Up to a temperature of ~830 °C the sound velocity increases with increasing temperature, up to a maximum, then, decreasing nonlinearly.

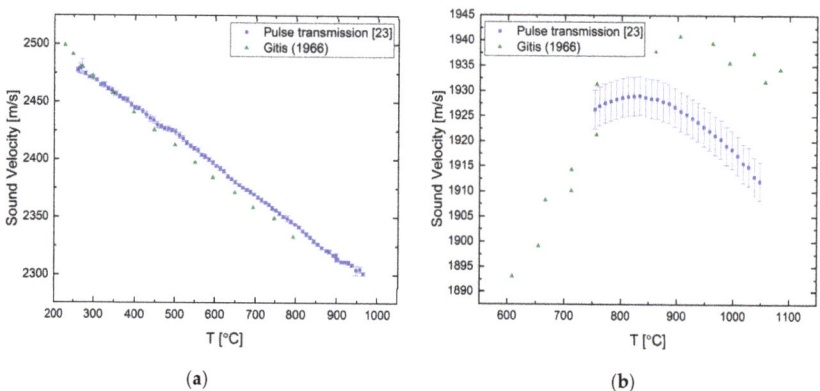

Figure 5. The temperature dependence of the sound velocity in (**a**) pure liquid tin [23] compared with data from [26] and in (**b**) pure liquid antimony [23] compared with data from [26].

The sound velocity of liquid gallium (99.99% purity) and liquid indium (99.999% purity) were measured using the pulse-echo setup and the results are presented in Figure 6 and detailed in the Supplementary Material. Both elements display normal behavior and a good agreement between the measured values and previously published data.

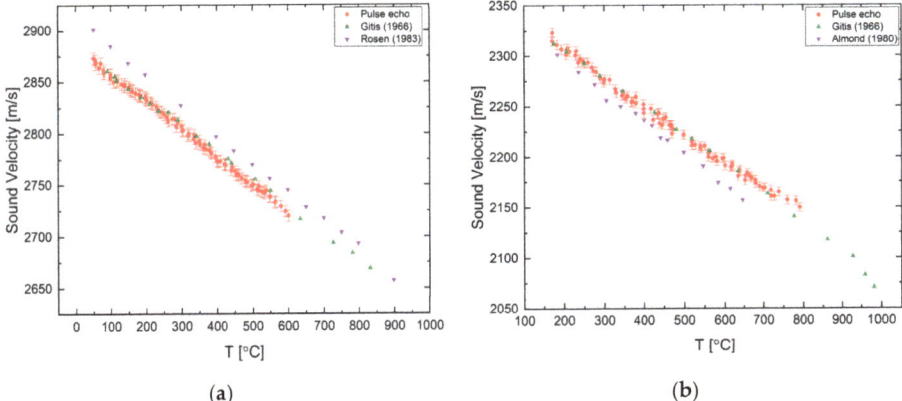

Figure 6. The temperature dependence of the sound velocity measured in (**a**) pure liquid gallium compared with data from [27,28] and in (**b**) pure liquid indium compared with data from [27,29].

3.2.2. Binary Systems

The sound velocities in liquid Pb-Sn and Bi-Sn were measured in [15] using the transmission method. Figure 7 presents the sound velocity as a function of temperature for the following four compositions of the Pb-Sn system up to 1000 °C: $Pb_{13}Sn_{87}$, $Pb_{26}Sn_{74}$, $Pb_{46}Sn_{54}$, and $Pb_{70}Sn_{30}$ (at.%). For all the compositions a normal behavior is observed, as in the two component elements.

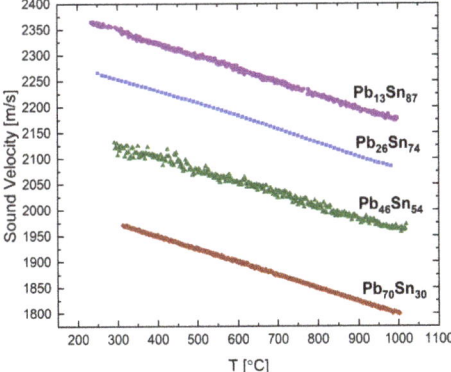

Figure 7. The temperature dependence of the sound velocity in the liquid Pb-Sn system at different alloy compositions adapted with permission from [15].

The sound velocity of binary systems Bi-Pb [25] and Bi-Sb [24] were measured by the pulse-echo technique. In Figure 8, we present some results of the Bi-Sb isomorphous binary alloy. Measurements in the liquid phase were carried up to temperatures of ca. 900 °C for the following selected compositions: $Bi_{13}Sb_{87}$, $Bi_{35}Sb_{65}$, $Bi_{53}Sb_{47}$, and $Bi_{70}Sb_{30}$ (all in at.%). In this system, the Sb-rich alloys, $Bi_{13}Sb_{87}$ and $Bi_{35}Sb_{65}$, display anomalous behavior similar to Sb, but with less significant trends. The temperature at which the sound velocity is maximal decreases from ~830 °C to ~700 °C at 13% Bi and ~520 °C at 35% Bi alloy composition. As the Bi concentration is increased the temperature dependence of the sound velocity becomes more linear, and for the Bi-rich alloy, $Bi_{70}Sb_{30}$, a semi-normal behavior at the low temperatures near the solidification is observed, similar to Bi.

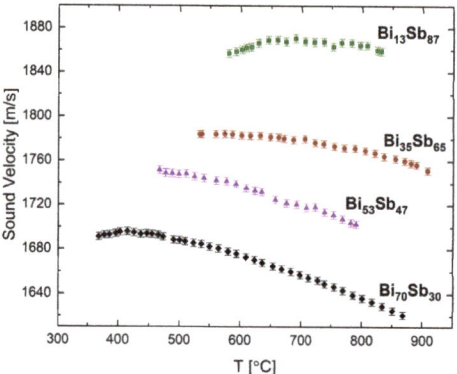

Figure 8. Sound velocity in the liquid Bi-Sb system as a function of temperature at selected alloy compositions, adapted with permission from [24].

4. Modeling Binary Phase Diagrams under Pressure

Phase diagrams of binary alloys are expected to vary with pressure. Measuring thermophysical properties under pressure is experimentally challenging in addition to the vast amount of data required to construct the diagram as a function of temperature, composition, and pressure. Therefore, we proposed to follow a different route, i.e., to calculate the pressure-dependent phase diagram with input from ambient measurements of sound velocity and density to obtain the variation of interaction parameters with pressure. Lastly, information on the temperature-pressure phase diagram of the elements constituting the binary system is required.

The equilibrium condition to determine phase line is an equality of the chemical potentials of the same component in the two different phases, calculated from the Gibbs free energy:

$$\mu_i = \left(\frac{\partial G}{\partial N_i}\right)_{P,T} \tag{2}$$

where μ_i is the chemical potential of component i, G is the Gibbs free energy, and N_i is the number of particles.

Most binary alloy systems do not behave as ideal solutions. In systems with an asymmetric miscibility gap, the Gibbs free energy can be expressed to the lowest order in composition in the form of a sub-regular solution:

$$G = X_A G_A + X_B G_B + RT(X_A \ln X_A + X_B \ln X_B) + J_0 X_A X_B + J_1 X_A X_B (X_A - X_B) \tag{3}$$

where X_A and X_B are the atomic fractions of each component, G_A and G_B are the partial Gibbs energy, and J_0 and J_1 are the regular and sub-regular interaction coefficients. The latter depend on pressure, and the variation of those parameters with pressure is a crucial input for the calculated phase diagram under pressure.

The pressure dependence of the interaction coefficient may be expanded to the second order and the deviation of the molar volume from its ideal value can be expressed in the following manner:

$$\delta V = \frac{\partial J_0}{\partial P} X_A X_B + \frac{\partial J_1}{\partial P} X_A X_B (X_A - X_B) \quad (4)$$

$$\frac{\partial \delta V}{\partial P} = \frac{\partial^2 J_0}{\partial P^2} X_A X_B + \frac{\partial^2 J_1}{\partial P^2} X_A X_B (X_A - X_B) \quad (5)$$

The sound velocity of the liquid at ambient pressure (C_s), is related to the molar volume and the adiabatic compressibility (K_S) by:

$$\frac{1}{C_s^2} = \rho K_s = -\frac{M}{V^2}\left(\frac{\partial V}{\partial P}\right)_S \quad (6)$$

Hence, the sound velocity of an ideal solution, which is the velocity of the elements weighted by the relative composition, can be related to the measured one using Equation (6) to obtain the relation:

$$\left(\frac{C_{id}}{C_s}\right)^2 - 1 = -\frac{2\delta V}{V_{id}} + \frac{\frac{\partial \delta V}{\partial P}}{X_A \frac{1}{C_{s,A}^2}\left(-\frac{V_A^2}{M_A}\right) + X_B \frac{1}{C_{s,B}^2}\left(-\frac{V_B^2}{M_B}\right)} + O(\delta V^2) \quad (7)$$

The pressure dependence of the interaction parameter, J(P), is derived from measurements of the sound velocity and density performed at ambient pressure to determine the deviation of the molar volume from its ideal values (δV) to estimate $\frac{\partial \delta V}{\partial P}$. Extending the CALPHAD methodology, we calculated the phase diagrams of several binary alloys under pressure, including both isomorphous and eutectic systems which included: Bi-Sb, Bi-Sn, Pb-Sn [15], and Bi-Pb [25]. The model cannot represent the formation of new high-pressure phases in the P-T diagram of the alloy. The limitation on the pressure range arises from the fact that the interaction parameters are expanded only up to the second order.

The phase diagram of the isomorphous binary system Bi-Sb was calculated in [15] up to a pressure of 1.7 GPa and is shown in Figure 9. The solidus decreases significantly with pressure, while the liquidus slightly decreases, mainly due to the anomaly in the melting temperature of the Bi with respect to pressure. The calculated phase diagram of this alloy is limited to a pressure of 1.7 GPa. Extension of this study to higher pressures is a subject for future study.

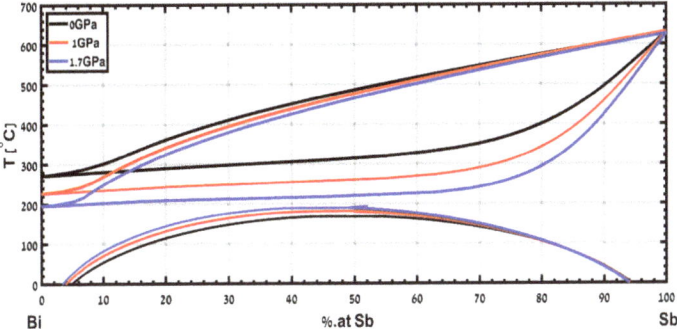

Figure 9. Calculated phase diagram of the isomorphic system Bi-Sb from ambient pressure up to pressure of 1.7 GPa adapted with permission from [15].

A different example is the eutectic phase diagram of Pb-Sn, which has been calculated in [15] up to a pressure of 1.25 GPa and is presented in Figure 10. The model captures the shifts in the eutectic composition and temperature. This useful information is hard to obtain experimentally. Note that the eutectic point shifts to a composition rich in Sn, and the eutectic temperature increases with increasing pressure.

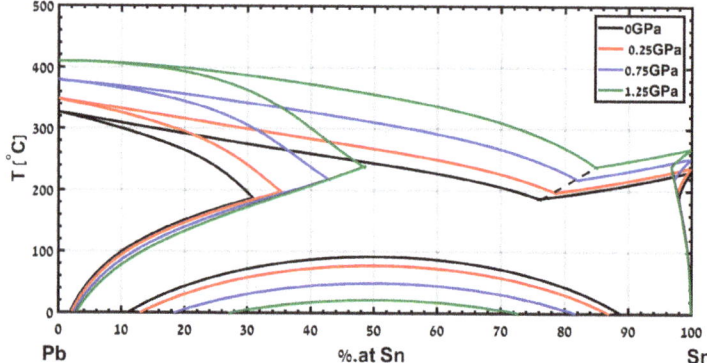

Figure 10. Calculated phase diagram of the eutectic system Pb-Sn from ambient pressure up to pressure of 1.25 GPa, adapted with permission from [15].

5. Electrical Resistivity

Electrical resistivity is one of the basic transport properties of a material. It is a useful experimental tool for studying phase transformations in solid and in liquid phases, for example, to identify melting point due to the abrupt change in resistivity resulting from the loss of the long-range order. In the literature, two classes of experimental techniques for resistivity measurements in liquid metals have been proposed, i.e., non-contact and contact methods. In this study, we applied a contact method implemented using a tabletop setup that was designed and built to be simple, modular, and accurate.

The experimental apparatus is based on an alternating current (AC) source with the four-point probe technique commonly used in the literature. The use of AC instead of direct current (DC) reduces the Seebeck effect and, by using a known reference frequency, enables better elimination of external noise [30]. In the present setup, the melt is held in a quartz test tube and is in direct contact with the immersed electrodes. Measurements are carried out under a protective gaseous argon atmosphere in a constant flow mode to avoid the enhanced reactivity of liquid metals at high temperature with the structural materials constructing the experimental chamber, which were chosen to have low reactivity. Figure 11 displays a schematic view of the apparatus.

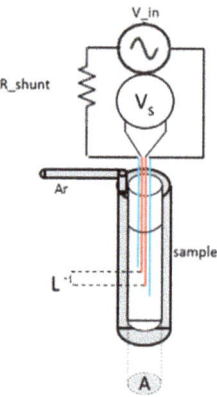

Figure 11. Schematic of the measuring system.

An alternating current is supplied to the sample, and the voltage drop across the sample is measured. The calculated resistivity (ρ_{sample}) from the measured voltage drop is as follows:

$$\rho_{sample} = V_{meas} \frac{R_{shunt}}{V_{in}} G \tag{8}$$

where G is the geometric constant of the cell, V_{meas} the measured voltage, V_{in} the input voltage, and R_{shunt} the shunt resistor. The precise cell dimensions are needed to convert the measured voltage to resistivity. However, if only the relative resistivity or the temperature coefficient is of interest, one can ignore the geometric constant.

The estimated error consists of statistical and systematic errors. The main contribution to systematic error arises from cell dimensions. The measured voltage is averaged over a temperature window of 5 °C, and the standard deviation of the statistical distribution of the voltage within this window is calculated. The error is, therefore, presented in Equation (9).

$$\frac{\Delta\rho}{\rho} = \sqrt{2\left(\frac{\Delta V_{meas}}{V_{meas}}\right)^2 + \left(\frac{\Delta R_{shunt}}{R_{shunt}}\right)^2 + \left(\frac{\Delta L_{elec}}{L_{elec}}\right)^2 + \left(\frac{\Delta A}{A}\right)^2} \tag{9}$$

where A and L are the sample's cross-section and length, determined by the voltage electrodes. We assume that only the voltage measurement contains both statistical and systematic errors. Other terms consist of systematic errors only.

The error in determining the absolute value of electrical resistivity is approximately 3%, a major part of which is derived from the uncertainty of the geometric dimensions of the cell. Consequently, the error in determining the relative values of the resistivity is only 0.1%.

5.1. Electrical Resistivity Results

5.1.1. Elemental Systems

The electrical resistivity was measured for elemental bismuth, tin, indium (99.999% purity), and gallium (99.99% purity). The results are presented in Figure 12 and in tabular form in the Supplementary Material. The resistance varies linearly with temperature in the liquid state, upon heating and cooling cycles. Deviation of the resistivity-temperature coefficient from published data can be seen for the temperature coefficient upon heating and cooling cycles for Bi and Ga.

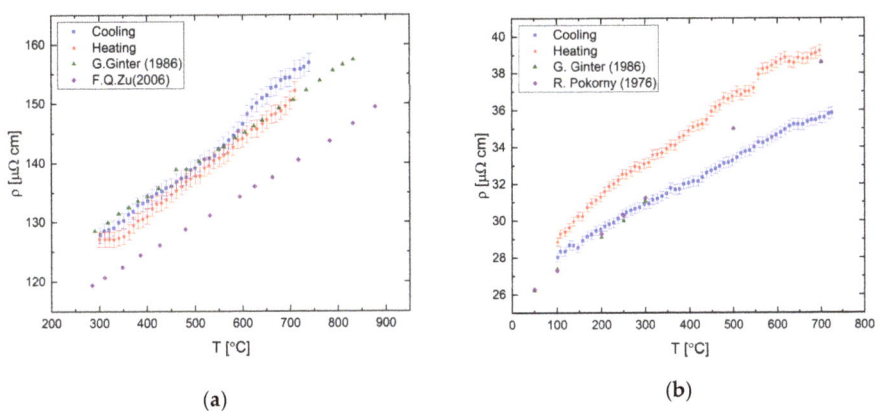

Figure 12. *Cont.*

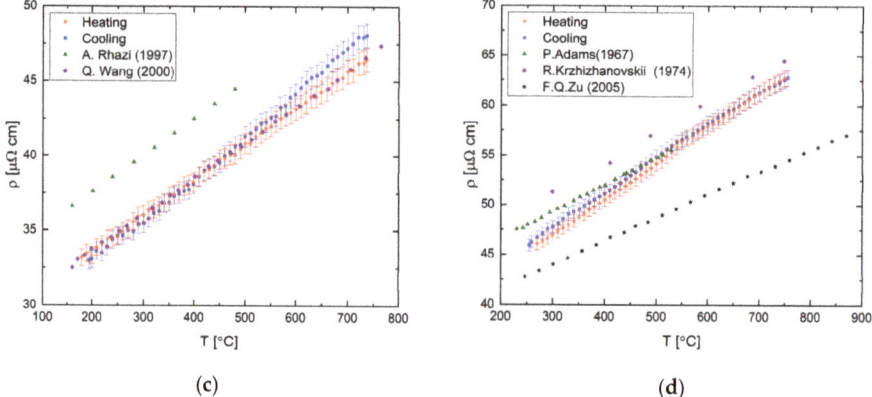

Figure 12. Resistivity of pure metals (**a**) bismuth and references [31,32]; (**b**) gallium and references [31,33]; (**c**) indium and references [34,35]; and (**d**) tin and references [36–38]. Cooling and heating rate of 60 °C/h.

5.1.2. Binary Systems

The electrical resistance of Bi-Ga and Ga-In alloys (prepared from the same sources as the elemental samples above) was measured as a function of temperature and composition (see Supplementary Material). To ensure reliable data, the resistivity of every composition was measured for two different samples, each undergoing at least three cooling and heating cycles.

The resistivity of the Bi-Ga system was measured for the following compositions: $Bi_{30}Ga_{70}$, $Bi_{33}Ga_{67}$, $Bi_{50}Ga_{50}$, $Bi_{67}Ga_{33}$, and $Bi_{70}Ga_{70}$ (in at.%) and the results are presented in Figure 13. A linear trend is found for all measured compositions, in which the resistivity increases with increasing temperature.

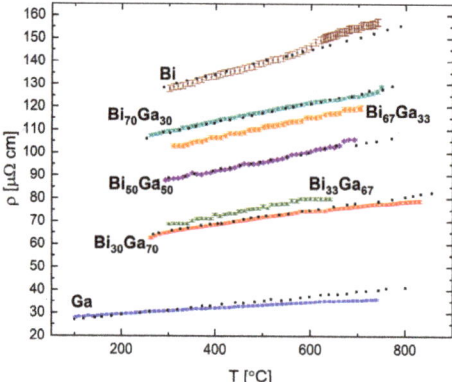

Figure 13. Electrical resistivity in the liquid Bi-Ga system as a function of temperature at selected alloy compositions. The uncertainty is smaller than the symbol size. The black dots are data from [31].

The temperature coefficients $\frac{d\rho}{dT}$ were calculated by a linear fit to the resistivity data and are presented as a function of Bi concentration in Figure 14. The coefficient values are calculated over the measured temperature range displayed in Figure 13. Our results suggest a possible correlation of the temperature coefficient in the Bi-Ga system to a second-order polynomial as a function of composition. Parabolic dependence is an expected behavior for alloys containing metals with mixed valences [39].

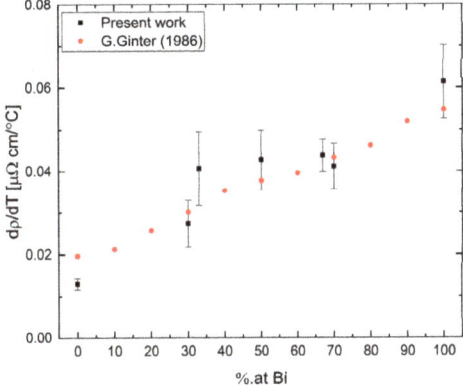

Figure 14. Resistivity coefficient of Bi-Ga alloy vs. Bi concentration.

The temperature dependence of the resistivity at different compositions is displayed in Figure 15. This isotherm plot shows a linear correlation between the absolute resistivity of the melt and the Bi concentration. The smooth dependence suggests that no obvious transitions are taking place in the melt with increasing Bi concentration. Furthermore, as the Bi concentration is increased, the slope of the resistivity curve increases as indicated from the distance between the isotherms.

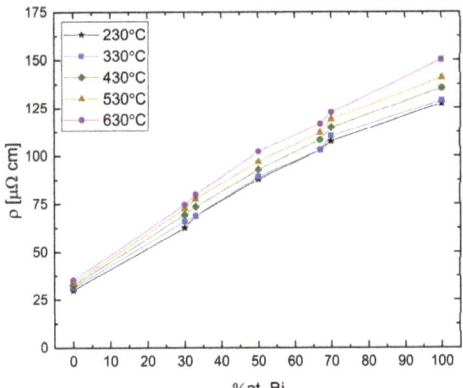

Figure 15. Absolute resistivity of Bi-Ga binary alloy vs. Bi concentration. The uncertainty is smaller than the symbol size.

A measurement of the electrical resistivity of liquid $Bi_{33}Ga_{67}$ (in at.%) alloy near the melting point obtained during slow cooling with a rate of 60 °C/h is presented in Figure 16. The resistivity-temperature curve presents an abrupt change at 260 °C, about 50 °C above the liquidus. This change is correlated with phase separation in the melt [40]. Following this shoulder, a drastic increase of resistivity is observed below 225 °C, which is characteristic of the two-phase zone that contains a mixture of liquid and solid states. The difference between the present results and the results reported by Wang et al. [40] may originate due to use of DC vs. AC measurements which may produce an out of phase signal near the solidification or due to degradation of the contacts.

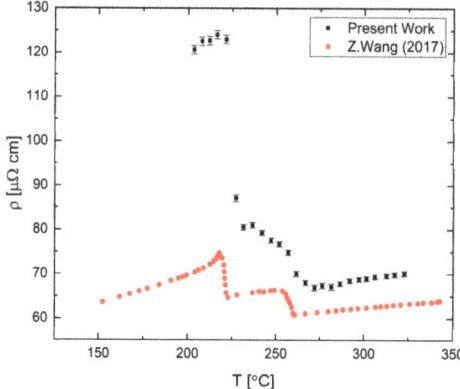

Figure 16. Resistivity of $Bi_{33}Ga_{67}$ liquid alloy compared to data from [40].

The resistivity of pure indium, gallium, and the selected binary alloys $Ga_{86}In_{14}$, $Ga_{70}In_{30}$, $Ga_{25}In_{75}$, and $Ga_{10}In_{90}$ (in at.%) were measured, and in Figure 17 we summarize the results for these compositions. The results present linear dependence of the resistivity with respect to temperature. No relation between the composition and the absolute resistivity values was found, in contrast to the Bi-Ga system. Two compositions exhibited the following outlying behaviors: the eutectic composition, $Ga_{86}In_{14}$, presented the lowest resistivity of the measured compositions; and the $Ga_{70}In_{30}$ had a significantly higher value of the resistivity-temperature slope.

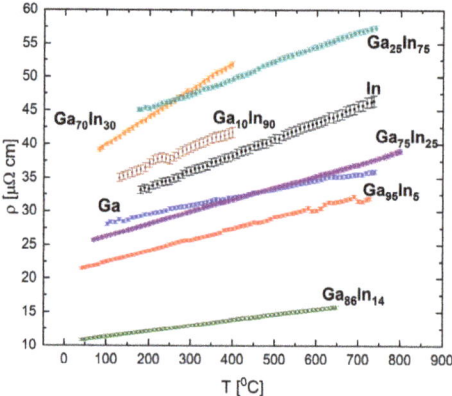

Figure 17. Electrical resistivity in the liquid Ga-In system as a function of temperature at selected alloy compositions. The uncertainty is smaller than the symbol size.

The eutectic alloy (Figure 18) displays abnormal behavior at 90 °C and 250 °C, which might indicate a possible transformation in the liquid phase. Structural transformation at 90 °C is seen upon heating and cooling cycles, which suggest a reversible process. A transformation at similar temperatures was reported previously based on XRD measurements [41].

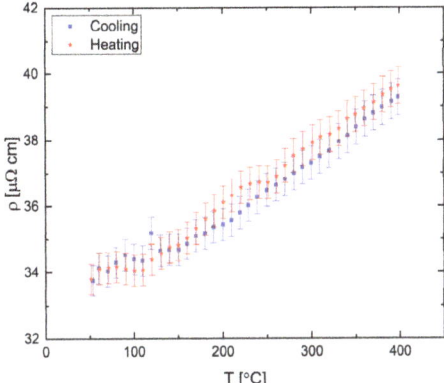

Figure 18. Electrical resistivity of eutectic Ga-In. Cooling and heating rate of 60 °C/h.

The temperature coefficient of the Ga-In system presents no clear tendency with indium concentration, as can be seen in Figure 19. This result is in contradiction to the theory [42] for alloys with an equivalent amount of valence electrons, in which the change in composition will maintain the electron density unchanged and the trend should be linear.

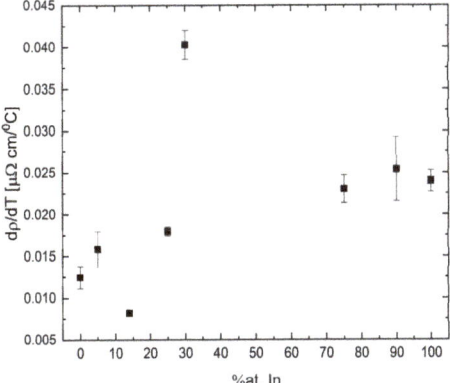

Figure 19. Resistivity coefficient of Ga-In as a function of indium concentration. Bars represent the uncertainty (symbols without bars have an uncertainty smaller than the symbol size).

The resistivity vallues as a function of composition for Ga-In binary alloys at several temperatures are presented in Figure 20. A unique phenomenon can be observed in these data, namely, that the hypereutectic area displays a general parabolic trend with a maximum resistivity between $Ga_{70}In_{30}$ and $Ga_{25}In_{75}$, and the hypoeutectic region shows a linear trend where the resistivity decreases until reaching a minimum at eutectic concentration, significantly below the resistivity of either elemental component.

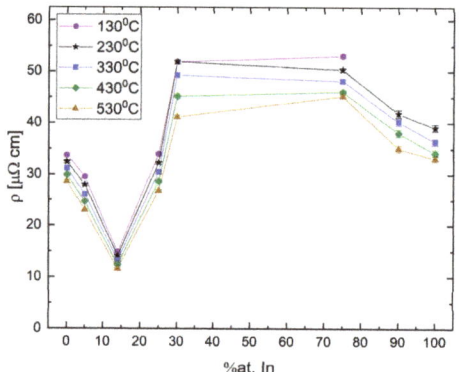

Figure 20. Absolute resistivity vs. indium concentration in Ga-In alloys.

6. Discussion

The present contribution presents an overview of experimental measurements and theoretical treatment to obtain physical and thermophysical properties of liquid metals and liquid binary alloys in the context of phase diagrams. We have carried out high-precision sound velocity, electrical resistivity and density measurements on several column III to V elements and their binary alloys.

Measurements were carried out as a function of temperature in the liquid state. We found that, where relevant, the present measurements stand in a good agreement with previously published data. These measurements were used to map phase transformations by direct measurements, i.e., electrical resistivity, or by incorporating sound velocity and density measurements into a thermodynamic model to calculate phase diagrams of binary alloys under pressure.

The measurements of the physical properties were also analyzed with respect to composition. Regarding sound velocity measurements, we observed that as the concentration of the element having a larger sound velocity is increased, the alloy sound velocity increases, respectively. Moreover, there is a connection between the elements' temperature coefficient and that of the alloys. For example, if the two elements have normal behavior, then the alloys of both the elements present normal behavior as well. In the Pb-Sn system, both the elements and the different compositions of the alloy display normal temperature coefficients. In the Bi-Sb binary system, the Sb rich alloy displays anomalous behavior similar to Sb. As the Bi concentration increases, the behavior becomes semi-normal similar to in Bi. With regard to electrical resistivity, we observe that the Bi-Ga system shows a parabolic correlation with composition, as expected from mixed valence alloys. The Ga-In system, in which the elements have the same valences, presents a complex dependence on composition and not a linear ratio as expected [42].

Electrical resistivity is sensitive to structural changes, for example, as manifested upon a transition from a solid to liquid phase where the resistivity in metals usually increases. Furthermore, the thermal coefficient of the resistivity is also strongly connected to the structural properties, and thus a change in the slope can indicate a change in the liquid's short-range order [32]. Such a change in the thermal coefficient is seen in the Bi-Ga system and is related to the region of the phase diagram presenting a phase separation in the liquid. The electrical resistivity measurements indicate, with high accuracy, these phase transformations from a homogenous liquid to the two liquids region and, then, to the zone of mixed phases of liquid and solid.

The use of CALPHAD allows us to exploit data measured at ambient pressure to calculate phase diagrams of binary alloys under pressure. The sound velocity and the density are used to calculate the pressure dependence of the interaction parameters on pressure to the second order. We apply this formalism for two examples, the isomorphic system Bi-Sb and the eutectic system Pb-Sn.

We demonstrated the use of experimental measurements to obtain nontraditional thermal analysis. These are thermophysical measurements that can shed some light on the dependence of the structure on temperature, and provide evidence on the presence or absence of a local change in the liquid state. From sound velocity and density measurements at ambient conditions, we are able to calculate the pressure dependence of the interaction parameters in the liquid state, which is required to calculate the phase diagram under pressure of binary systems. This provides an innovative path to predict properties of the material under pressure and the dependence of eutectic point and liquidus temperature as a function of composition.

Supplementary Materials: The following are available online at http://www.mdpi.com/1996-1944/12/23/3999/s1.

Author Contributions: Conceptualization, J.L., E.Y., and G.M.; methodology, E.Y., G.M., and Y.G.; investigation, Y.K., S.B.S., M.E., and Y.G.; writing—original draft preparation, all authors; writing—review and editing, S.B.S., E.Y., and G.M.

Funding: This research was partially funded by the Pazy Research Foundation and by the Korea–Israel Joint Research Program, Nano & Pressure Phase Diagram of Alloys (NADIA), by the National Research Foundation of Korea (NRF) grant funded by the Ministry of Science and ICT (MSIT) (NRF-2016K1A3A1A31913031) and by a grant from the Ministry of Science, Technology & Space, Israel.

Acknowledgments: The authors acknowledge the technical assistance of Aviram Berko.

Conflicts of Interest: The authors declare no conflict of interest. The funders had no role in the design of the study; in the collection, analyses, or interpretation of data; in the writing of the manuscript, or in the decision to publish the results.

References

1. Zhao, J.-C. *Methods for Phase Diagram Determination*; Elsevier B.V.: Oxford, UK, 2011.
2. Lukas, H.; Fries, S.G.; Sundman, B. *Computational Thermodynamics: The Calphad Method*; Cambridge University Press: New York, NY, USA, 2007.
3. Saunders, N.; Miodownik, A.P. *CALPHAD: Calculation of Phase Diagrams A Comprehensive Guide*; Elsevier Ltd.: Oxford, UK, 1998.
4. Chang, Y.A.; Chen, S.; Zhang, F.; Yan, X.; Xie, F.; Schmid-Fetzer, R.; Oates, W.A. Phase Diagram Calculation: Past, Present and Future. *Prog. Mater. Sci.* **2004**, *49*, 313–345. [CrossRef]
5. Dinsdale, A.T. SGTE Data for Pure Elements. *Calphad* **1991**, *15*, 317–425. [CrossRef]
6. Heuze, F.E. High-Temperature Mechanical, Physical and Thermal Properties of Granitic Rocks—A Review. *Int. J. Rock Mech. Min. Sci.* **1983**, *20*, 3–10. [CrossRef]
7. Brosh, E.; Makov, G.; Shneck, R.Z. Thermodynamic analysis of high-pressure phase equilibria in Fe–Si alloys, implications for the inner-core. *Phys. Earth Planet. Inter.* **2009**, *172*, 289–298. [CrossRef]
8. *High Pressure Process Technology: Fundamentals and Applications*; Bertucco, A.; Vetter, G. (Eds.) Elsevier Science B.V.: Amsterdam, The Netherlands, 2001.
9. Dalladay-Simpson, P.; Howie, R.T.; Gregoryanz, E. Evidence for a New Phase of Dense Hydrogen above 325 Gigapascals. *Nature* **2016**, *529*, 63–67. [CrossRef]
10. Dewaele, A.; Stutzmann, V.; Bouchet, J.; Bottin, F.; Occelli, F.; Mezouar, M. High Pressure-Temperature Phase Diagram and Equation of State of Titanium. *Phys. Rev. B* **2015**, *91*, 134108. [CrossRef]
11. Young, D.A. *Phase Diagrams of the Elements*; University of California Press: Berkeley, CA, USA, 1991.
12. Makov, G.; Yahel, E. Liquid-Liquid Phase Transformations and the Shape of the Melting Curve. *J. Chem. Phys.* **2011**, *134*, 204507. [CrossRef]
13. Zu, F.-Q. Temperature-Induced Liquid–Liquid Transition in Metallic Melts: A Brief Review on the New Physical Phenomenon. *Metals* **2015**, *5*, 395–417. [CrossRef]
14. Mayo, M.; Yahel, E.; Greenberg, Y.; Makov, G. Short Range Order in Liquid Pnictides. *J. Phys. Condens. Matter* **2013**, *25*, 505102. [CrossRef]
15. Emuna, M.; Greenberg, Y.; Hevroni, R.; Korover, I.; Yahel, E.; Makov, G. Phase Diagrams of Binary Alloys under Pressure. *J. Alloys Compd.* **2016**, *687*, 360–369. [CrossRef]
16. Makov, G.; Emuna, M.; Yahel, E.; Kim, H.G.; Lee, J. Effect of Pressure on the Interactions and Phase Diagrams of Binary Alloys. *Comput. Mater. Sci.* **2019**, *169*, 109103. [CrossRef]

17. Iida, T.; Roderick, I.L.G. *The Thermophysical of Metallic Liquids*, 1st ed.; Oxford University Press: New York, NY, USA, 2015.
18. Strauss, S.W.; Richards, L.E.; Brown, B.F. The Density of Liquid Lead and of Dilute Solutions of Nickel in Lead. *Nucl. Sci. Eng.* **1960**, *7*, 442–447. [CrossRef]
19. Greenberg, Y.; Yahel, E.; Caspi, E.N.; Benmore, C.; Beuneu, B.; Dariel, M.P.; Makov, G. Evidence for a Temperature-Driven Structural Transformation in Liquid Bismuth. *EPL* **2009**, *86*, 36004. [CrossRef]
20. Cahill, J.A.; Kirshenbaum, A.D. The Density of Liquid Bismuth from Its Melting Point to Its Normal Boiling Point and an Estimate of Its Critical Constants. *J. Inorg. Nucl. Chem.* **1963**, *25*, 501–506. [CrossRef]
21. Papadakis, E.P. The Measurement of Small Changes in Ultrasonic Velocity and Attenuation. *C R C Crit. Rev. Solid State Sci.* **1973**, *3*, 373–418. [CrossRef]
22. Calder, C.A.; Wilcox, W.W. Acoustic Velocity Measurement Across the Diameter of a Liquid Metal Column. *Exp. Mech.* **1978**, *19*, 171–174. [CrossRef]
23. Greenberg, Y.; Yahel, E.; Ganor, M.; Hevroni, R.; Korover, I.; Dariel, M.P.; Makov, G. High Precision Measurements of the Temperature Dependence of the Sound Velocity in Selected Liquid Metals. *J. Non-Cryst. Solids* **2008**, *354*, 4094–4100. [CrossRef]
24. Emuna, M.; Greenberg, Y.; Yahel, E.; Makov, G. Anomalous and Normal Dependence of the Sound Velocity in the Liquid Bi-Sb System. *J. Non-Cryst. Solids* **2013**, *362*, 1–6. [CrossRef]
25. Okavi, S.; Emuna, M.; Greenberg, Y.; Yahel, E.; Makov, G. Interactions in Liquid Bismuth-Lead from Sound Velocity Studies. *J. Mol. Liq.* **2016**, *220*, 788–794. [CrossRef]
26. Gitis, M.B.; Mikhailov, I.G. Velocity of Sound and Compressibility of Certain Liquid Metals. *Sov. Phys. Acoust.* **1966**, *11*, 372–375.
27. Gitis, M.B.; Mikhailov, I.G. Correlation of the Velocity of Sound and Electrical Conductivity in Liquid Metals. *Sov. Phys. Acoust.* **1966**, *12*, 14–17.
28. Rosen, M.; Salton, Z. Temperature Dependence of the Sound Velocity and Ultrasonic Attenuation in Liquid Bi-Ga and Bi-Sn Alloys. *Mater. Sci. Eng.* **1983**, *58*, 189–194. [CrossRef]
29. Almond, D.P.; Blairs, S. Ultrasonic Speed, Compressibility, and Structure Factor of Liquid Cadmium and Indium. *J. Chem. Thermodyn.* **1980**, *12*, 1105–1114. [CrossRef]
30. Tupta, M.A. *AC Versus DC Measurement Methods for Low-Power Nanotech and Other Sensitive Devices*; Keithley Instruments: Cleveland, OH, USA, 2007.
31. Ginter, G.; Gasser, J.G.; Kleim, R. The Electrical Resistivity of Liquid Bismuth, Gallium and Bismuth-Gallium Alloys. *Philos. Mag. B* **1986**, *54*, 543–552. [CrossRef]
32. Li, X.F.; Zu, F.Q.; Ding, H.F.; Yu, J.; Liu, L.J.; Xi, Y. High-Temperature Liquid-Liquid Structure Transition in Liquid Sn-Bi Alloys: Experimental Evidence by Electrical Resistivity Method. *Phys. Lett. Sect. A Gen. Atmotic Solid State Phys.* **2006**, *354*, 325–329. [CrossRef]
33. Pokorny, M.; Astrom, H.U. Temperature Dependence of the Electrical Resistivity of Liquid Gallium between Its Freezing Point (29.75 Degrees C) and 752 Degrees C. *J. Phys. F Met. Phys.* **1976**, *6*, 559–565. [CrossRef]
34. Rhazi, A.; Auchet, J.; Gasser, J.G. Electrical Resistivity of Ni-In Liquid Alloys. *J. Phys. Condens. Matter* **1997**, *9*, 10115–10120. [CrossRef]
35. Wang, Q.; Chen, X.; Lu, K. Electrical Resistivity and Absolute Thermopower of Liquid GaSb and InSb Alloys. *J. Phys. Condens. Matter* **2000**, *12*, 5201–5207. [CrossRef]
36. Adams, P.D.; Leach, J.L. Leach. Resistivity of Liquid Lead-Tin Alloys. *Phys. Rev.* **1967**, *156*, 178–183. [CrossRef]
37. Krzhizhanovskii, R.E.; Sidorova, N.P.; Bogdanova, I.A. Experimental Investigation of the Electrical Resistivity of Some Molten Bismuth-Tin Binary Alloys and of the Thermal Conductivity of Bismuth, Tin, and a Eutectic Bismuth-Tin Alloys. *Inzhenerno-Fizicheskii Zhurnal* **1974**, *26*, 33–36. [CrossRef]
38. Li, X.F.; Zu, F.Q.; Ding, H.F.; Yu, J.; Liu, L.J.; Li, Q.; Xi, Y. Anomalous Change of Electrical Resistivity with Temperature in Liquid Pb-Sn Alloys. *Physica B* **2005**, *358*, 126–131. [CrossRef]
39. Verhoeven, J.D.; Lieu, F.Y. Resistivity in the Molten Bi-Sn System. *Acta Metall.* **1965**, *13*, 927–929. [CrossRef]
40. Wang, Z.; Sun, Z.; Wang, X.; Zhang, H.; Jiang, S. Effects of Element Addition on Liquid Phase Separation of Bi-Ga Immiscible Alloy: Characterization by Electrical Resistivity and Coordination Tendency. *JMADE* **2017**, *114*, 111–115. [CrossRef]
41. Yu, Q.; Wang, X.D.; Su, Y.; Cao, Q.P.; Ren, Y.; Zhang, D.X.; Jiang, J.Z. Liquid-to-Liquid Crossover in the GaIn Eutectic Alloy. *Phys. Rev. B* **2017**, *95*. [CrossRef]

42. Faber, T.E.; Ziman, J.M. A Theory of the Electrical Properties of Liquid Metals III. The Resistivity of Binary Alloys. *Philos. Mag.* **1964**, *11*, 153–173. [CrossRef]

 © 2019 by the authors. Licensee MDPI, Basel, Switzerland. This article is an open access article distributed under the terms and conditions of the Creative Commons Attribution (CC BY) license (http://creativecommons.org/licenses/by/4.0/).

Article

Robust Interferometry for Testing Thermal Expansion of Dual-Material Lattices

Weipeng Luo *, Shuai Xue, Cun Zhao, Meng Zhang and Guoxi Li *

College of Intelligent Science, National University of Defense Technology, Changsha 410073, China; shuaixue1991@163.com (S.X.); zhaocun18@163.com (C.Z.); z.mengdr@gmail.com (M.Z.)
* Correspondence: luoweipeng09@nudt.edu.cn (W.L.); lgx2020@sina.com (G.L.)

Received: 15 November 2019; Accepted: 7 January 2020; Published: 9 January 2020

Abstract: Dual-material lattices with tailorable coefficients of thermal expansion have been applied to a wide range of modern engineering systems. As supporting techniques for fabricating dual-material lattices with given coefficients of thermal expansion, the current existing methods for measuring the coefficient of thermal expansion have limited anti-interference ability. They ignore the measuring error caused by micro-displacement between the measurement sensor and the test sample. In this paper, we report a robust interferometric test method which can eliminate the measurement error caused by micro-displacement between the measurement sensor and the test sample. In the presented method, two parallel plane lenses are utilized to avoid the measurement error caused by translation, and the right lens is utilized as an angle detector to eliminate the measurement error caused by rotation. A robust interferometric testing setup was established using a distance measuring set and two plane lenses. The experiment results indicated that the method can avoid the measurement error induced by translation and has the potential to eliminate the measurement error induced by rotation using the rotational angle. This method can improve the anti-interference ability and accuracy by eliminating the measurement error. It is especially useful for high-precision thermal expansion measurement of dual-material lattices.

Keywords: dual-material lattices; interferometry; measurement; thermal analysis; thermal expansion

---

1. Introduction

Dual-material lattices with tailorable coefficients of thermal expansion (CTE) have been widely used in many applications [1–3]. Various dual-material lattices have been proposed to obtain tailorable CTEs [4–8]. In order to guide their design and machining, the equivalent CTEs of the dual-material lattices must be accurately measured ahead. Usually, the CTE measuring process takes a long time due to the slow heating. During this long process, vibrations from the environment and the thermal deformation of the measuring device will cause micro-displacement between the measurement sensor and the test sample. The micro-displacement will generate unacceptable measurement error for high-precision measurement. Hence, developing a robust measurement method for testing the CTEs of dual-material lattices is meaningful to enhance measurement accuracy. However, the existing thermal expansion measurement CTE technologies are sensitive to the micro-displacement between the measurement sensor and the test sample.

Various techniques are available to measure the CTE of materials, including strain gauge technique [9,10], capacitance method [11,12], interferometric technique [13,14], and digital image correlation (DIC) method [1,15,16]. Among these, laser interferometric techniques and the DIC method are the most used to measure the CTEs of dual-material lattices. The laser interferometric technique uses interference fringe variation to measure thermal deformation with high accuracy. However, the laser interferometric fringe pattern is very sensitive to the vibration of the environment, and this method cannot eliminate the measurement error caused by the micro-displacement [13,14]. We measured the CTE of a dual-material lattice with negative thermal expansion using laser interferometry [17], but the measurement error was large, and we did not consider the measurement error caused by the micro-displacement. Digital image correlation provides a full-field thermal deformation by comparing the images captured before and after deformation [18]. It uses a high-resolution camera to capture the digital images of the test sample. This method can eliminate part of the measurement error caused by the micro-displacement via data processing. However, this method is sensitive to temperature variation, air turbulence, and out-of-plane displacement of the sample [19,20]. Therefore, developing a robust measurement method for testing the CTEs of dual-material lattices which can eliminate the measurement error caused by micro-displacement is still challenging and has never been reported (according to the authors' knowledge).

In this paper, we report on a robust laser interferometric measurement method for dual-material lattices to overcome the measurement error caused by micro-displacement. The presented method has high anti-interference capability by using a distance measuring system. The distance measuring system consists of a distance measuring set and two parallel plane lenses. The two parallel lenses can avoid the measurement error caused by the translational component of the micro-displacement. The right lens, working as a micro-rotation angle indicator, can measure the rotational angle of the micro-displacement. The measurement error caused by the rotational component of the micro-displacement is eliminated with the rotational angle. With this method, the measurement error caused by micro-displacement can be eliminated completely.

2. Thermal Expansion Measurement System

2.1. Principle of the Measurement System

The schematic diagram of the experimental system for CTE measurement is shown in Figure 1. It includes a temperature control system and distance measurement system. The temperature control system consists of a thermostat, an electric heating plate, a temperature controller, and a thermocouple thermometer. The thermostat is used to provide uniform ambient temperatures for the sample to reduce temperature inhomogeneity. The electric heating plate is heated by the electric resistance. It uses the temperature controller to realize the temperature control. This electric heating plate is used for rapid heating of the sample. The thermocouple thermometer has two thermocouple probes to monitor the temperature at different locations of the sample. It can indicate whether the temperature of the test sample is uniform.

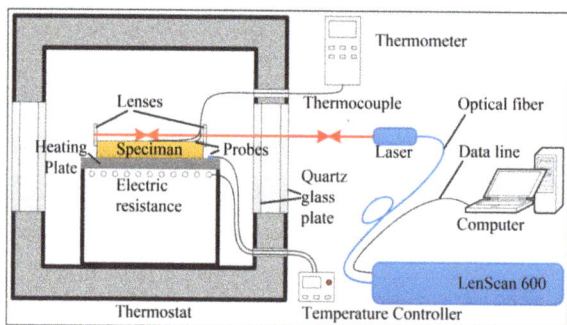

Figure 1. Schematic diagram of the measurement system. It consists of a temperature control system and distance measurement system.

The distance measuring system is composed of a distance measuring set and two plane optical lenses. The two lenses are mounted on both ends of the sample to reflect the laser beam. The right lens, near the laser, is used to measure the rotational angle between the sample and the laser beam.

2.2. Establishment of the Experimental System

The actual experimental setup for the CTE measurement is shown in Figure 2. In the actual experimental setup, an oven (Lichen-101BS, Shanghai, China) with temperature control was used as the thermostat. The temperature controller (HS-618F, Shanghai, China), with an accuracy of ±1 °C, was used to control the temperature of the heating plate. It allows non-contact temperature settings via an infrared remote controller. The thermocouple thermometer (UT320, Dongguan, China), with two thermocouple probes, was used to monitor the temperature at different locations of the sample.

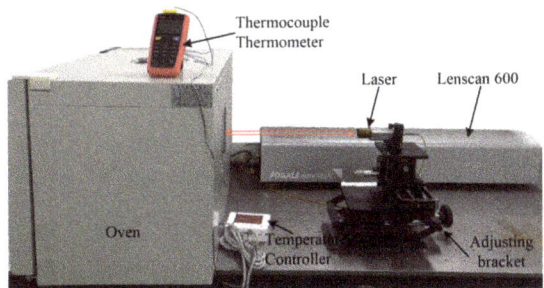

Figure 2. Actual experimental setup. The oven, thermocouple thermometer heating plate, and temperature controller form the heating system. The Lenscan 600 (Nimes, France) was used to measure the distance.

The distance measuring set (LS600, Nimes, France), with an absolute accuracy of ±1 μm, was used to measure the thermal deformation of the sample. It can measure the length of the air gap between the two lenses and the thickness of the right lens based on low coherence interferometry [21]. The two plane lenses (N-BK7, Shanghai, China) were fixed on the sample by two mounting brackets. Each mounting bracket had three angle adjusting screws. Thus, the angles of the lenses could be adjusted to allow the reflected laser to coincide with the incident laser beam.

2.3. Measurement Steps

The following steps were the experimental measurement procedures as shown in Figure 3.

2.3.1. Fixation of the Sample, Lenses, and Thermocouple Probes

The sample was fixed on the surface of the heating plate. The two thermocouple probes were fixed at different positions on the sample by high-temperature rubberized fabric. Then, the two lenses were fixed on the sample using the two mounting brackets.

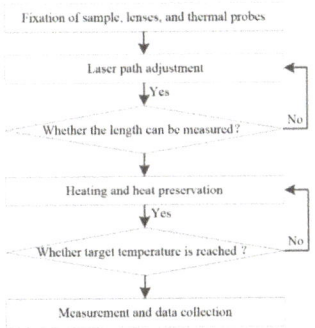

Figure 3. Flowchart of the steps for coefficient of thermal expansion (CTE) measurement.

2.3.2. Laser Path Adjustment

First, the height of the adjusting bracket was adjusted to ensure the center height of the laser probe was consistent with the center height of the two lenses. Second, the pitch angle and yaw angle of the laser probe were adjusted to ensure that the laser beam could pass through the center of the two lenses. Third, the angle adjusting screws of the two lenses were adjusted to reflect the light point to coincide with the incident point. Each lens had to be adjusted independently. Then, the length of the air gap between the two lenses was measured to determine whether the measurement quality was acceptable; if it was not, the laser path was readjusted.

2.3.3. Heating and Heat Preservation

First, the temperature of the thermostat and the heating plate were set at the target temperatures and the sample was heated. When the thermostat and hot plate both reached their target temperatures, they were kept warm for approximately half an hour. Second, the thermocouple thermometer was used to monitor the temperature at different positions of the sample. The readings of the two thermocouple probes were observed until the difference between the two readings was less than 0.5 °C.

2.3.4. Measurement and Data Collection

The sample was measured five times at each temperature point and the results recorded. Then, the temperature setting was changed, and Sections 2.3.3 and 2.3.4 were repeated until the measurements were completed.

3. Measurement Error Analysis

During the heating process, in order to guarantee the uniformity and accuracy of the sample temperature, the sample was heated slowly. Thus, the whole measurement process needed a long time. Over such a long period of time, the vibration of the environment and the thermal deformation of the measuring system caused uncertain micro-displacement between the sample and the measurement sensor. This micro-displacement can be decomposed into a translational component and a rotational component as shown in Figure 4. First, we assumed that these plane lenses were rigidly connected to the sample: the micro-displacement of the sample and the micro-displacement of the lenses were the same. Second, we assumed that the reflected laser coincided with the incident laser in the initial conditions: the initial angle between the sample and the laser was zero. Third, we assumed that the sample did not warp during the measurements. The following analyzes the measurement errors caused by these two components and how to eliminate these errors.

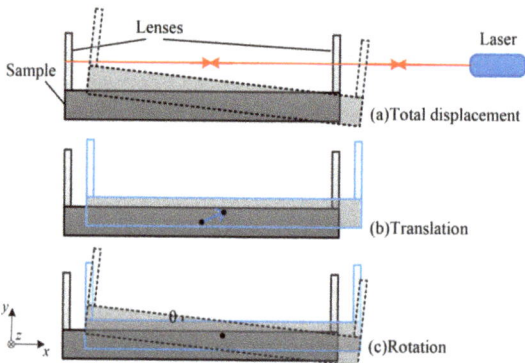

Figure 4. Schematic diagram of the micro-displacement decomposition. (**a**) Total displacement. (**b**) The translational component. (**c**) The rotational component.

During the measurement process, the temperature dependence of the refractive index must be considered. The distance measuring set measures the optical path based on a linear optical encoder [21]. Then, the actual air gap and thickness of the lens are calculated by the measurement software automatically. By setting the refractive index of each material in the initial measurement model, the measuring software can obtain the actual distance or thickness through calculation. However, in the process of thermal expansion measurement, the refractive indexes of the materials change with the temperature increase. This will lead to an inaccurate result. In order to improve the measurement accuracy, the influence of temperature on the refractive index should be considered.

3.1. Influence of the Translational Component

The translational component is one of the sources of measurement errors. In this presented CTE measurement system, it measures the CTE by measuring the optical path of the air gap between the two lenses. The optical path of the air gap was obtained by the difference of the optical paths between the two lenses and the laser source as shown in Figure 5b. It can be expressed as:

$$D_{AB} = D_{LA} - D_{LB} \tag{1}$$

In Equation (1), D_{AB} is the optical path of the air gap, D_{LA} is the optical path of the left lens and the laser source, and D_{LB} is the optical path between of the right lens and the laser source. As shown in Figure 5b, the translational component can cause changes in the optical path between the lenses and the laser source.

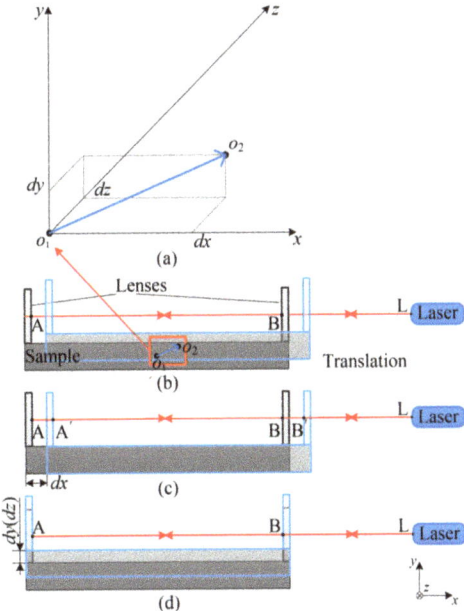

Figure 5. Schematic diagram of the translation displacement decomposition. (**a**) Translation displacement decomposition along the axis. (**b**) Total translation displacement. (**c**) The x component. (**d**) The y and z component.

The translational component can be decomposed along the axis into three components: dx, dy, and dz as shown in Figure 5a. Through the laser path adjustment, the lens' surfaces are perpendicular to the laser beam. The two lenses are parallel to each other. Thus, dy and dz will not change distances between the lenses and the laser source as shown in Figure 5d. Therefore, dy and dz will not change the optical path of the air gap. When the sample moves along the x-axis, as shown in Figure 5c, the optical path of D_{LA} and D_{LB} change to $D_{LA'}$ and $D_{LB'}$. However, the lengths of D_{LA} and D_{LB} decrease by the same value, dx. Thus:

$$\frac{\partial D_{AB}}{\partial x} = 0 \tag{2}$$

In summary, the three translational components (i.e., dx, dy, and dz) will not change the optical path of the air gap. The translational component of the micro-displacement will not generate measurement errors for the CTE measurement. Thus, the measurement errors caused by the translational component can be avoid by the two parallel lenses.

3.2. Measurement Error Analysis and Elimination of Rotational Component

3.2.1. Influence of Rotational Component

The rotational component is another source of measurement error. An excessive rotation angle will prevent the laser receiver from receiving reflected light. It will lead to measurement failure. The rotational component can be decomposed along the axis into three components in cartesian coordinates: around the x-axis, around the y-axis, and around the z-axis. The rotational component around the x-axis only makes the sample rotate around the laser beam. It does not cause extra optical path changes. Thus, it will not make measurement errors. Considering the rotational component around the y-axis and the rotational component around the z-axis, each component will cause the angle change between the sample and the laser beam. The situation will be more complex when the two components both occur. In order to simplify the calculation, the two components can be described in the polar coordinates. The total rotational component has just one direction and one angle to the laser beam. The direction of the total rotational component is circular symmetric. Thus, we can simplify the complicated situation into rotation around only one transverse axis. Take the rotation around the y-axis as an example for a small rotation angle; the incident and reflected rays no longer coincide as shown in Figure 6.

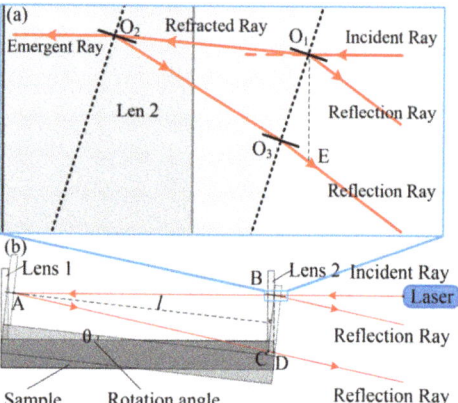

Figure 6. Schematic diagram of the laser transmission after the rotation. (**a**) The laser transmission at the right lens. (**b**) The laser transmission of the measurement system after rotation.

When the sample rotates for angle θ, the lenses are no longer perpendicular to the laser beam as shown in Figure 6b. According to the laws of reflection and refraction, the reflected rays will be deflected, and transmitted light will be refracted. Then, the optical path of the air gap changes. It can be expressed as:

$$D_{gap} = \frac{D_{AB} + D_{AC} + D_{CD}}{2} = n_1 l / \cos\theta + n_1 l \sin^2\theta / \cos(2\theta) \tag{3}$$

where D_{gap} is the total optical path of the air gap between the two lenses, D_{AB} is the optical path of the two points A and B, D_{AC} is the optical path of the two points A and C, D_{CD} is the optical path of the two points C and D, n_1 is the refractive index of the air, l is the vertical distance of the two lenses, and θ is the rotation angle. If the rotation angle θ is zero, the optical path is minimized. Then:

$$D_{gap} = n_1 l \tag{4}$$

According to Equation (3), we can find the measurement result after the rotation is slightly larger. In Equation (3), considering the rotation angle θ is very small, the second term is:

$$n_1 l \sin^2\theta / \cos(2\theta) \approx n_1 l \sin^2\theta = o(\theta) \tag{5}$$

To simplify Equation (3), we ignore high-order small quantities. Thus:

$$D_{gap} = n_1 l / \cos\theta \tag{6}$$

$$\frac{\partial D_{gap}}{\partial \theta} = \frac{n_1 l \sin\theta}{\cos^2\theta} \tag{7}$$

According to Equation (7), when θ is zero, the derivative of D_{gap} is zero. Thus, the total optical path of the air gap reaches a minimum. In summary, no matter if the rotation angle is positive or negative, as long as rotation occurs, the optical path D_{gap} will become larger than the initial length after this rotation. This will cause the measurement result to be slightly larger and generate measurement error. To improve the accuracy of measurement, the measurement error caused by rotation must be reduced.

3.2.2. Measurement Error Elimination

According to Equation (6), if we obtain the value of the rotation angle through measurement, the measurement error caused by the rotational component can be compensated. In the measurement system reported in this paper, considering the right lens is fixed on the sample rigidly, the rotational angle of the right lens is equal to the rotational angle of the sample.

$$t = \frac{D_{lens}}{n_2} = \frac{D_{O_1 O_2} + D_{O_2 O_3} + D_{O_3 E}}{n_2} = t_0 / \cos\theta + \frac{n_1 t_0 \sin^2\theta / \cos(2\theta)}{n_2} \tag{8}$$

where D_{lens} is the optical path of the right lens, t_0 is the original thickness of the right lens, n_1 is the refractive index of the air, n_2 is the refractive index of the right lens, and θ is the rotation angle. If the rotation angle θ is zero, the thickness of the right lens is $t = t_0$.

In Equation (8), considering the rotational angle θ is very small, thus the second term is:

$$\frac{n_1 t_0 \sin^2\theta / \cos(2\theta)}{n_2} \approx \frac{n_1 t_0 \sin^2\theta}{n_2} = o(\theta) \tag{9}$$

To simplify Equation (8), we ignore high-order small quantities. Thus:

$$t = t_0 / \cos\theta \tag{10}$$

Then, according to Equation (10), the rotational angle is:

$$\theta = \arccos(\frac{t}{t_0}) \tag{11}$$

If the value of t is obtained, the rotational angle can be calculated according to Equation (11). In this measurement system, the thickness of the right lens and the air gap can be measured simultaneously using Lenscan 600. Thus, it can guarantee the rotational angle of the right lens, and the distance of the air gap is measured synchronously. When the sample rotates, the rotational angle of the sample can be obtained by comparing the measured thickness of the right lens with the initial thickness. Thus, the measurement error of the rotational component can be eliminated by using Equation (6).

3.2.3. Calculation of CTE

Generally, the variation in the sample's length during the heating process can be obtained by comparing the optical path of the air gap measured at different temperatures T_1 and T_2. To improve the accuracy of calculation, the change in the refractive index of the air and lens with the temperature must be considered. The CTE of the lens should be considered too.

$$\Delta d = \frac{D_{gapT2} - D_{gapT1}}{n_{1T2}} \cos\theta \tag{12}$$

where Δd is the deformation of the sample along the x-axis from temperature T_1 to T_2, n_{1T_2} is the refractive index of the air at temperature T_2, D_{gapT1} and D_{gapT2} are the optical paths of the air gap between the two lenses at different temperatures, and the rotation angle θ is the parameter to be measured.

The rotational angle of the sample can be obtained by comparing the measured thickness of the right lens with the initial thickness if the temperature does not change. However, the temperature increases during CTE measurement. The measured thickness of the right lens will include thermal expansion deformation and optical path variation induced by refractive index of the right lens. These factors must be removed from the measurement result. Thus, the pure thickness of the right lens is:

$$t = \frac{D_{lens}}{n_{2T2}} - \alpha_{lens} t_0 (T_2 - T_1) \tag{13}$$

where D_{lens} is the optical path of the right lens, n_{2T_2} is the refractive index of the right lens at temperature T_2, and α_{lens} is the CTE of the right lens. The rotational angle can be calculated according to Equation (11).

Then, combining Equations (11)–(13), the pure deformation of the sample can be obtained. Thus, the thermal expansion coefficient is:

$$\alpha = \frac{\Delta d}{l(T_2 - T_1)} \tag{14}$$

4. Experiment

In order to verify the effectiveness of the experimental measurement system, we used a truss structure as a reference. It was manufactured by electric discharge machining from a 5 mm thick plate of aluminum. The properties of the glass and air are provided in Table 1. The refractive index of the air at different temperatures was calculated according to the equation of Rüeger [22]. The effects of the translational component and rotational component on the measurement result were verified by the experiment. Then, the CTE along the length of the aluminum sample was measured to verify the validity of this experimental system.

Table 1. Properties for the glass [23] and air [22].

Material	Refractive Index	Temperature Coefficient of the Refractive Index ($\times 10^{-6}$ K^{-1})	Coefficient of Thermal Expansion α ($\times 10^{-6}$ K^{-1})
Glass N-BK7	1.5035829	2.4	8.3
Air	1.000271	-	-

4.1. Translational Experiment

The x direction displacement of the laser was adjusted using the adjusting bracket. The y and z direction displacements of the laser were adjusted by a 2D precision mobile platform as shown in

Figure 7a. The position of the laser was detected by a micrometer (Shahe5313-02, Zhejiang, China) with an accuracy of ±2 μm. According to the measurement steps described in Section 2.3, the temperature was kept at room temperature (i.e., 20 °C); the laser was moved respectively along the x, y, and z directions from −50–50 μm; the interval was 10 μm; and each position was measured five times. The effects of the translational component on the length of the air gap are shown in Figure 7, and the measured lengths of the air gap after translating in the x, y, and z directions are provided in Figure 7b–d, respectively. "L_x" is the length of the air gap after x-axis translation. "L_y" is the length of the air gap after y-axis translation. "L_z" is the length of the air gap after z-axis translation. "L_{Ix}", "L_{Iy}", and "L_{Iz}" are the initial lengths of the air gap represented by the dashed lines in Figure 7b–d, respectively. According to the results, the lengths of the air gap fluctuate around the mean length. The amount of fluctuation was less than the accuracy of the measurement (±1 μm). Thus, the lengths of the air gap had no obvious trend with translation. This proves that translation had little effect on the length measurement. This is consistent with the previous analysis.

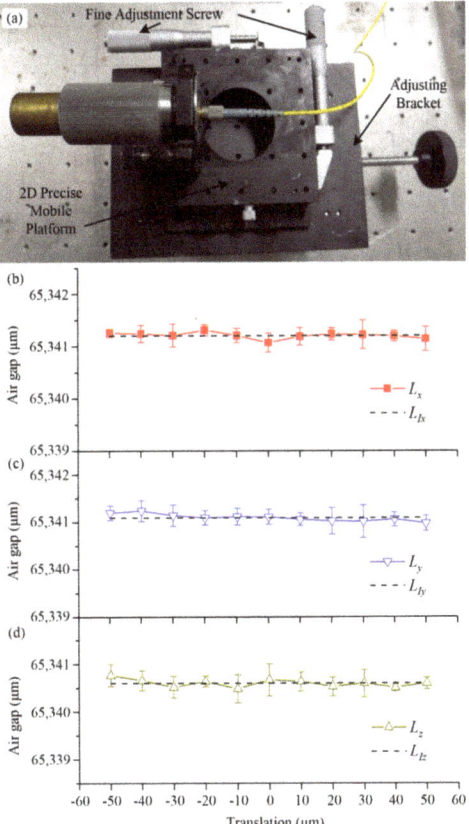

Figure 7. Measurement results after artificial additional translation. (**a**) The translation adjustment device. (**b**) Measurement results after x-axis translation. "L_x" is the length of the air gap after x-axis translation. (**c**) Measurement results after y-axis translation. "L_y" is the length of the air gap after y-axis translation. (**d**) Measurement results after z-axis translation. "L_z" is the length of the air gap after z-axis translation. The dashed lines are the initial length of the air gap before translation.

4.2. Rotational Experiment

The rotation of the sample was realized by a rotation platform (KSP-656M, Guangdong, China) with an accuracy of ±52.2″ as shown in Figure 8a. The temperature was kept at room temperature (i.e., 20 °C); the fine adjustment screw was adjusted to rotate the sample around the y-axis from 0 to 0.6°; the interval was 0.058°; each position was measured five times. Figure 8b shows the effect of the rotation on the sample length measurement at 20 °C. "L_r" is the measurement length of the air gap after rotation. "L_{Th}" is the theoretical length of the air gap after rotation. "L_e" is length of the air gap after eliminating the measurement error of rotation. The dashed line "L_I" is the initial length of the air gap before rotation. According to the measurement result, the measurement length of the air gap increased when the rotation angle increased. This proves that the measurement length of the air gap after rotation was larger than the actual length. The rotational component generates extra measurement error. It will reduce the measurement accuracy of the thermal expansion coefficient. In Figure 8b, the red line provides the length of the air gap after eliminating the measurement error of rotation. This proves that the elimination was effective in reducing the measurement errors caused by rotation. Therefore, in order to improve the measurement accuracy, it is necessary to take measures to eliminate the measurement errors caused by rotation.

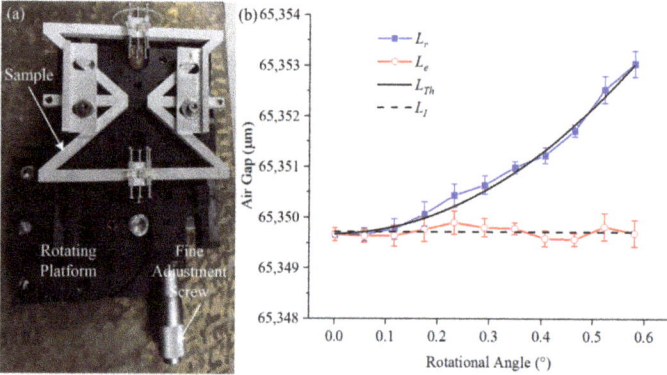

Figure 8. Measurement results after artificial additional rotation. (**a**) The experimental device of the sample rotation. (**b**) Measurement results after rotation. "L_r" is the measurement length of the air gap after rotation. "L_{Th}" is the theoretical length of the air gap after rotation. "L_e" is length of the air gap after elimination of the measurement error of rotation. The dashed line is the initial length of the air gap before rotation.

4.3. Measurement of the CTE of the Sample

According to the measurement steps described in Section 2.3, with interval of 20 °C, the sample was heated to 120 °C. The length of the air gap was measured five times at each temperature. The sample was heat-treated to reduce the residual stress. This reduces the warping deformation of the sample during measurement.

As shown in Figure 9, The "Air Gap" is the distance between the two lenses fixed on the sample. The length of the air gap between the two lenses increased with the increase in temperature. The blue scale bar represents 0.1 µm in length of the error bar of the air gap. "α" is the equivalent CTE of the sample. The average CTE of the sample was $22.99 \pm 0.54 \times 10^{-6}$ K^{-1}. Compared with the recommended value (23.1×10^{-6} K^{-1}) in the existing literature [24], this proves that this experimental system is valid and can be used for measuring the thermal expansion of truss structures.

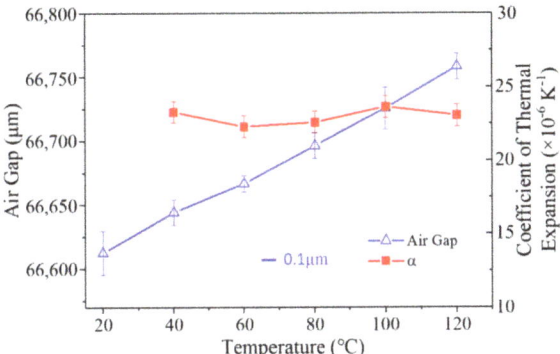

Figure 9. CTE measurement of the aluminum truss structure. The air gap is the distance between the two lenses fixed on the sample. It increased with an increase in the temperature. The blue scale bar represents 0.1 μm in length of the measurement error of the air gap. "α" is the CTE of the sample.

5. Discussion

Generally, the measurement accuracy of the thermal expansion coefficient depends on the accuracy of the temperature control and length measurement. The accuracy of the temperature control can be improved by advanced temperature control technology and high precision temperature sensor. In addition, the accuracy of the length measurement by laser interferometry has reached the nanometer scale. Thus, the measurement error will become the key factor to improving measurement precision. Vibration isolation is a good error control method in most cases. However, with temperature changes, it will bring many uncertain micro-displacements which cannot be suppressed by vibration isolation. This method can effectively remove the measurement error caused by the micro-displacements. It will reduce the isolation requirements for the environment and enhance the anti-interference ability. In addition, with the development of modern grating processing and high precision rail technology, interferometers with higher precision will be manufactured. The measurement accuracy of this method can be further improved.

This method has the potential to eliminate the measurement error caused by micro-displacement, and its effectiveness was verified by the experiment in Section 4. However, limited by the length measurement accuracy of the Lenscan 600, it cannot be used to eliminate the rotation error in actual measurements. In this presented measurement system, the rotational angle of the sample was obtained by detecting the thickness change of the right lens. Considering that the measurement accuracy of the Lenscan 600 is ±1 μm, the minimum detection angle of the lens with a thickness of 0.75 mm is 3°. This angle is beyond the angular tolerance of Lenscan 600 and far more than the actual rotational angle. One possible approach to decrease the minimum detection angle is to increase the thickness of the right lens. If the thickness of the right lens is 10 mm, the minimum detection angle is 0.81°. If the thickness of the lens is 656 mm, the minimum angle that can be measured is 0.1°. This method requires a large thickness of lens in order to achieve a sufficient angular detection accuracy. Hence, limited to the length measurement accuracy in the experimental system, this method is difficult to apply in actual measurement.

With the development of high-precision measuring instruments, especially with the accuracy of interferometry reaching 1 nm, this method could achieve sufficient angular detection accuracy. Assuming the measurement accuracy of the distance measuring system is 1 nm, if the thickness of the lens is 1 mm, the minimum detection angle is 0.08°. If the thickness of the lens is 20 mm, the minimum angle that can be measured is 1.5′. Then, the angular detection accuracy would be sufficient. The measurement error caused by rotation can be eliminated by this method. This method can effectively improve the accuracy of thermal expansion measurement. This is meaningful for the design and fabrication of zero thermal expansion lattice. Furthermore, it can be determined from theoretical analysis that rotational error can be eliminated as long as the rotational angle is obtained. There are several other approaches that can also achieve high-precision angle measurement such as high-precision electronic autocollimator and lens surface measurement. With these methods, we expect that the measurement error can be eliminated accurately. This method could greatly improve the anti-interference ability and accuracy by eliminating the measurement error.

The lattices with tailorable coefficients of thermal expansion are realized using two or three kinds of materials to form complex truss structures. Some of the structures are symmetric and the thermal expansion of the structures are axial elongation or shrinkage, without a rotational component. Considering a more general situation where the truss structures are asymmetric, these would have different stiffnesses in each direction. Then, the thermal expansion of the designed structures would have an axial component and a rotational component. Even though this method can eliminate the measurement error, it still has some restrictions. It cannot be used to measure the complex thermal deformation when there is an axial component and a rotational component. This method tests the CTE by measuring the length variation of the lattice. It can only measure the change in distance between two points along the optical axis without rotation. If thermal deformation of the dual-material lattice contains both of the two components, the rotational component of the CTE will mix with the rotational error. This will lead to a wrong result. Thus, this method cannot be used to measure the CTE of the axial component and the rotational component simultaneously. One possible solution is to measure the CTE of each component alone using this method and then to obtain the total CTE though calculation.

6. Conclusions

In this paper, we reported on a robust interferometric method that can eliminate the measurement error induced by micro-displacement. This method can greatly enhance the anti-interference ability and accuracy by eliminating the measurement error. This method is especially useful for CTE measurement of non-standard samples like dual-material truss structures with high precision requirements. The theoretical analysis indicated that translation did not affect the measurement results and that rotation made the measurement results lager. The robust experimental system was designed and built using two plane lenses and a distance measuring interferometer. The experimental results of the aluminum truss structure proved that this method can avoid the translation error and has the potential to eliminate the measurement error caused by rotation. The CTE measurement results of the sample successfully demonstrated the validity of the measurement system.

Author Contributions: Conceptualization and methodology, W.L., S.X. and C.Z.; formal analysis, W.L.; experiment and investigation, W.L. and S.X.; writing—original draft preparation, W.L.; writing—review and editing, S.X., M.Z. and G.L.; project administration, G.L. All authors have read and agreed to the published version of the manuscript.

Funding: This research received no external funding.

Acknowledgments: The authors would like to thank Kai Zhang, Renxiu Han, and Tian Lan for their support. The authors also would like to thank Qiang Xu of National University of Defense and CTEhnology Microsystem Laboratory for his support.

Conflicts of Interest: The authors declare no conflict of interest.

References

1. Parsons, E.M. Lightweight cellular metal composites with zero and tunable thermal expansion enabled by ultrasonic additive manufacturing: Modeling, manufacturing, and testing. *Compos. Struct.* **2019**, *223*, 110656. [CrossRef]
2. Wei, K.; Peng, Y.; Wang, K.; Duan, S.; Yang, X.; Wen, W. Three dimensional lightweight lattice structures with large positive, zero and negative thermal expansion. *Compos. Struct.* **2018**, *188*, 287–296. [CrossRef]
3. Dong, K.; Peng, X.; Zhang, J.; Gu, B.; Sun, B. Temperature-dependent thermal expansion behaviors of carbon fiber/epoxy plain woven composites: Experimental and numerical studies. *Compos. Struct.* **2017**, *176*, 329–341. [CrossRef]

4. Lakes, R.S. Cellular solid structures with unbounded thermal expansion. *J. Mater. Sci. Lett.* **1996**, *15*, 475–477. [CrossRef]
5. Lakes, R. Cellular solids with tunable positive or negative thermal expansion of unbounded magnitud. *Appl. Phys. Lett.* **2007**, *90*, 221905. [CrossRef]
6. Miller, W.; Mackenzie, D.S.; Smith, C.W.; Evans, K.E. A generalised scale-independent mechanism for tailoring of thermal expansivity: Positive and negative. *Mech. Mater.* **2008**, *40*, 351–361. [CrossRef]
7. Steeves, C.A.; Lucato, S.L.D.S.; He, M.; Antinucci, E.; Hutchinson, J.W.; Evans, A.G. Concepts for structurally robust materials that combine low thermal expansion with high stiffness. *J. Mech. Phys. Solids* **2007**, *55*, 1803–1822. [CrossRef]
8. Grima, J.N.; Farrugia, P.S.; Gatt, R.; Zammit, V. A system with adjustable positive or negative thermal expansion. *Proc. R. Soc. A* **2007**, *463*, 1585–1596. [CrossRef]
9. Grossinger, R.; Muller, H. New device for determining small changes in length. *Rev. Sci. Instrum.* **1981**, *52*, 1528–1535. [CrossRef]
10. Walsh, R.P. Use of strain gages for low temperature thermal expansion measurements. In Proceedings of the 16th International Cryogenic Engineering Conference/International Cryogenic Materials Conference (ICEC/ICMC), Kitakyushu, Japan, 20–24 May 1996; pp. 661–664.
11. Bijl, D.; Pullan, H. A new method for measuring the thermal expansion of solids at low temperatures; the thermal expansion of copper and aluminium and the Gruneisen rule. *Physica* **1954**, *21*, 285–298. [CrossRef]
12. Subrahmanyam, H.N.; Subramanyam, S.V. Accurate measurements of thermal expansion of solids between 77 K and 350 K by 3-terminal capacitance method. *Pramana* **1986**, *27*, 647–660. [CrossRef]
13. Pirgon, O.; Wostenholm, G.H.; Yates, B. Thermal expansion at elevated temperatures IV carbon-fibre composites. *J. Phys. D Appl. Phys.* **1973**, *6*, 309–321. [CrossRef]
14. Wolff, E.G.; Savedra, R.C. Precision interferometric dilatometer. *Rev. Sci. Instrum.* **1985**, *56*, 1313–1319. [CrossRef]
15. Steeves, C.A.; Mercer, C.; Antinucci, E.; He, M.Y.; Evans, A.G. Experimental investigation of the thermal properties of tailored expansion lattices. *Int. J. Mech. Mater. Des.* **2009**, *5*, 195–202. [CrossRef]
16. Wei, K.; Chen, H.; Pei, Y.; Fang, D. Planar lattices with tailorable coefficient of thermal expansion and high stiffness based on dual-material triangle unit. *J. Mech. Phys. Solids* **2016**, *86*, 173–191. [CrossRef]
17. Luo, W.; Xue, S.; Zhang, M.; Zhao, C.; Li, G. Bi-material negative thermal expansion inverted trapezoid lattice based on a composite rod. *Materials* **2019**, *12*, 3379. [CrossRef] [PubMed]
18. Strycker, M.D.; Schueremans, L.; Paepegem, W.V.; Debruyne, D. Measuring the thermal expansion coefficient of tubular steel specimens with digital image correlation techniques. *Opt. Lasers Eng.* **2010**, *48*, 978–986. [CrossRef]
19. Pan, B.; Xie, H.; Hua, T.; Asundi, A. Measurement of coefficient of thermal expansion of films using digital correlation method. *Polym. Test.* **2009**, *28*, 75–83.
20. Pan, B. Thermal error analysis and compensation for digital image/volume correlation. *Opt. Lasers Eng.* **2018**, *101*, 1–15. [CrossRef]
21. FOGALE Nanotech. Sensors & Systems Documentations. Available online: http://www.fogale.fr/brochures.html (accessed on 10 December 2017).
22. Rüeger, J.M. *Refractive Indices of Light, Infrared and Radio Waves in the Atmosphere*; University of New South Wales: Sydney, Australia, 2002.
23. SCHOTT Glass Made of Ideas. SCHOTT Optical Glass Data Sheets. Available online: https://www.schott.com/english/download/index.html (accessed on 22 October 2019).
24. Yan, M.G. Aluminium alloy. In *China Aeronautical Materials Handbook*, 2nd ed.; China Standard Press: Beijing, China, 2001; Volume 3, pp. 308–318.

© 2020 by the authors. Licensee MDPI, Basel, Switzerland. This article is an open access article distributed under the terms and conditions of the Creative Commons Attribution (CC BY) license (http://creativecommons.org/licenses/by/4.0/).

Article

Spark Plasma Sintering Apparatus Used for High-temperature Compressive Creep Tests

Barak Ratzker, Sergey Kalabukhov and Nachum Frage *

Department of Materials Engineering, Ben-Gurion University of the Negev, P.O.B. 653, Beer-Sheva 84105, Israel; ratzkerb@post.bgu.ac.il (B.R.); kalabukh@bgu.ac.il (S.K.)
* Correspondence: nfrage@bgu.ac.il; Tel.: +972-8-646-1468; Fax: +972-8-647-9441

Received: 10 November 2019; Accepted: 13 January 2020; Published: 15 January 2020

Abstract: Creep is a time dependent, temperature-sensitive mechanical response of a material in the form of continuous deformation under constant load or stress. To study the creep properties of a given material, the load/stress and temperature must be controlled while measuring strain over time. The present study describes how a spark plasma sintering (SPS) apparatus can be used as a precise tool for measuring compressive creep of materials. Several examples for using the SPS apparatus for high-temperature compressive creep studies of metals and ceramics under a constant load are discussed. Experimental results are in a good agreement with data reported in literature, which verifies that the SPS apparatus can serve as a tool for measuring compressive creep strain of materials.

Keywords: spark plasma sintering apparatus; compressive creep test; stress exponent; electric current

1. Introduction

Creep is the continuous deformation of a material subjected to a constant load or stress, often lower than its yield strength. It is a process that is very sensitive to temperature, generally considered as high-temperature deformation, which occurs at roughly ≥ 0.5 T/T_m (where T/T_m in K is the homologous temperature) [1]. It is an imperative issue, which could lead to failure of engineering materials that are under stress and exposed to high-temperature environments. Compressive creep tests are a convenient method for investigating high-temperature deformation, especially for strong and brittle refractory materials such as ceramics [2–4]. Creep generally includes three different stages: the primary, secondary and tertiary stage (Figure 1). The primary and tertiary are the initial and final stages, in which there is a transient strain rate that decelerates to a steady rate or accelerates up to failure by rupture, respectively. Meanwhile, during the second stage, termed steady-state creep the strain rate is constant (or nearly constant). The creep rate in this stage is used to evaluate high-temperature deformation and creep behavior of materials.

Measuring steady-state creep rate is important for prediction of the service lifetime for structural components, as well as understanding the micromechanics and metallurgical aspects of high-temperature deformation. Accurate strain rate measurements make it possible to determine creep parameters and identify the operating deformation mechanisms [5,6]. Creep rate generally depends on either external conditions, such as temperature and applied stress, or material properties, such as grain size and presence of precipitates or dopants.

Creep strain rate is most sensitive to the temperature and applied stress and can be described by an Arrhenius exponential and power-law dependency, respectively. The basic power-law creep equation is generally presented as:

$$\dot{\varepsilon} = A\sigma^n \exp\left(-\frac{Q}{RT}\right) \qquad (1)$$

where A is a constant, σ is the applied stress, n is the stress exponent, Q is the apparent activation energy, R is the gas constant and T is the temperature. The experimental creep rates can then be analyzed according to their dependency on stress and temperature which allows to determine values of Q and n.

$$\ln \dot{\varepsilon} = \ln A + n \ln \sigma - \frac{Q}{RT} \qquad (2)$$

The creep apparent activation energy (Q), which is inherent to the material, reflects the diffusion processes taking place at elevated temperatures and determines the creep temperature dependence. Diffusion also acts as an accommodating process to grain boundary sliding and dislocation creep mechanisms at lower temperatures [6,7]. The stress exponent (n) reflects the sensitivity to the applied stress and creep dependence on the load. According to Equation (2), Q and n can be determined by either measuring creep at different temperatures under same applied load, or at the same temperature under different applied loads, respectively. A proper creep test setup must allow precise gauging of the temperature and stress applied to the sample with accurate measurements of the axial displacement [8].

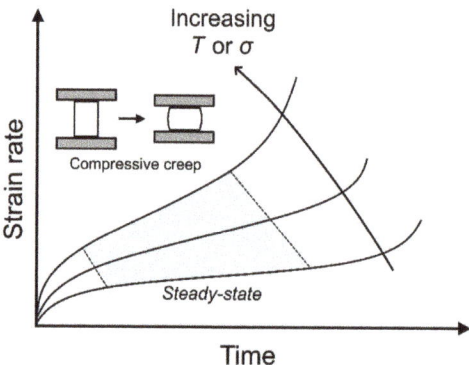

Figure 1. Schematic illustration of creep curves; strain rate increases with temperature or applied stress.

Spark plasma sintering (SPS) is an advanced pressure-assisted sintering technique, which utilizes an electric current for heat generation within conductive tooling and powder compacts [9,10]. This combination makes it possible to achieve excellent sintering capabilities of many metallic, ceramic and composite materials [11]. To track the densification progress, the SPS apparatus is equipped with a strain gauge and built in LVDT. It records the punch displacement every second with an accuracy of ~1 μm (depending on the SPS system). Consequently, the SPS apparatus encompasses all the necessary components (including temperature and load control) to perform compressive creep tests. Therefore, it was suggested that the SPS apparatus could be used as an accurate creep testing tool [12]. Such capabilities have already been demonstrated for both ceramics [12,13] and metals [14,15]. It is worthy to note that SPS creep test results have already shown good agreement with previously reported data obtained by conventional testing methods for the same materials at similar temperature/pressure ranges [12–15]. Moreover, it has been proposed that the SPS apparatus' ability to apply an electric current to the sample makes it possible to investigate to some extent electro-plastic effects in conductive materials during high temperature deformation [14,15]. This is particularly important for SPS, because it could also be directly connected to the enhanced sintering behavior of conducting materials [16,17].

In the present study, we describe in detail the use of an SPS apparatus for creep investigation of metals and ceramics, highlight the advantages and disadvantages of this method, and provide prominent experimental examples of creep tests performed by an SPS apparatus applied as a creep testing device.

2. Test Setup and Procedure

2.1. SPS Apparatus Technical Details

The SPS apparatus requires practically no modification to serve as a creep testing device in compliance with ASTM technical standards [18]. The following description is based on an FCT system SPS (FCT Systeme GmbH, Rauenstein, Germany), but would generally be the same for other SPS machines from other manufacturers. The system allows to easily set the testing parameters (i.e., load, temperature) and track them continuously in 1 s intervals, along with many other parameters derived from them (e.g., punch displacement, current, voltage). The specifications of the SPS apparatus used in this study are given in Table 1.

Table 1. Lab-scale HP-D 10 FCT system SPS apparatus specifications relevant to creep testing.

Temperature Range, °C	Pressing Force, kN	Displacement Resolution, mm	Programmable Test Segments	Electric Current Applied to Sample	Chamber Atmosphere
Up to 2400	3–100 (stress depends on sample cross-section)	0.001	Yes	Possible for conductive samples	Vacuum (10^{-2} mbar) with argon flow

To perform a creep test, a columnar sample is set at the center between the punches. To ensure that the sample is placed in an unconstrained manner, the initial and final sample dimensions should be considered prior to the test. The relative punch displacement (RPD) is monitored with an accuracy of ~1 μm for an HP-D10 FCT System (this may vary for other machines). The measured RPD can be converted to strain, simply by dividing it by the initial sample height while taking in account the thermal expansion of the material. The corresponding creep rate can then be determined from the slope or derivative of the strain curve [12,14].

2.2. Test Configurations and Temperature Considerations

The temperature in the SPS apparatus is typically measured using the built-in system pyrometer or thermocouples. For the best accuracy during creep tests, it was suggested to place a thermocouple (C, K or S type for our system) in direct contact with the sample surface (see Figure 2). Temperature distribution in SPS is a known issue which also depends on the tooling configuration [19,20]. If there is only resistive heating of the sample during creep tests of conducting materials (i.e., tooling without a surrounding die as was used in our previous studies [14,15]) the temperature deviations may be relatively large due to significant heat loss from the sample surface. To mitigate this, it is suggested in the present study to apply an electric current while using the graphite die, like the configuration discussed in [21]. The minor disadvantage in this case is that the electric current value applied by the SPS apparatus splits between the die and sample. The actual current applied to the sample (I_s) can then only be estimated according to the relative electrical resistance of the tooling and the sample according to the Kirchhoff's law. Thus, the different heating configurations for SPS apparatus creep tests are as depicted in Figure 2. In which conductive materials can be resistively heated by passage of an electric current as well as by heat convection and radiation from the graphite punches and die, respectively. The electric current can be avoided by separating the sample from the graphite by the means of a ceramic insulator, such as alumina. When tests are conducted at relatively high temperatures a graphite felt should be placed around the die to further mitigate heat dissipation.

To estimate possible temperature distributions in samples tested with different tooling configuration, a special set of temperature measurements was conducted. The measurements were performed on a cylindrical copper sample with several 1.5 mm holes drilled 6 mm deep into the center, top and bottom (Figure 3). Using two thermocouples simultaneously, multiple temperature measurements (in the 450–550 °C temperature range) were performed. Each measurement was taken when the temperature was stabilized. The results for both tooling configurations are summarized in Table 2. The temperature was defined according to the thermocouple located on the sample surface. When testing the difference between the center and top/bottom regions the temperature was defined according to the thermocouple located in the center of the sample.

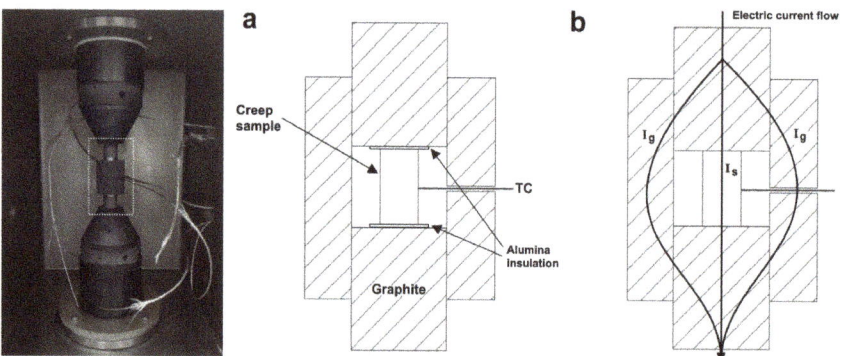

Figure 2. Schematics and images of configurations creep testing (**a**) without and (**b**) with electric current applied to the sample. The electric current flow is portrayed schematically.

The tests revealed the presence of a temperature gradient within the sample. In the case of a sample insulated from the electric current (Figure 2a), the radial temperature difference, between the surface and center is about ~15 °C. This difference in temperature would exists in any conventional creep test apparatus, since the sample is heated from the outside. As for the case with the electric current, the measurements showed a larger difference of roughly ~20–30 °C between the center and surface. However, we believe that these values should be taken with a grain of salt. The temperature measured in the center may be inaccurate due to higher current density developing around the hole. Even a small addition of current would cause significant extra localized heating around the thermocouple, making these measurements erroneous. In fact, considering the rate of heat loss from the surface, we expect that the real temperature gradient is roughly the same as was observed for the insulated sample (or even lower since the whole sample is heated by the current), but in the reverse direction. Additionally, there is also a difference between the top and the bottom of the sample This difference can be attributed to the SPS apparatus design in which the upper punch is the positive electrode and usually hotter. As shown by Sweidan et al. [21], approaches can be taken to minimize temperature deviations and achieve more accurate testing.

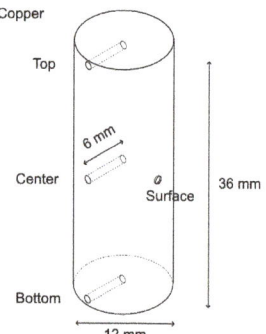

Figure 3. Schematic illustration of copper sample and the holes used for temperature measurements.

The uncertainty of the temperature measurements may be problematic to directly compare results of creep tests with and without an electric current and to discuss the effect of electro-plasticity.

Furthermore, it should be considered that for such creep tests at elevated temperatures joule heating effects also contribute to deformation and stress relaxation and cannot be easily de-coupled from the electro-plastic effect [22]. This issue will be further considered in the next section.

Table 2. Temperatures measured at various regions of copper sample for the different testing configurations with or without passage of an electric current.

Test Configuration	Temperature at the Surface, °C	Temperature at the Center, °C	Temperature at the Top, °C	Temperature at the Bottom, °C
With electric current	450	473 *		
	500	524 *	515 *	501 *
	550	579 *		
Insulated from the electric current	450	438		
	500	486		
	550	536		

* Suspected to be inaccurate due to high current density causing a temperature rise around the thermocouple locations.

2.3. Creep Testing Procedure

The creep test itself is performed in a relatively simple manner, by setting the designated temperature and load (pressure is calculated according to sample cross-section) and tracking RPD. The test can be conducted under constant temperature or load, but also with various pressure or temperature steps [15], to obtain multiple creep rate measurements from a single sample. An example for a testing procedure of alumina (at 1250 °C under 80 MPa), including all relevant experimental data necessary for creep evaluation, is presented in Figure 4. The heating stage **I** (Figure 4), is conducted prior to the creep test, while a minimal or designated test force is applied. At this stage, the negative RPD indicates the thermal expansion of the graphite tooling system and stainless-steel pistons. In segments **II–III** (Figure 4), the recorded displacement reflects only sample deformation after a mechanical and thermal equilibrium have been reached. In segment **II** there is a rapid decrease in the strain rate, while in segment **III** the strain rate is practically constant. Thus, segments **II–III** are considered primary and steady-state creep, respectively. It should be noted that the steady-state mentioned above (segment **III**) is sometimes a quasi-steady-state, due to the continuous reduction of true stress during creep (Figure 4) as well as concurrent grain growth [23,24]. This would be more of an issue at high strain rates under testing conditions of relatively high temperatures or applied stress [12].

Nevertheless, using an SPS apparatus for creep tests has several technical limitations. A minimal load of 3 kN must be applied during the test in order to receive displacement recording. This limits the sample size and determines the minimal applied stress. Furthermore, the SPS system cannot be set for a certain stress or constant strain rate and thus the accurate measurement of high-temperature compressive strength cannot be performed. Since the SPS apparatus typically only allows to apply a constant load, the actual stress on the sample continuously decreases during the test as the sample cross-section expands with increasing strain (Figure 5). This issue is treated by calculating the true strain (for any given moment) [25].

$$\varepsilon_t(t) = \ln \frac{l_t}{l_0} \qquad (3)$$

where ε_t is the true strain, l_t is the current specimen height at a given point in time and l_0 is the initial specimen height. Considering volume conservation (before excessive cavitation at final stages of creep), the true stress σ_t can then be calculated.

$$\sigma_t = \frac{F}{a_0^2}[\exp(\varepsilon_t)] \qquad (4)$$

where F is the applied load and a_0^2 is the initial sample cross-section area.

Thus, the strain rate during compressive creep tests can be summarized as the following equation.

$$\dot{\varepsilon} = A\left(\frac{F}{a_0^2}[\exp(\varepsilon_t)]\right)^n \exp\left(\frac{-Q}{RT}\right) \qquad (5)$$

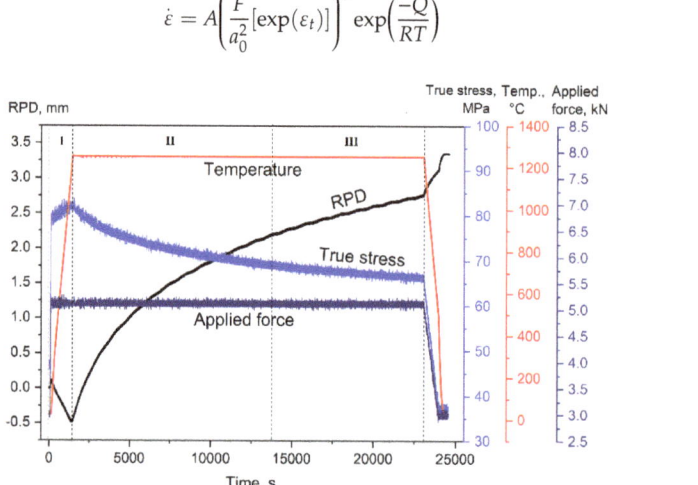

Figure 4. Creep test variables recorded by the SPS system (true stress is calculated). The different test stages (segments I, II and III) related to conducting a creep test are depicted [12]. Reproduced with permission of Elsevier.

Figure 5. Example of AlSi$_{10}$Mg samples before and after creep test (~34% strain) at 225 °C under an initial applied pressure of 130 MPa.

3. Results and Discussion

3.1. Creep Testing of Metals

As discussed in the previous section, metals (and other conductive materials) can be tested by heating generated from the graphite tooling while insulated from the punches or including resistive heating within the sample by allowing passage of the electric current through the sample. In our previous study on copper [14], it was shown that SPS-measured creep rates without the current agree quite well with results of conventional tensile creep tests that have been performed on copper under at the same temperature range (400–600 °C). The slope against the reciprocal of temperature is similar, which means that the creep apparent activation energy Q is the same. In this case, Q was equal to about 110 kJ/mol, which corresponds to vacancy migration (dislocation motion in vacancy saturation) [26].

This validated the SPS apparatus as an accurate creep testing tool. Nevertheless, for the case with applied electric current we used only resistive heating without surrounding die and the obtained results regarding electro-plastic effect may be questionable.

Therefore, in the present study, we performed additional creep experiments on similar copper samples with a surrounding die, which allows to lower the temperature gradient. These isothermal creep tests were conducted with load increments of 30, 40 and 50 N, while maintaining a constant temperature of 500 °C. This makes it possible to investigate the stress dependence and determine the stress exponent n. Each of these tests was performed in both possible configurations, with and without an electric current. The obtained creep curves are presented in Figure 6a, and the calculated creep rates are presented as a function of stress in Figure 6b. It was found that the value of n without the current was close to 4 which agrees with the known values for stress exponent of copper at relatively low stress (<100 MPa) and intermediate homologues temperatures [27]. While the value with the applied current was significantly lower, at around 2.3 (which is very low for copper). It has to be pointed out that stress exponent does not depend a relatively small temperature difference. See for instance a comparison with reported data for copper creep (tensile) at various temperatures (Figure 6c). Thus, the considerable difference of in the stress exponent could only be attributed to some electro-plastic effect which may affect the creep mechanism. Nevertheless, it is quite difficult to de-couple between the additional heating and electro-plastic effect, but the latter clearly has a contribution to susceptibility to creep [28,29]. Such results help to explain enhanced densification during SPS of conducting materials under influence of an electric current [16]. It perhaps may be possible to gain a deeper understanding on the matter by performing creep tests with different electric currents by altering the SPS tooling [30].

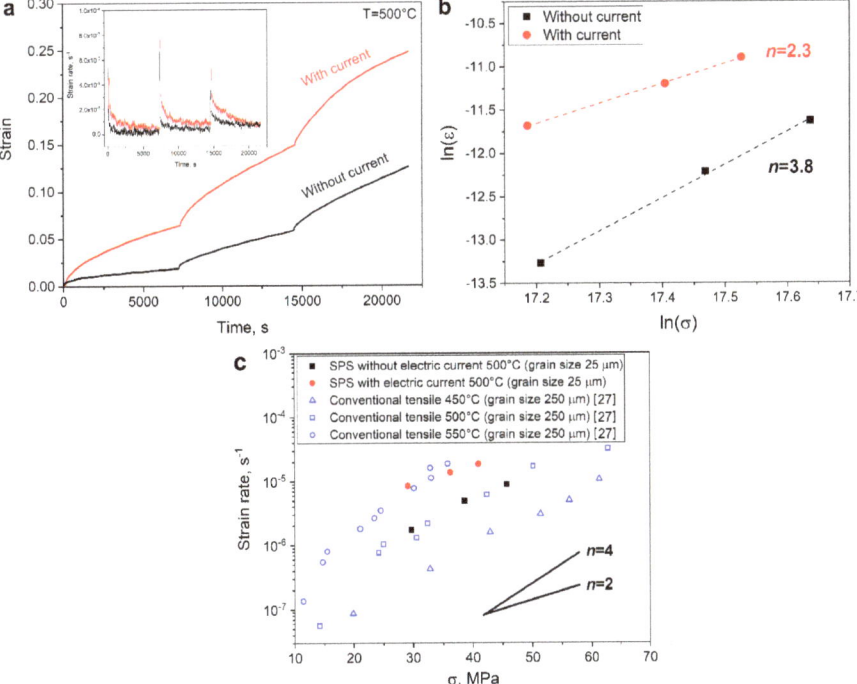

Figure 6. (**a**) Example of copper creep test at 500 °C with three load steps with and without an applied current, the corresponding strain rate in the insert; (**b**) the creep rates as a function of stress (natural log) with the slope value n indicated and (**c**) as a function of stress compared with other reported results [27] for tensile creep of copper at various temperatures (notice the different grain size).

3.2. Creep Testing of Ceramics

In the case of insulating ceramics, a configuration involving the graphite die must be used (Figure 2). However, it should be taken into account that some ceramics are still conductive, such as ZrN, and would involve an effect of electric current within the sample [17]. Also, under strong electric fields there may also be field effects in ceramics that can influence the creep behavior [22]. SPS was applied for a creep study of polycrystalline $MgAl_2O_4$ (for the first time) to clarify the deformation mechanisms at high temperature under relatively high stress [13]. For instance, it was found that the apparent activation energy decreases with increased applied stress. To further validate the creep measurements obtained by an SPS apparatus, it was used to examine alumina, perhaps the most widely researched ceramic material. Alumina single stage creep curves from a previous study [12], performed at various temperatures (in the range of 1125–1250 °C) and under an applied stress of 100 MPa, are presented in Figure 7a. Corresponding strain rates (according to the derivative over time) are presented in Figure 7b. The dramatic effect of the temperature on the total strain and strain rate (slope) can be easily observed. Furthermore, alumina creep rate values obtained at 1200 °C under varying loads are presented as a function of stress, alongside values from a study conducted in compression by Bernard-granger et al. [31] (Figure 7c). Both studies were performed on alumina with a similar fine grain size (0.5 and 0.42 µm, respectively). Here, as well, there is a good agreement between creep measurements by SPS and data reported in literature. The slope reflects the sensitivity to the applied stress and corresponds to a stress exponent n of about 1.8 which is close to 2 and established for fine-grained alumina [12,32].

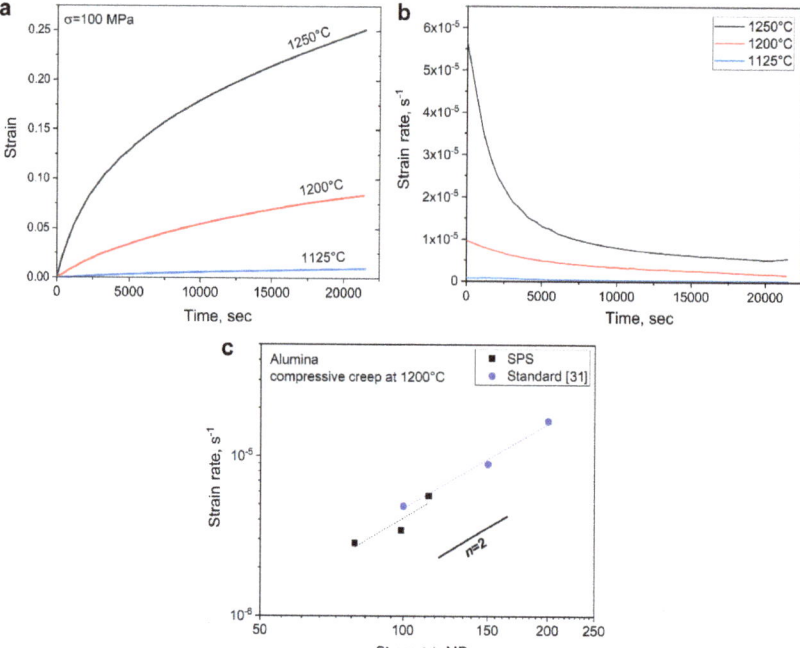

Figure 7. (**a**) Example of alumina creep tests strain under an applied stress of 100 MPa at the 1125–1250 °C temperature range and (**b**) corresponding strain rates; (**c**) comparison of creep rates (at 1200 °C and varying stress) measured for alumina by an SPS apparatus in compression (black squares) and conventional compression testing equipment [31] (blue circles).

3.3. High-Pressure Creep Tests

In some cases, such as for Ni-based superalloys, there is a lot of interest in creep properties under high-stress conditions [33–35]. Typically, it is difficult to examine high-temperature deformation under high applied stress. However, one of the advantages of the SPS apparatus as a creep testing device is that it can allow tests under stresses of hundreds of MPa, using proper high-pressure tooling such as SiC punches [12]. The materials that can be investigated will depend on properties of the high-pressure tooling. For instance, SiC has significantly higher resistance to creep [36] compared to oxide ceramics and can be applied for such tests under high pressure. This was demonstrated for alumina, which was tested under an applied pressure of 400 MPa (sample dimensions 5 × 5 × 10 mm), as presented in Figure 8. These types of tests can clarify creep mechanisms and unique mechanical behavior of materials subjected to a combination of high temperature and stress.

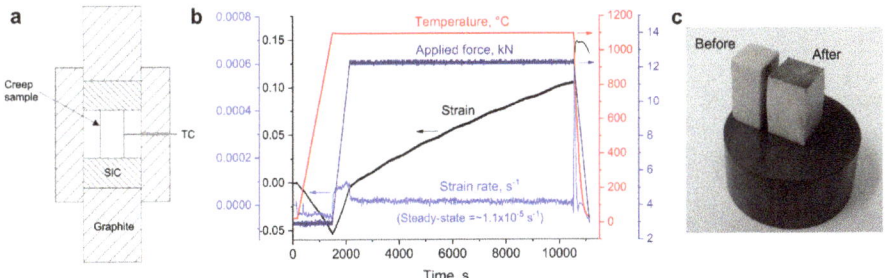

Figure 8. (**a**) Schematic illustration of hybrid SiC-graphite tooling for high-pressure experiments; (**b**) an example for a high-pressure creep test of alumina (sample size 5 × 5 × 10 mm, initial applied stress 400 MPa); (**c**) appearance of samples before and after the high-stress creep test.

4. Conclusions

An SPS apparatus can be used for studying high temperature mechanical properties, particularly compressive creep of metals and ceramics. In this apparatus, a wide range of temperatures and pressures can be applied, and all the necessary data for creep tests can be easily acquired. However, the SPS apparatus, as a tool for mechanical testing, has some technical limitations, including the mandatory minimal load of 3 kN, the lack of a possibility for tensile testing, and the fact that only a constant load regime can be applied. Nevertheless, different tooling configurations may be used so that conductive materials can be tested with or without an applied electric current. This affects the different temperature gradients which exist in the sample, especially when an electric current is applied. Nevertheless, the current that the SPS apparatus utilizes can make it possible to investigate electro-plastic effects to some extent. Furthermore, creep tests under relatively high applied stress (in the range of few hundreds MPa) can also be realized by the SPS apparatus. Several creep test results obtained by an SPS apparatus were presented, and experimental creep results for metals (copper) and ceramics (alumina) proved the accuracy of this device for creep testing of engineering materials. Thus, the SPS apparatus can serve as a relatively simple and convenient method for a wide range of creep testing of both metals and ceramics.

Author Contributions: This study was conceptualized by B.R. and N.F.; experiments were performed by S.K.; data curation by S.K. and B.R.; data analysis by B.R. and S.K.; writing by B.R.; supervised by N.F. All authors have read and agreed to the published version of the manuscript.

Funding: This research received no external funding.

Conflicts of Interest: The authors declare no conflict of interest.

References

1. Kassner, M.E. *Fundamentals of Creep in Metals and Alloys*, 3rd ed.; Butterworth Heinemann imprint of Elsevier: Waltham, MA, USA, 2015.
2. Carroll, D.F.; Wiederhorn, S.M. High temperature creep testing of ceramics. *Int. J. High Technol. Ceram.* **1988**, *4*, 227–241. [CrossRef]
3. Yoon, K.J.; Wiederhorn, S.M.; Luecke, W.E. Comparison of tensile and compressive creep behavior in silicon nitride. *J. Am. Ceram. Soc.* **2004**, *83*, 2017–2022. [CrossRef]
4. Longhin, M.E.; Shelleman, D.L.; Hellmann, J.R. A methodology for the accurate measurement of uniaxial compressive creep of refractory ceramics. meas. *J. Int. Meas. Confed.* **2017**, *111*, 69–83. [CrossRef]
5. Cannon, W.R.; Langdon, T.G. Review creep of ceramics-Part 1 mechanical characteristics. *J. Mater. Sci.* **1983**, *18*, 1–50.
6. Langdon, T.G. Identifying creep mechanisms in plastic flow. *Z. Für Met.* **2005**, *96*, 522–531. [CrossRef]
7. Raj, R.; Ashby, M.F. On grain boundary sliding and diffusional creep. *Metall. Trans.* **1971**, *2*, 1113–1127. [CrossRef]
8. Gibeling, J.C. *ASM Handbook Volume 8: Mechanical Testing and Evaluation*; ASM International: Novelty, OH, USA, 2000.
9. Guillon, O.; Gonzalez-Julian, J.; Dargatz, B.; Kessel, T.; Schierning, G.; Räthel, J.; Herrmann, M. Field-assisted sintering technology/spark plasma sintering: mechanisms, materials, and technology developments. *Adv. Eng. Mater.* **2014**, *16*, 830–849. [CrossRef]
10. Cavaliere, P. *Spark Plasma Sintering of Materials*; Springer Nature: Cham, Switzerland, 2019.
11. Orru, R.; Licheri, R.; Locci, A.M.; Cincotti, A.; Cao, G. Consolidation/Synthesis of materials by electric current activated/assisted sintering. *Mater. Sci. Eng. R Rep.* **2009**, *63*, 127–287. [CrossRef]
12. Ratzker, B.; Sokol, M.; Kalabukhov, S.; Frage, N. Using a spark plasma sintering apparatus as a tool in a compressive creep study of fine-grained alumina. *Ceram. Int.* **2017**, *43*, 9369–9376. [CrossRef]
13. Ratzker, B.; Sokol, M.; Kalabukhov, S.; Frage, N. Creep of polycrystalline magnesium aluminate spinel studied by an SPS apparatus. *Materials* **2016**, *9*, 493. [CrossRef]
14. Ratzker, B.; Sokol, M.; Kalabukhov, S.; Frage, N. compression creep of copper under electric current studied by a spark plasma sintering (SPS) apparatus. *Mater. Sci. Eng. A* **2018**, *712*, 424–429. [CrossRef]
15. Uzan, N.E.; Ratzker, B.; Landau, P.; Kalabukhov, S.; Frage, N. Compressive creep of AlSi$_{10}$Mg parts produced by selective laser melting additive manufacturing technology. *Addit. Manuf.* **2019**, *29*, 100788. [CrossRef]
16. Li, W.; Olevsky, E.A.; McKittrick, J.; Maximenko, A.L.; German, R.M. Densification mechanisms of spark plasma sintering: Multi-step pressure dilatometry. *J. Mater. Sci.* **2012**, *47*, 7036–7046. [CrossRef]
17. Lee, G.; Olevsky, E.A.; Manière, C.; Maximenko, A.; Izhvanov, O.; Back, C.; McKittrick, J. Effect of electric current on densification behavior of conductive ceramic powders consolidated by spark plasma sintering. *Acta Mater.* **2018**, *144*, 524–533. [CrossRef]
18. E209-18 *Standard Practice for Compression Tests of Metallic Materials at Elevated Temperatures with Conventional or Rapid Heating Rates and Strain Rates*; ASTM E209-18; ASTM International: West Conshohocken, PA, USA, 2018.
19. Achenani, Y.; Saâdaoui, M.; Cheddadi, A.; Fantozzi, G. Finite element analysis of the temperature uncertainty during spark plasma sintering: Influence of the experimental procedure. *Ceram. Int.* **2017**, *43*, 15281–15287. [CrossRef]
20. Manière, C.; Durand, L.; Brisson, E.; Desplats, H.; Carré, P.; Rogeon, P.; Estournès, C. Contact resistances in spark plasma sintering: From in-situ and ex-situ determinations to an extended model for the scale up of the process. *J. Eur. Ceram. Soc.* **2017**, *37*, 1593–1605. [CrossRef]
21. Sweidan, F.B.; Kim, D.H.; Ryu, H.J. Minimization of the sample temperature deviation and the effect of current during high-temperature compressive creep testing by the spark plasma sintering apparatus. *Materialia* **2020**, *9*, 100550. [CrossRef]
22. Conrad, H. Thermally activated plastic flow of metals and ceramics with an electric field or current. *Mater. Sci. Eng. A.* **2002**, *322*, 100–107. [CrossRef]
23. Chokshi, A.H.; Porter, J.R. Analysis of concurrent grain growth during creep of polycrystalline alumina. *J. Am. Ceram. Soc.* **1986**, *37*, 36–37. [CrossRef]
24. Xue, L.A.; Chen, I. Deformation and grain growth of low-temperature-sintered high-purity alumina. *J. Am. Ceram. Soc.* **1990**, *73*, 3518–3521. [CrossRef]

25. Venkatachari, K.R.; Raj, R. Superplastic flow in fine-grained alumina. *J. Am. Ceram. Soc.* **1986**, *38*, 135–138. [CrossRef]
26. Feltham, P.; Meakin, J.D. Creep in face-centred cubic metals with special reference to copper. *Acta Metall.* **1959**, *7*, 614–627. [CrossRef]
27. Raj, S.V.; Langdon, T.G. Creep behavior of copper at intermediate temperatures-I. Mechanical characteristics. *Acta Metall.* **1989**, *37*, 843–852. [CrossRef]
28. Klypin, A.A. Plastic deformation of metals in the presence of electric influences. *Strength Mater.* **1975**, *7*, 810–815. [CrossRef]
29. Cao, W.D.; Conrad, H. Effect of Stacking Fault Energy and Temperature on the Electroplastic Effect in FCC Metals. In *Micromechanics of Advanced Materials: A Symposium in Honor of Professor James C.M. Li's 70th Birthday*; Minerals, Metals & Materials Society: Warrendale, PA, USA, 1995.
30. Giuntini, D.; Olevsky, E.A.; Garcia-cardona, C.; Maximenko, A.L.; Yurlova, M.S.; Haines, C.D.; Martin, D.G.; Kapoor, D. Localized overheating phenomena and optimization of spark-plasma sintering tooling design. *Materials* **2013**, *6*, 2612–2632. [CrossRef]
31. Bernard-Granger, G.; Guizard, C.; Duclos, R. Compressive creep behavior in air of a slightly porous as-sintered polycrystalline α-alumina material. *J. Mater. Sci.* **2007**, *42*, 2807–2819. [CrossRef]
32. Ruano, O.A.; Wadsworth, J.; Sherby, O.D. Deformation of fine-grained alumina by grain boundary sliding accommodated by slip. *Acta Mater.* **2003**, *51*, 3617–3634. [CrossRef]
33. Xu, Q.; Hayhurst, D.R. The evaluation of high-stress creep ductility for 316 stainless steel at 550 °C by extrapolation of constitutive equations derived for lower stress levels. *Int. J. Press. Vessel. Pip.* **2003**, *80*, 689–694. [CrossRef]
34. Izuno, H.; Yokokawa, T.; Harada, H. An applicability of a creep constitutive equation for Ni-base superalloys on low temperature high stress condition. *J. Jpn. Inst. Met.* **2006**, *70*, 674–677. [CrossRef]
35. Wu, X.; Dlouhy, A.; Eggeler, Y.M.; Spiecker, E.; Kostka, A.; Somsen, C.; Eggeler, G. On the nucleation of planar faults during low temperature and high stress creep of single crystal Ni-base superalloys. *Acta Mater.* **2018**, *144*, 642–655. [CrossRef]
36. Munro, R.G. Material properties of a sinterd α-SiC. *J. Phys. Chem. Ref. Data.* **1997**, *26*, 1195–1203. [CrossRef]

© 2020 by the authors. Licensee MDPI, Basel, Switzerland. This article is an open access article distributed under the terms and conditions of the Creative Commons Attribution (CC BY) license (http://creativecommons.org/licenses/by/4.0/).

Article

Thermal Analysis of High Entropy Rare Earth Oxides

Sergey V. Ushakov [1,*], Shmuel Hayun [2,*], Weiping Gong [3,*] and Alexandra Navrotsky [1,*]

1. School of Molecular Sciences, and Center for Materials of the Universe, Arizona State University, Tempe, AZ 85287, USA
2. Department of Materials Engineering at the Ben-Gurion University of the Negev, Beer-Sheva 84105, Israel
3. Guangdong Provincial Key Laboratory of Electronic Functional Materials and Devices, Huizhou University, Huizhou 516001, China
* Correspondence: sushakov@asu.edu (S.V.U.); hayuns@bgu.ac.il (S.H.); weiping_gong@csu.edu.cn (W.G.); anavrots@asu.edu (A.N.)

Received: 5 June 2020; Accepted: 6 July 2020; Published: 14 July 2020

Abstract: Phase transformations in multicomponent rare earth sesquioxides were studied by splat quenching from the melt, high temperature differential thermal analysis and synchrotron X-ray diffraction on laser-heated samples. Three compositions were prepared by the solution combustion method: $(La,Sm,Dy,Er,RE)_2O_3$, where all oxides are in equimolar ratios and RE is Nd or Gd or Y. After annealing at 800 °C, all powders contained mainly a phase of C-type bixbyite structure. After laser melting, all samples were quenched in a single-phase monoclinic B-type structure. Thermal analysis indicated three reversible phase transitions in the range 1900–2400 °C, assigned as transformations into A, H, and X rare earth sesquioxides structure types. Unit cell volumes and volume changes on C-B, B-A, and H-X transformations were measured by X-ray diffraction and consistent with the trend in pure rare earth sesquioxides. The formation of single-phase solid solutions was predicted by Calphad calculations. The melting point was determined for the $(La,Sm,Dy,Er,Nd)_2O_3$ sample as 2456 ± 12 °C, which is higher than for any of constituent oxides. An increase in melting temperature is probably related to nonideal mixing in the solid and/or the melt and prompts future investigation of the liquidus surface in Sm_2O_3-Dy_2O_3, Sm_2O_3-Er_2O_3, and Dy_2O_3-Er_2O_3 systems.

Keywords: high entropy oxides; rare earth oxides; laser melting; aerodynamic levitation; phase transition; melting; thermodynamics

---

1. Introduction

Alloys often contain tens of elements in strictly defined ratios with one element as "the base" of the alloy (e.g., all steels have more than 70 at.% Fe). Recently, a new design approach had emerged, which is focused on "baseless" or multi-principle element alloys (MPEAs) with the concentration of each element no more than 35 at.% but not less than 5 at.% [1]. The reports on complex, concentrated alloys (CCAs) appeared in the literature since the 1960s [2]; however, the new research direction took off in 2004 after the discovery of remarkable hardness, yield strength, and resistance to annealing softening in several MPEAs made by Taiwanese metallurgists [1,3]. They also introduced the term "high entropy alloys (HEA)," arguing that in these complex compositions, the gain in configurational entropy is responsible for the formation of simple single-phase solid solutions, rather than intermetallic compounds which would have a deleterious effect on the properties.

The high entropy design approach was recently applied to carbides [4,5] borides [6–8], and oxides [9–18] for high temperature and battery-related applications. Most of the high entropy compositions that were successfully prepared as single phases are within the 15% limit of atomic radii differences known as a Hume-Rothery [19] rule to metallurgists and as a Goldschmidt [20] limit for isomorphic mixtures to mineralogists (the majority of mineral species meet HE definition! [21]). While the argument about the role of configurational entropy is highly contentious [2,22], the name has its merits and rightfully attracts attention to thermodynamic controls, and we use the high entropy (HE) term to refer to five component rare earth oxides studied in this work.

It soon will be a century since Goldschmidt et al. [23] published the first research on rich polymorphism in rare earth sesquioxides (R_2O_3, where R is a lanthanide, Y or Sc). They originally divided quenchable polymorphs into A, B, and C types (Figure 1). The A-type is trigonal (*P-3m*1), typical for sesquioxides of the large lanthanides, and also called La_2O_3-type; the B-type is monoclinic (*C2/m*), typical for lanthanides in the middle of series and also called Sm_2O_3-type [24]; the C-type is cubic (*Ia-3*), typical for small lanthanides, Y and Sc, and also called bixbyite-type after the naturally occurring $(MnFe)_2O_3$ mineral.

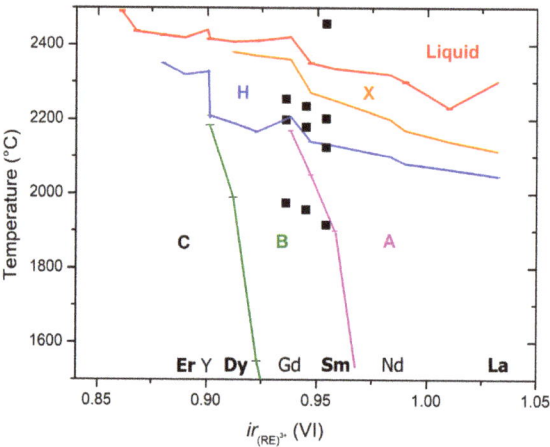

Figure 1. Phase transformations in rare earth and yttrium sesquioxides vs. ionic radii for octahedral coordination. Lines connect the best values for pure sesquioxides. The data points represent temperatures of phase transitions from thermal analysis of three $(La_{0.2}Sm_{0.2}Dy_{0.2}Er_{0.2}RE_{0.2})_2O_3$ compositions studied in this work, where RE–Nd, Gd, or Y, plotted vs. average ionic radius.

Most rare earth oxides can be obtained in more than one structure type (polymorph) at ambient conditions: normally an A-type La_2O_3 and Nd_2O_3 can be synthesized in a C-type structure [25] at temperatures below ~500 °C (and C-type was predicted [26] to be their ground state structure); while normally C-type oxides from Dy to Yb were obtained in B-type structure in nanoparticles [27]. The oxides of trivalent actinides also found in these structure types [28]. Two high temperature structures were first identified by Foex and Traverse [29,30]. The H-type is hexagonal (*P6$_3$/mmc*) [31] and was reported for all rare earth and Y sesquioxides except Lu and Yb [32]. For oxides from La to Dy, the transformation of the hexagonal phase to the cubic X-type (*Im-3m*) structure was detected before melting [33]. The X-type structure was also reported to be formed in Tm_2O_3 and Lu_2O_3 after irradiation with Xe and Au ion beams [34].

The systematic research on phase equilibria in rear earth oxides has mostly been focused on pure oxides and several binary systems. There are only a few systematic investigations of ternaries and they are limited to studies on quenched samples [35,36]. All the studied systems of trivalent rare earth oxides are characterized by wide ranges of solid solutions in the structures identified in pure oxides. Eleven interlanthanide perovskites are known to form in several systems combining large and small rare earths: $LaRO_3$ (R = Y, Ho-Lu), $CeRO_3$ (R = Tm-Lu), and $PrRO_3$ (R = Yb-Lu) [37–39]. They all show an orthorhombic (*Pnma*) distortion and do not melt congruently, but decompose at 800–2000 °C into solid solutions of one of rear earth oxide structure types [39]. $LaGdO_3$ in a B-type structure attracted attention for application as high-k gate dielectric [40] and as an optical temperature sensor when doped with Er/Yb [41,42].

Mixed three-four valent Ce, Tb, and Pr oxides with cubic defect fluorite related structures have been studied for gas sensor and catalyst applications [43]. Following the high entropy approach, Tseng et al. [44] studied thermal expansion and magnetic susceptibility of $(Gd,Tb,Dy,Ho,Er)_2O_3$ composition, which formed solid solution in a C-type structure. Djenadic et al. [11] reported that the presence of Ce^{4+} in several HE rare earth oxide compositions produced defect fluorite solid solutions.

In this work, we studied three compositions containing five rare earth sesquioxides in equiatomic ratios: $(La,Sm,Dy,Er,RE)_2O_3$, with RE either Nd or Gd or Y. All chosen rare earths are trivalent in the solid state, and their sesquioxides represent all polymorphs: A-type (La, Nd), B-type (Sm, Gd, Dy), and C-type (Er, Gd). However, they all form a H-type structure at high temperatures (Figure 1) with very intriguing properties, such as fast oxygen ion conductivity and superplasticity [45], but was never quenched to room temperature.

We performed laser melting, splat quenching, and annealing of the samples and characterized their high temperature phase transformations and thermal expansion by a combination of in situ differential thermal analysis and synchrotron diffraction on laser-heated samples. An unexpected and surprising finding was the substantial (>100 °C) increase in melting temperatures compared to those expected from consideration of melting points of constituent oxides.

2. Materials and Methods

The intimately mixed rare rear earth oxides of desired stoichiometry were first synthesized by the solution combustion method [46] and characterized by X-ray diffraction (XRD). Then, samples were laser melted in the hearth and in aerodynamic levitator and used for high temperature synchrotron XRD, differential thermal analysis (DTA), splat quenching, and prolonged annealing experiments. The experiment flow chart is provided in the Supplementary Materials (Figure S1).

2.1. Sample Synthesis

Aqueous solutions of rare earth nitrates (Sigma-Aldrich 99.9% metals base) were mixed in desired stoichiometry. Ethylene glycol and citric acid were mixed at a molar ratio of 1 to 2 and added to the nitrate water solution. The mixed nitrate–citrate solution was evaporated at 150 °C under agitation by magnetic stirring until a highly viscous foam-like colloid was formed. This colloid was annealed in air at 800 °C for 96 h. An additional heat treatment was performed at 1100 °C for 12 h. The samples were analyzed by room temperature powder X-ray diffraction after each treatment.

2.2. Laser Melting and Splat Quenching

Powders after heat treatment at 1100 °C were laser melted in air on the copper hearth with 400 W CO_2 laser and remelted in an argon flow in the aerodynamic levitator. The resulting samples were oblate spheroids 2.6–2.9 mm in diameter, with a flattening of ~0.1. The structure and phase transformations in obtained samples were studied by XRD and DTA. Sample composition and homogeneity were characterized by electron microprobe analysis. Laser-melted spheroids were further processed by splat quenching using a splittable nozzle aerodynamic levitator. The employed device is part of a drop-and-catch (DnC) calorimeter, described in detail earlier [47]. For quenching experiments, solid copper plates were installed in place of the calorimeter sensors (Figure S2). The samples produced by splat quenching were analyzed by room temperature XRD.

2.3. Microprobe Analysis

A Cameca SX-100 electron microprobe was used for imaging and analysis of the chemical composition of laser-melted samples. Energy dispersive spectroscopy and backscattered electron imaging (BSE) were used for the characterization of sample homogeneity. Quantitative chemical analysis was performed by wavelength dispersive spectroscopy (WDS) using synthetic rare earth orthophosphate crystals for calibration standards for all rare earths except Y, for which synthetic $Y_3Al_5O_{12}$ (YAG) was used due to flux originated Pb contamination detected in YPO_4 standard.

2.4. Room Temperature X-ray Diffraction

Room temperature powder XRD was used to characterize samples after precipitation, laser melting, splat quenching, differential thermal analysis, and synchrotron diffraction experiments. The measurements were performed using Bruker D8 Advance diffractometer (Bruker, Madison, WI, USA) with CuK_a radiation and a rotating sample holder. The operating parameters were 40 kV and 40 mA, with a step size of 0.01° and dwell 3 s/step. Lattice parameters, phase fractions, and crystallite sizes of powders after annealing were refined using whole profile refinement as implemented in MDI Jade 2010 software package (Materials Data, Livermore, CA, USA). GSAS-II [48] was used for Rietveld [49] refinement of lattice parameters and phase fractions in laser-melted samples.

2.5. High Temperature X-ray Diffraction

High-temperature X-ray diffraction experiments were performed on an aerodynamic levitator at beamline 6-ID-D at the Advanced Photon Source (APS), Argonne National Laboratory. The levitator at the beamline provided by Materials Developments, Inc. (Evanston, IL, USA) and described in detail elsewhere [50]. The samples 63–70 mg in weight, were prepared by laser melting as described above.

The diffraction experiments were performed in transmission geometry with X-ray wavelength 0.1236 Å (100.3 keV energy). The beam was collimated in a "letterbox" shape, 500 µm wide and 200 µm tall. The samples were levitated in argon flow and heated from the top with a 400-W CO_2 laser. The levitator software provided manual control of the levitation gas flow rate and manual or automated laser power control for sample heating. Diffraction data were collected in 100-°C increments based on the surface temperature of the levitated bead, which was monitored with a single color pyrometer (875–925 nm spectral band, IR-CAS3CS, Chino Co., Tokyo, Japan) with emissivity set to 0.92. Emissivities for rare earth oxides above 2000 °C are unknown [51], and thermal gradient in laser-heated aerodynamically levitated bead exceeds 100 °C [52–54]. In this work, the temperatures of diffracted volume were assigned based on phase transformation temperatures obtained from DTA measurements.

The diffraction images were recorded with a Perkin-Elmer XRD 1621 area detector positioned at a distance 1099 mm from the sample. The exposure time was set to 0.1 s to avoid detector saturation; 100 exposures were summed and recorded into a single image used for further processing with GSAS-II software [48]. The sample to detector distance, detector tilt, and beam center coordinates was calibrated using NIST CeO_2 powder standard available at the beamline and with Y_2O_3 bead prepared by laser melting. The images from area detector were integrated from 1 to 7° 2Θ at 70–120 ° azimuth into diffraction patterns with 1600 points (0.00375 steps in 2Θ) (see Figure S3). Room temperature diffraction images were collected from every bead before and after laser heating. During the processing of diffraction data from the levitator, sample displacement was refined at room temperature from known cell parameters and kept constant during further refinements. Pawley [49,55] method, as implemented in GSAS-II, was used for refinement of unit cell parameters at high temperatures.

2.6. Differential Thermal Analysis

Differential thermal analysis was performed with a Setaram Setsys 2400 instrument modified to enable excursions to 2500 °C. The experiments were conducted in Ar flow at heating and cooling rates 20 °C/min using WRe differential heat flow sensor and thermocouple for furnace temperature control.

Laser-melted beads, 100–140 mg in weight were placed in tungsten crucibles and sealed under Ar atmosphere to avoid the possibility of sample and standards contamination with carbon vapor from vitreous carbon protection tube. Multiple measurements were performed on each sample. Temperatures of phase transformations were determined as average from the onset [56] of endothermic peaks on heating. Enthalpies of phase transformation were calculated as the averages of absolute values of endothermic heat effects on heating and exothermic heat effects on cooling. The instrument and methodology were described in detail elsewhere [53,57–59]. Temperature and sensitivity calibrations were performed using melting and phase transition temperatures and enthalpies of Au (1064 °C), Al_2O_3 (2054 °C), Nd_2O_3 (A-H, H-X, and X-L at 2077, 2201 and 2308 °C, respectively), and Y_2O_3 (C-H and H-L at 2348 °C and 2439 °C, respectively). It must be noted that international temperature scale ITS-90 [60] defines no fixed points above the freezing point of gold, albeit alumina melting temperature 2054 ± 6 °C was recommended to be included as a secondary reference point on the ITS [61,62].

2.7. Calphad Modeling

Calphad [63–66] modeling was performed to compare with experimental results. Calphad-type thermodynamic database for rare earth sesquioxides was created by Zinkevich [67]. He critically reviewed all relevant experimental data available before 2006 and evaluated missing fusion enthalpies values based on measured enthalpy of fusion and volume change on melting for Y_2O_3. In a more recent evaluation by Zhang and Jung [68] evaluation, missing data were estimated from liquidus in RE_2O_3-Al_2O_3 phase diagrams. However, new measurements for fusion enthalpies [53] are in better agreement with Zinkevich's assessment. Zhang and Jung's evaluation varies largely with the values proposed by Konings et al. [69]. Zinkevich [67] database for rare earth sesquioxides is openly available at the NIST website [70] and was used in this work without any modifications. Thermo-Calc (Stockholm, Sweden) software was used for calculations.

3. Results

3.1. Chemical Composition

The composition was determined by microprobe analysis on laser-melted samples recovered after high temperature diffraction experiments. The measured ratios were close to nominal, indicating no preferential loss of any component on melting and laser heating during diffraction experiments. The microprobe results are reported in Table 1, with variations given as two standard deviations of the mean of 12 analyses per sample. Backscattered electron micrographs are included in Supplementary Materials (Figures S4–S6). The following stoichiometries of synthesized rare earth sesquioxides were obtained: $(La_{0.18}Sm_{0.20}Dy_{0.18}Er_{0.18}Y_{0.26})_2O_3$, $(La_{0.19}Sm_{0.21}Dy_{0.21}Er_{0.20}Gd_{0.19})_2O_3$, and $(La_{0.20}Sm_{0.20}Dy_{0.21}Er_{0.20}Nd_{0.19})_2O_3$. For the sake of brevity, we will refer to these compositions as HE-Y, HE-Gd, and HE-Nd, respectively, where "HE" stands for "high entropy" and element symbol is for rare earth element in $(La,Sm,Dy,Er,RE)_2O_3$ nominal composition.

Table 1. Atomic percent of rare earth cations in laser-melted rare earth sesquioxides from the results of wavelength dispersive microprobe analysis.

Rare Earth	HE-Nd	HE-Gd	HE-Y
La	19.5 ± 0.2	18.7 ± 0.4	17.7 ± 0.8
Sm	20.1 ± 0.1	20.7 ± 0.1	19.8 ± 0.1
Dy	20.8 ± 0.1	21.1 ± 0.2	18.0 ± 0.3
Er	20.3 ± 0.1	20.3 ± 0.2	18.4 ± 0.3
Y	-	-	26.0 ± 0.4
Gd	-	19.3 ± 0.2	-
Nd	19.3 ± 0.1	-	-
MW g/mol	350.89	358.84	323.09
Ave radii [1] Å	0.954	0.945	0.936

[1] The average ionic radius of rare earth after Shannon [71] for RE^{+3} in octahedral coordination.

3.2. Phases after Solution Combustion Synthesis and Annealing

After 800 °C annealing of powders from solution combustion synthesis, the cubic C-type phase with crystallite size 20–40 nm was a major phase in all samples. The B-type phase was also detected in HE-Nd and HE-Gd samples (Table 2, Figures S7 and S8). After annealing of the powders at 1100 °C, only the B-type phase was identified in HE-Nd sample; the amount of B-type phase in HE-Gd sample increased to 70 wt.%. The C-type phase was retained in HE-Y composition, and its crystallite size increased to ~65 nm. The decrease in volume on C-B transformation was calculated from XRD data in HE-Nd and HE-Gd samples as 8% on average.

Table 2. The phase composition of powders after calcination at 800 °C and heat treatment at 1100 °C.

Experiment	Phase	HE-Nd	HE-Gd	HE-Y
	C-Type (Cubic, Bixbyite-Type) $Ia\text{-}3$, $Z = 16$			
Air	a, Å	10.903(1)	10.863(2)	10.814(3)
800 °C	V, Å^3/z	81.0(1)	80.1(1)	79.0(1)
96 h	Size	38 ± 1 nm	21 ± 1 nm	21 ± 1 nm
	wt.%	~85 wt.%	~90 wt.%	100 wt.%
	B-Type (Monoclinic, Sm_2O_3-Type) $C2/m$, $Z = 6$			
	a, Å	14.259(5)	13.90(4)	-
	b, Å	3.620(1)	3.53(5)	
	c, Å	8.862(3)	9.00(4)	
	β, °	100.67(1)	96.8(1)	
	V, Å^3/z	74.9 ± 0.1	73 ± 2	
	Size	32 ± 2 nm	13 ± 2 nm	
	wt.%	~15 wt.%	~10 wt.%	
	V (C→B) %	−7.5 ± 0.1%	−9.7 ± 0.3%	
	C-type (cubic, bixbyite-type) $Ia\text{-}3$, $Z = 16$			
Air	a, Å		10.825(1)	10.804(1)
1100 °C	V, Å^3/z	-	79.0(1)	78.8(1)
12 h	Size		49 ±1 nm	65 ± 1 nm
	wt.%		~30 wt.%	100 wt.%
	B-type (monoclinic, Sm_2O_3-type) $C2/m$, $Z = 6$			
	a, Å	14.242(1)	14.227(1)	
	b, Å	3.6152(2)	3.601(1)	
	c, Å	8.857(1)	8.833(1)	
	β, °	100.62(4)	100.66(1)	
	V, Å^3/z	74.7(1)	74.1(1)	-
	Size	73 ± 1 nm	68 ± 1 nm	
	wt.%		~70 wt.%	
	V (C→B) %	−8.7 ± 0.1%	−6.8 ± 0.1%	

The numbers in brackets indicate uncertainty in the last digit from the refinement of XRD data.

3.3. Phases after Laser Melting, Splat Quenching, and Annealing

Melt processing yielded a B-type phase in all studied compositions. Room temperature XRD patterns are shown in Figure 2; the example of the whole profile refinement plot is included in Figure S9. The unit cell parameters and crystallite sizes of B-phase measured after splat quenching from ~3000 °C and after laser melting and 60 days annealing, are listed in Table 3.

In splat-quenching experiments, samples were heated in oxygen flow to the temperature several hundred degrees above the melting point to allow for cooling during ~100 ms drop time from the splittable nozzle to the splat-quenching plates. Sixty days annealing at 1100 °C was performed on the laser-melted samples, recovered after thermal analysis. The crystallite sizes of splat-quenched samples were about 80 nm. The crystallite sizes of the laser-melted samples after thermal analysis and annealing were in the range of 95–150 nm. The decrease in volume of B-type phase after annealing compared to splat-quenched samples was calculated from measured cell parameters as 0.6–0.8%. This variation is consistent with the retention of thermally induced defects in splat-quenched samples.

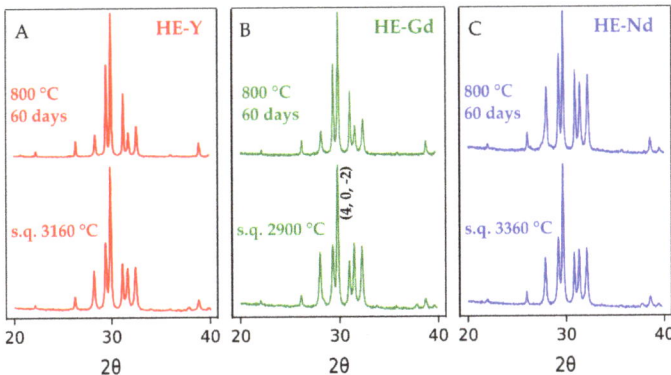

Figure 2. (**A**), (**B**), (**C**): powder X-ray diffraction patterns of three $(La_{0.2}Sm_{0.2}Dy_{0.2}Er_{0.2}X_{0.2})_2O_3$ compositions, where X–Y, Gd, or Nd, respectively. Patterns were collected at room temperature using CuK radiation (λ = 1.54056 Å) on samples obtained by splat quenching (s. q.) of the melts from indicated temperatures and after annealing at 800 °C for 60 days. All patterns identified as monoclinic (B-type) phases. Table 3 lists the results of cell parameters refinements. The typical profile refinement plot is presented in supporting information (Figure S9 in Supplementary Materials).

Table 3. Room temperature unit cell parameters and crystallite sizes of HE samples after splat quenching from ~3000 °C, and after laser melting and annealing at 800 °C for 60 days. All samples were indexed in a monoclinic B-type structure (S.G. $C2/m$, Sm_2O_3-type).

Unit Cell Parameters	HE-Nd * Splat Quench	800 °C/60 d	HE-Gd Splat Quench	800 °C/60 d	HE-Y Splat Quench	800 °C/60 d
a, Å	14.245(1)	14.244(1)	14.194(1)	14.180(1)	14.159(2)	14.139(1)
b, Å	3.6150(1)	3.6025(1)	3.5956(1)	3.5840(1)	3.5741(2)	3.5617(1)
c, Å	8.857(1)	8.839(1)	8.818(1)	8.797(1)	8.781(1)	8.758(1)
β, °	100.63(1)	100.69(1)	100.59(1)	100.59(1)	100.61(1)	100.65(1)
V, Å^3/z	74.72(1)	74.28(1)	73.73(1)	73.24(1)	72.79(1)	72.24(1)
Cryst. size	76 ± 1 nm	95 ± 1 nm	80 ± 1 nm	101 ± 1 nm	86 ± 1 nm	147 ± 3 nm

* Cell parameters refined with internal standard on HE-Nd sample after HTXRD experiments (a, b, c, β): 14.244(1) Å, 3.610(2) Å, 8.852(2) Å, 100.611(5)°.

All samples after melting show an increase in the intensity of (4, 0, −2) peak of B-phase compare with calculated from an ideal B-type structure. This variation is not related to the complex composition of the studied sample since we observed a similar effect in pure Sm_2O_3 (Figure S10) and it is likely due to twinning [72].

3.4. Temperatures and Enthalpies of Phase Transformations from DTA Experiments

Thermal analysis was performed on laser-melted samples to enable sealing W crucibles; thus, the initial structure for all samples was B-type. The transition temperatures detected in the samples by differential thermal analysis are plotted in Figure 1; the data are summarized in Table 4.

On heating to 2400 °C, three reversible heat effects were observed in all samples, which were assigned to B-A, A-H, and H-X phase transformations (Figure S11). Enthalpies of transformations obtained from endothermic peaks on heating and exothermic peaks on cooling were generally consistent and were averaged to obtain the values listed in Table 4. The range of undercooling increased with transition temperature, with maximum observed values 14, 43, and 63 °C for B-A, A-H, and H-X transformations, respectively. For the A-H transition, the width of DTA peaks on heating in studied samples was similar to that observed for pure Nd_2O_3 at the same heating rate [59]. However, the peaks corresponding to B-A and H-X transformations were substantially wider than those for A-H in HE compositions and for H-X in pure Y_2O_3 and Nd_2O_3 (e.g., 41–44 °C for H-X transition in HE samples vs. 12 °C for pure Nd_2O_3).

Table 4. Results of thermal analysis of $(La_{0.20}Sm_{0.20}Dy_{0.21}Er_{0.20}Nd_{0.19})_2O_3$ (HE-Nd), $(La_{0.19}Sm_{0.21}Dy_{0.21}Er_{0.20}Gd_{0.19})_2O_3$ (HE-Gd), and $(La_{0.18}Sm_{0.20}Dy_{0.18}Er_{0.18}Y_{0.26})_2O_3$ (HE-Y) samples.

	HE-Nd	HE-Gd	HE-Y
T_{B-A} °C	1916 ± 9(5) *	1957 ± 5(4)	1975 ± 13(4)
H_{B-A} J/g	57 ± 3(10)	56 ± 7(8)	56 ± 8(8)
H_{B-A} kJ/mol	19.8 ± 1.0	20.3 ± 2.7	18.0 ± 2.7
S_{B-A} J/mol/K	9.0 ± 0.1	9.1 ± 0.1	8.0 ± 0.2
T_{A-H} °C	2125 ± 3(5)	2180 ± 2(4)	2199 ± 4(6)
H_{A-H} J/g	22 ± 3(10)	23 ± 1(7)	28 ± 1(9)
H_{A-H} kJ/mol	7.7 ± 0.9	8.3 ± 0.5	9.2 ± 0.3
S_{A-H} J/mol/K	3.2 ± 0.1	3.4 ± 0.1	3.7 ± 0.1
T_{H-X} °C	2202 ± 4(2)	2235 ± 5(2)	2254 ± 8(4)
H_{H-X} J/g	79 ± 1(2)	85 ± 23(3)	126 ± 7(4)
H_{H-X} kJ/mol	27.8 ± 0.2	30.6 ± 8.3	40.6 ± 2.4
S_{H-X} J/mol/K	11.2 ± 0.1	12.2 ± 0.4	16.1 ± 0.1
T_m °C	2456 ± 12		

* The uncertainties are reported as two standard deviations of the mean with the number of experiments given in parentheses.

Temperatures and enthalpies of transitions increased with decreasing average ionic radius, from HE-Nd to HE-Gd to HE-Y, consistent with the trend among pure rare earth sesquioxides. The exception was the enthalpy of B-A transition in HE-Y sample, which, although the same within calculated uncertainties, appeared ~2 kJ/mol smaller than that for B-A transition in HE-Gd sample.

Transition entropies were calculated from $_{trs}S = {}_{trs}H/T$, where T is the temperature in Kelvin at the transition onset on heating. Entropies of transitions are nearly constant between compositions, with average values ~9 J/mol/K for B-A transformation; ~3.4 J/mol/K for A-H transformation, and ~11.5 J/mol/K for H-X transformation, except for ~16 J/mol/K value for H-X transformation of HE-Y at 2254 °C. This deviation might be related to the fact that pure Y_2O_3 does not undergo H-X transformation.

During DTA experiments above 2300 °C, the failure of the sensor is frequent, and the maximum achievable temperature is limited by magnitude and direction of temperature drift of the control WRe thermocouple. For the calibration of the DTA thermocouple in this temperature range, pure Y_2O_3 (Tm 2439 ± 12 °C) [73] was used as a standard. All studied compositions were heated to temperatures above 2450 °C; however, the peak corresponding to melting onset was registered only for HE-Nd sample at temperature 17 °C above the melting point of Y_2O_3 (Figure 3). Due to the significant uncertainties in baseline choice and sensitivity calibration at this range, the enthalpy and entropy of fusion were not evaluated.

3.5. Volume Changes and Thermal Expansion from High-Temperature XRD

In diffraction experiments on levitated samples, powder-like diffraction patterns are obtained by ensuring the rotation of the solid sample in the gas flow [53,54,58]. When sample spheroids are prepared by melt solidification, as in this study, the exact shape of each bead depends on the surface tension of the particular composition, volume change on melting, and stochastic nature of nucleation. Due to these variations and crystal growth at high temperature, the rotation does not always produce the required random orientation of crystallites. In some cases, data are amenable to structure refinement [32,52]; however, in the present study, variations in intensities only allowed unambiguous identification of B-A and H-X transformations and refinement of the unit cell parameters of corresponding phases.

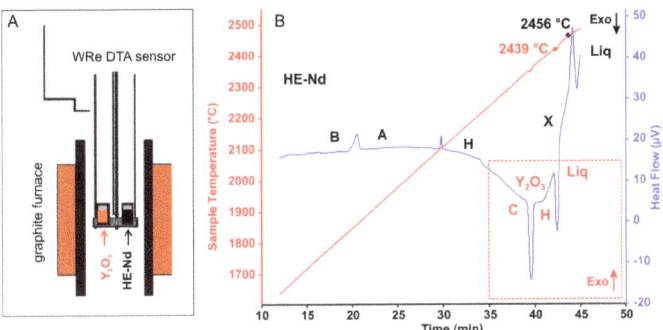

Figure 3. (**A**) The schematic of differential thermal analyzer and samples placement. (**B**) Heat flow trace (no baseline subtraction), showing heat effects on heating from HE-Nd ((La$_{0.20}$Sm$_{0.20}$Dy$_{0.21}$Er$_{0.20}$Nd$_{0.19}$)$_2$O$_3$) and Y$_2$O$_3$ samples. Endothermic B-A-H-X-Liquid transformations for HE-Nd are labeled in black. Endothermic C-H-Liquid transformations for Y$_2$O$_2$ are labeled in red. The assignment of the exothermic direction of the heat flow signal for HE-Nd and Y$_2$O$_3$ is the opposite due to the placement of the samples. Melting temperatures for Y$_2$O$_3$ and HE-Nd are marked on the temperature trace. The Y$_2$O$_3$ melting point (2439 °C) was used for calibration.

The A-H transformation shows well pronounced peaks in DTA measurements (Figure S11); however, the diffraction patterns of A and H phases are very similar (Figure 4). In our experiments, the data quality did not allow us to unambiguously identify the A-H transition from diffraction on levitated samples. Volume changes on B-A and H-X transformations were calculated from unit cell parameters at phase transformation temperatures. The data are presented in Table 5, and examples of refinements are included in Supplementary Materials (Figures S12 and S13). Both transitions are accompanied by volume contraction. The volume change on B-A transition was refined for all samples and found to be −2.5 ± 0.1% for HE-Nd and HE-Gd and −3.1 ± 0.1% for HE-Y (Table 5).

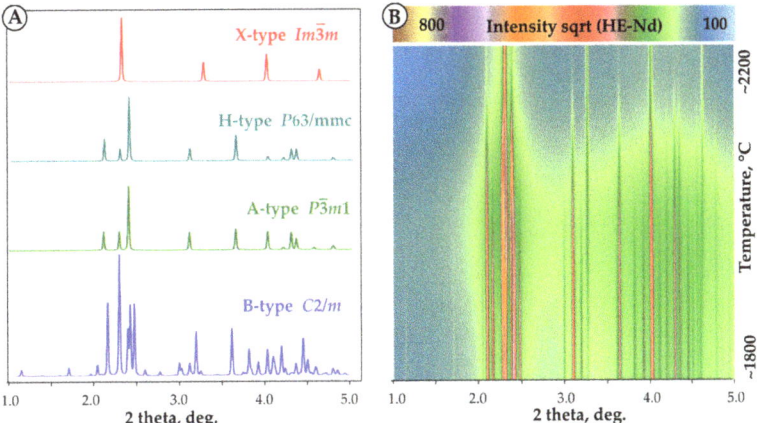

Figure 4. (**A**) Calculated X-ray diffraction patterns for different structure types for (La, Sm, Dy, Er, Nd)$_2$O$_3$ composition and instrument parameters corresponding to the experimental condition (λ = 0.1236 Å). Note the similarity of diffraction patterns for A and H structures. (**B**) GSAS-II contour plot of experimental diffraction patterns collected at 1800–2200 °C on laser-heated HE-Nd bead. See Figures S11 and S12 for refinement plots.

Table 5. High-temperature unit cell parameters and volume changes on B-A and H-X transformation in HE samples from synchrotron diffraction on laser-heated levitated samples.

Structure/Sample		HE-Nd	HE-Gd	HE-Y
T$_{tr}$ (DTA)	T$_{B-A}$, °C	1916 ± 9	1957 ± 5	1975 ± 13
B-type monoclinic C2/m, Z = 6	a, Å	14.433(8)	14.403(5)	14.367(7)
	b, Å	3.711(1)	3.697(1)	3.674(1)
	c, Å	9.026(5)	8.987(2)	8.965(3)
	β, °	101.35(2)	101.18(1)	101.09(2)
	V, Å^3/z	79.0(2)	78.2(1)	77.4(1)
A-type trigonal P-3m1, Z = 1	a, Å	3.885(1)	3.874(1)	3.864(1)
	c, Å	6.199(1)	6.175(2)	6.171(1)
	V, Å^3/z	81.03(2)	80.24(2)	79.78(1)
	V (B→A), %	−2.5 ± 0.1	−2.5 ± 0.1	−3.1 ± 0.1
T$_{tr}$ (DTA)	T$_{H-X}$, °C	2202 ± 4	2235 ± 5	2254 ± 8
H-type P6$_3$/mmc, Z = 1	a, Å	3.898(1)		3.869(1)
	c, Å	6.216(1)		6.168(1)
	V, Å^3/z	81.81(1)		79.96(1)
X-type Im-3m, Z = 1	a, Å	4.324(1)		4.2989(1)
	V, Å^3/z	80.85(2)		79.44(1)
	V (H→X), %	−1.2 ± 0.1		−0.6 ± 0.1

The thermal expansion of the B-type phase from room temperature to the B-A transition was calculated from room temperature cell parameters of annealed samples (Table 3) and cell parameters at the transition temperature (Table 5). The thermal expansion is anisotropic and similar for all three compositions. The smallest expansion is observed in a parameter $(7.8 ± 0.7) × 10^{−6}$/K, linear thermal expansion coefficients along b and c directions are $(1.6 ± 0.2) × 10^{−5}$/K and $(1.2 ± 0.5) × 10^{−5}$/K, respectively. The average volume thermal expansion coefficient for all studied compositions in the B-type structure is $(3.5 ± 0.2) × 10^{−5}$/K.

Due to the narrow temperature range of stability of A and H phases, the accurate calculation of thermal expansion is not possible. For HE-Nd sample, the average volume thermal expansion of A and H phases appears to be similar to that for the B phase, but for HE-Y an increase in thermal expansion is observed. The volume change on H-X transformation was refined for HE-Nd and HE-Y compositions as −1.2 ± 0.1% and −0.6 ± 0.1%, respectively.

4. Discussion

4.1. Experiment vs. Calphad Predictions

The Calphad approach is used widely in the high entropy alloys and ceramics field [74–76]. Senkov et al. demonstrated [74] that for high entropy alloys, Calphad computations of type and number of phases agreed well with experimental results only when more than half of binary systems were fully assessed; however, even when there are enough data on binary and ternary systems, Calphad predictions of transformation temperatures, phase compositions and fractions are challenging. Compared with metals, the databases for oxides are much more limited, especially for the temperature range addressed in this study. Nevertheless, it is instructive to compare our experimental results with Calphad predictions.

We used an open-access database created by Zinkevich [67]. It is based on his 2007 assessments of thermodynamic properties of pure rare earth sesquioxides from room temperature to the melting point. This database is often used as a starting point for the creation of new thermodynamic assessments for systems with rare earth oxides [77,78]. Without any modifications, the database allows modeling of phase equilibria using the ideal solution model (which assumes zero enthalpies of mixing and the largest configurational entropy gains). A set of interaction parameters for the regular solution model (which accounts for mixing enthalpies) are included in the Zinkevich review [67], but they are based on a limited number of selected binary systems, not included in the database [70] and were not used in the present study. The phase fractions as a function of temperature from Calphad modeling are shown in Figure S14.

4.1.1. C-B Transition

After calcination at 800 °C, C-type was a major phase in all compositions and was the only phase detected in the HE-Y sample (Table 2). Calphad computations correctly predicted the formation of the C-type single-phase solid solution in HE-Y and as a major phase in HE-Gd and HE-Nd. After annealing at 1100 °C, a single-phase C-type solid solution was retained in the HE-Y sample, B-type solid solution became a major phase in He-Gd sample He-Nd completely transformed to B-type structure. The results for HE-Nd and HE-Gd are consistent with Calphad predictions; however, in HE-Y the B-type phase was predicted to appear but was not observed experimentally.

Laser melting and splat quenching produced B-type solid solutions in all samples. Calphad calculations predicted B-type single-phase field for all compositions, with low-temperature boundary shifting from 900 to 1400 °C with decreasing average ionic radius of RE from HE-Nd to HE-Y. The heating and cooling of laser-melted samples in DTA and two months of annealing at 800 °C did not reverse the transformation. The B-C transformation below 1000 °C is known to be kinetically hindered in pure oxides as well [79]; thus, retaining the B phase in our experiments does not indicate slower diffusion in multicomponent compositions.

4.1.2. B-A-H-X Transitions

Calphad calculations correctly predicted B-A transformation in HE-Nd and HE-Gd samples. DTA experiments indicated that transformation proceeds over a 40–70 °C range; however, from Calphad computations, B-A transformation proceeds over a temperature range of several hundred degrees with a change in the fractions and compositions of phases at equilibrium. For HE-Y composition, B-type was predicted to transform directly to the H phase over a narrow biphase region. In contrast, experimental results indicate the B-A transformation in all samples. In agreement with the experiment, the formation of single-phase solid solutions in H and X structure types was predicted for all compositions. H-X transformation was predicted to proceed over an indistinguishably narrow temperature range, with effectively congruent melting at circa ~2400 °C. The narrow biphasic regions on H-X transition and congruent melting from Calphad modeling is consistent with single peaks in DTA.

4.1.3. Biphasic Fields

In a five-component system at constant pressure and temperature, the phase rule allows for up to six phases. In the experiments, we never observed more than two phases at any temperature. Calphad modeling showed mostly single-phase or biphasic fields and very narrow temperature ranges of three-phase co-existence (A, B, C and B, A, H). The absence of large biphasic fields for the B-A transformation from DTA measurements compared with Calphad computations might be attributed to not reaching equilibrium in scanning experiments below 2000 °C. However, the very narrow temperature ranges for H-X transformation and melting are predicted from Calphad and observed in DTA and are likely to be real. In experimental binary phase diagrams of rare earth oxides, the biphasic fields on melting and high temperature transformations are usually not resolved and often added as suggested dotted lines with expectation for them to occur [33,80,81]. Calphad modeling indicates that biphasic fields on A-H-X-L transitions in most binary rare earth oxide systems are often extremely small (~20 °C or less) [67].

It is tempting to assume that observed shrinkage of two-phase fields is the effect of the entropy. Indeed, in many phase diagrams, multiphase fields usually shrink with temperature as solid solution and melt ranges increase. However, that is not always the case. For example, the liquidus loop in the

Os-Re system at 3100 °C is as wide as in Cu-Ni system at 1300 °C and shows 20 at.% difference in composition between solid and liquid phases. The narrow two-phase fields for high temperature phase transformations and melting are peculiar to intra-rare earth systems. It was demonstrated in rare earth metal binaries, which melt below 1600 °C and studied more extensively and with higher accuracy than oxide systems [82,83]. Apparent congruent melting across intra-rare earth binaries was discussed in detail by Okatomo and Massalski [83]. Spedding et al. [84] concluded the study of Er-Y and Tb, Dy, Ho, Er binaries with the statement "the isothermal arrests observed for the melting and transformation temperatures show that there is no appreciable enrichment of one component over the other during these processes, which is what would be expected from the general experience encountered in trying to separate rare earths."

4.2. Thermal Expansion and Volume Change on Mixing

A peculiar feature of rare earth oxides is the negative volume change on temperature-induced C-B, B-A, and H-X transformations. In this work, volumes of C- and B-type solid solutions at room temperature were obtained from the analysis of quenched samples, and volumes of B-, A-, H-, and X-type solid solutions were obtained at temperatures of B-A and H-X transitions. This allowed calculation of volume changes on C-B transformation at room temperature, and for B-A and H-X transformation at transition temperatures. Axial thermal expansion of the monoclinic B-type phase was derived from unit cell parameters measured at B-A transition and on samples quenched to room temperature. In the sections below, we compare the behavior of high entropy compositions with pure oxides.

4.2.1. Volume Change on C-B Transition

All rare earth sesquioxides can be obtained in their ground-state C-type structure at room temperature. The B-type structure is both a high pressure and a high temperature phase for sesquioxides from Sm to Ho, and high pressure phase for sesquioxides from Er to Lu, Y and Sc [85–88]. In Figure 5, volumes of C and B phases for samples studied in this work are plotted vs. average ionic radius together with volumes of pure sesquioxides. The volumes of B-type phase in HE samples show no deviation from the trend. For C-type solid solution, only HE-Y sample was obtained as single-phase (in HE-Nd and HE-Gd samples C-type phase coexisted with B-type phase). Nevertheless, the volumes of C-type phase in HE samples show good agreement with the trend.

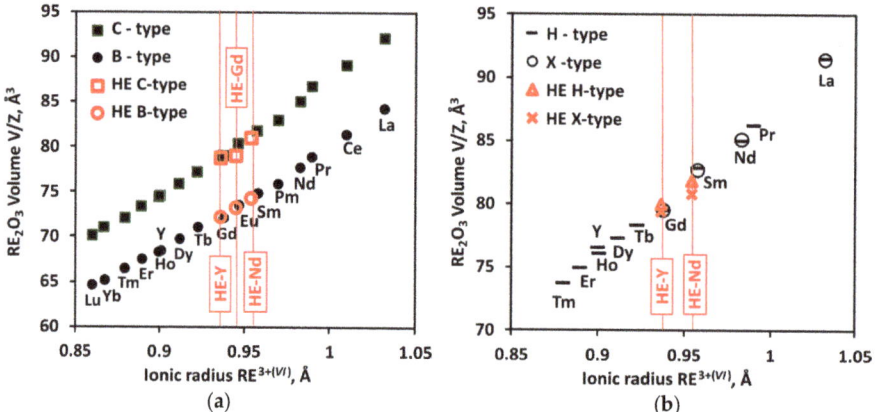

Figure 5. Volumes of C and B phases (**a**) at room temperature and H and X phases (**b**) at transition temperature in pure oxides [30,67] compared with $(La_{0.2}Sm_{0.2}Dy_{0.2}Er_{0.2}RE_{0.2})_2O_3$ HE-RE compositions studied in this work, where RE–Nd, Gd, or Y, plotted vs. average ionic radius.

The volume contraction on C-B transition in HE samples (~8%, Table 2) is indistinguishable from those in constituent oxides, and the B-type phase is both a high temperature and high pressure phase for the studied compositions. This comparison gives no indication of strong deviations from ideal mixing in the solid solution in the B-type structure. For ideal solid solutions, Gibbs energy of mixing does not depend on pressure and there is no excess volume of mixing. For the equiatomic compositions studied in this work, the volume of an ideal solid solution is an average of corresponding volumes of constituent oxides, thus they would follow the trend of volume change vs. average ionic radius.

4.2.2. Thermal Expansion of B-Type Solid Solutions

Taylor [89] reviewed thermal expansion data for pure rare earth sesquioxides and found no data on axial expansion for B-type phases. Ploetz et al. [90] studied the linear expansion of B-type Gd, Eu, and Sm sesquioxides by interferometry and reported the values $(10.0-10.8) \times 10^{-6}$ /K for 30 to 850 °C range. Since the Ploetz et al. measurements were performed on polycrystalline samples, volumetric thermal expansion can be estimated as three times the linear expansion and corresponds to $(3.0-3.2) \times 10^{-5}$/K. These values are in good agreement with average volume thermal expansion for multicomponent rare earth oxides measured in this work: $(3.5 \pm 0.2) \times 10^{-5}$/K from room temperature to the B-A transformation temperatures (1916–1975 °C).

4.2.3. Volume Changes on B-A and H-X Transformations

The volume change on B-A transformation was refined for HE-Nd and HE-Gd samples as -2.5 ± 0.1% and for HE-Y sample as -3.1 ± 0.1% The volume change on H-X transformation was refined for He-Nd and HE-Y samples as -1.2 ± 0.1 and -0.6 ± 0.1%, respectively. Since A, H, and X phases were not quenchable in these compositions, our values were determined from diffraction experiments at the respective transition temperatures. The volume change on B-A and H-X transformation reported for pure oxides are about -2% and -0.5%, respectively [30,67]. Thus, within the resolution of the data, we do not observe any anomalies in volume change on B-A, and H-X transformation in the studied solid solutions compared with pure oxides.

4.2.4. Volumes of H-Type Solid Solution vs. Pure Oxides

In Figure 5, the volumes of the H-type phase for HE-Nd and HE-Y are overlaid with values reported by Foex and Traverse [30] for pure oxides. Foex and Traverse's values correspond to the temperatures from 2120 °C for La_2O_3 to 2330 °C for Ho_2O_3 and Y_2O_3. The volume for the H phase for Sm_2O_3 and Gd_2O_3 refers to the temperatures 2200 and 2250 °C, respectively. Our data on the volume of H phase in HE-Nd and He-Gd from diffraction on levitated samples were assigned to temperatures 2202 ± 4 °C and 2254 ± 8 °C based on co-existence with X-type phase (Table 5) and DTA results. Thus, we can compare volumes of solid solutions vs. pure oxides at a similar temperature. The observed agreement in volume is within 2%. In contrast with the calculation of volume change on transition in coexisting phases, XRD measurement of the absolute values for the volumes at high temperature are affected by sample shift and temperature calibration. Thus, within the resolution of the data, we do not observe deviations from ideal volumetric mixing in the H-type solid solution for the studied compositions.

4.3. Increase of Melting Temperature in HE-RE

The measured melting temperature for HE-Nd is 2456 ± 12 °C. The error is assigned based on uncertainty in melting temperature of Y_2O_3 used for calibration, 2439 ± 12 °C [73]. If the solid solution were to melt congruently, one might estimate, to a very crude first approximation, its melting point as a weighted average of the endmember melting points [91], namely 2357 ± 16 °C. The uncertainties assigned to melting points of constituent oxides in HE-Nd (La,Sm,Dy,Er,Nd)$_2$O$_3$ composition are similar to Y_2O_3 (± 15 vs. ± 12 °C), however, Y_2O_3 melting temperature was established from independent measurements in ten laboratories [73], while values for Dy_2O_3 and Er_2O_3 are based on results of one group and regarded only as "probable" [91].

This observation of apparent increase of melting temperature compare with constituent oxides may be attributed to a combination of kinetic and thermodynamic factors. If the observed behavior does represent equilibrium, then the solidus–liquidus relations must be strongly perturbed by nonideal mixing behavior in the solid, liquid, or both.

To raise the melting temperature, nonideality must favor the solid over the liquid. This can occur by one or both of the following: negative deviations from ideality in the solid phase or positive deviations in the liquid phase. Negative deviations in the solid solution are generally associated with local ordering, hinting at a tendency toward compound formation, and resulting in negative heats of mixing. The volumetric behavior described above shows no evidence for ordering, but volume is not necessarily a good proxy for energetics. The other possibility is strongly positive deviations from ideal mixing in the molten oxides, leading, in the extreme, to liquid–liquid phase separation. We know very little about the structure and thermodynamics of molten binary or multicomponent rare earth oxide systems, except for a recent report from Nakanishi and Allanore [92] on non-ideal mixing in La_2O_3-Y_2O_3 melt, so this possible scenario cannot be assessed at present.

Increases in melting temperature in multicomponent systems compared with end members is not common in oxides, however, it is known to occur, most notably on the ZrO_2-rich side of binaries with Dy-Yb and Y sesquioxides, where melting temperature increase over 100 °C compared with pure ZrO_2 for the solid solution with 25 mol. % of Yb_2O_3 [93]. This almost certainly relates to the formation of favorable short-range order and defect clustering involving tetravalent and trivalent ions and oxygen vacancies. Such a short-range order probably is less important in the trivalent oxide systems studied here. In our system, it is probable that the melting temperature increase should be manifested at least in one of the related binary systems. No increase in melting temperatures was reported in studied binaries with La_2O_3 [33,94,95]. Sm, Dy, and Er are a common constituent in all HE compositions studied in this work. The study of liquidus in the binary systems, currently unknown, is needed to further identify the extent and source of the increase of melting temperatures observed in HE-RE compositions.

5. Conclusions

In this work, we studied three five-component compositions of trivalent rare earth sesquioxides: $(La,Sm,Dy,Er,RE)_2O_3$, with all oxides in equimolar ratios and RE either Nd or Gd or Y. All studied compositions demonstrated a C-B-A-H-X transformation sequence into structure types typical for rare earth sesquioxides. Monoclinic B-type phase was obtained in all compositions by laser melting and splat quenching and was retained after prolonged annealing at 800 °C. The experimentally observed phases are in good agreement with Calphad calculations performed using thermodynamic data for pure sesquioxides and the ideal solution model.

Compared with constituent oxides, A-type and X-type phases occur at wider temperature ranges in the studied compositions. The measured room temperature volumes of C and B phases and volume changes on C-B, B-A, and H-X transitions are in good agreement with those predicted from constituent oxides. No anomalies in thermal expansion of B-type solid solution and in volumes of H-type phases were detected. The obtained data on temperatures, enthalpies, and entropies of transitions can be used for benchmarking the next generation of thermodynamic databases for rare earth oxides. The observed increase in melting temperature compared with constituent oxides invites experimental and theoretical investigations of Sm_2O_3-Dy_2O_3, Dy_2O_3-Er_2O_3, Er_2O_3-Sm_2O_3 systems for which no data on melting temperatures are available.

Supplementary Materials: The following are available online at http://www.mdpi.com/1996-1944/13/14/3141/s1, Figure S1: The flow chart of performed experiments, Figure S2: Aerodynamic levitator with splittable nozzle and copper plates for splat quenching. Figure S3: Integration of area detector diffraction images of aerodynamically levitated bead, Figures S4–S6: Back-scattered electron micrograph of the laser-melted samples, Figure S7: Room-temperature X-ray diffraction patterns of HE-Y, HE-Gd and HE-Nd samples from solution combustion synthesis, Figure S8: Rietveld refinement plot of HE-Nd sample after calcination in air at 800 °C for 96 h, Figure S9: Rietveld refinement plot of HE-Nd sample after splat quenching from melt, Figure S10: Room-temperature powder XRD patterns on Sm_2O_3 sample after annealing at 800 °C and after laser melting, Figure S11: Heat flow trace vs. sample temperature for HE-Gd sample, Figure S12: Pawley refinement of unit cell parameters for B and A phases of HE-Gd sample at transition temperature, Figure S13: Pawley refinement of unit cell parameters for H and X phases of HE-Y sample at transition temperature, Figure S14: Calphad modeling of phase fractions in HE-Y, HE-Gd and HE-Nd samples.

Author Contributions: Conceptualization, S.V.U., S.H. and A.N.; samples synthesis and characterization, S.H. and S.V.U.; thermal analysis S.V.U.; Calphad computations, W.G.; writing—original draft preparation, S.V.U. and S.H.; writing—review and editing, W.G. and A.N.; visualization, S.V.U. and S.H. All authors have read and agreed to the published version of the manuscript.

Funding: This research was funded by National Science Foundation under the award NSF-DMR 1835848 (changed to NSF-DMR 2015852 on funding moved from UC Davis to ASU). Use of the Advanced Photon Source (APS, beamline 6-ID-D), an Office of Science User Facility operated for the DOE Office of Science by Argonne National Laboratory, was supported by the DOE under Contract No. DEACO2-06CH11357.

Acknowledgments: The authors gratefully acknowledge Matvei Zinkevich, Maren Lepple and Vladislav Gurzhiy for helpful discussions. The high temperature diffraction experiments would not be possible without Chris Benmore and Richard Weber ensuring operation and upgrades of aerodynamic levitator at beamline 6-ID-D at APS. Microprobe analysis was performed by Nicolas Botto.

Conflicts of Interest: The authors declare no conflict of interest.

References

1. Yeh, J.-W.; Chen, S.-K.; Lin, S.-J.; Gan, J.-Y.; Chin, T.-S.; Shun, T.-T.; Tsau, C.-H.; Chang, S.-Y. Nanostructured High-Entropy Alloys with Multiple Principal Elements: Novel Alloy Design Concepts and Outcomes. *Adv. Eng. Mater.* **2004**, *6*, 299–303. [CrossRef]
2. Miracle, D.B.; Senkov, O.N. A critical review of high entropy alloys and related concepts. *Acta Mater.* **2017**, *122*, 448–511. [CrossRef]
3. Yeh, J.-W.; Lin, S.-J.; Chin, T.-S.; Gan, J.-Y.; Chen, S.-K.; Shun, T.-T.; Tsau, C.-H.; Chou, S.-Y. Formation of simple crystal structures in Cu-Co-Ni-Cr-Al-Fe-Ti-V alloys with multiprincipal metallic elements. *Metall. Mater. Trans. A* **2004**, *35*, 2533–2536. [CrossRef]
4. Castle, E.; Csanadi, T.; Grasso, S.; Dusza, J.; Reece, M. Processing and Properties of High-Entropy Ultra-High Temperature Carbides. *Sci. Rep.* **2018**, *8*, 1–12. [CrossRef] [PubMed]
5. Feng, L.; Fahrenholtz, W.G.; Hilmas, G.E.; Zhou, Y. Synthesis of single-phase high-entropy carbide powders. *Scr. Mater.* **2019**, *162*, 90–93. [CrossRef]
6. Gild, J.; Zhang, Y.; Harrington, T.; Jiang, S.; Hu, T.; Quinn, M.C.; Mellor, W.M.; Zhou, N.; Vecchio, K.; Luo, J. High-Entropy Metal Diborides: A New Class of High-Entropy Materials and a New Type of Ultrahigh Temperature Ceramics. *Sci. Rep.* **2016**, *6*, 37946. [CrossRef]
7. Tallarita, G.; Licheri, R.; Garroni, S.; Orrù, R.; Cao, G. Novel processing route for the fabrication of bulk high-entropy metal diborides. *Scr. Mater.* **2019**, *158*, 100–104. [CrossRef]
8. Liu, D.; Wen, T.; Ye, B.; Chu, Y. Synthesis of superfine high-entropy metal diboride powders. *Scr. Mater.* **2019**, *167*, 110–114. [CrossRef]
9. Rák, Z.; Maria, J.P.; Brenner, D.W. Evidence for Jahn-Teller compression in the (Mg, Co, Ni, Cu, Zn)O entropy-stabilized oxide: A DFT study. *Mater. Lett.* **2018**, *217*, 300–303. [CrossRef]
10. Rost, C.M.; Sachet, E.; Borman, T.; Moballegh, A.; Dickey, E.C.; Hou, D.; Jones, J.L.; Curtarolo, S.; Maria, J.-P. Entropy-stabilized oxides. *Nat. Commun.* **2015**, *6*, 8485. [CrossRef]
11. Djenadic, R.; Sarkar, A.; Clemens, O.; Loho, C.; Botros, M.; Chakravadhanula, V.S.K.; Kübel, C.; Bhattacharya, S.S.; Gandhi, A.S.; Hahn, H. Multicomponent equiatomic rare earth oxides. *Mater. Res. Lett.* **2017**, *5*, 102–109. [CrossRef]
12. Chen, K.; Pei, X.; Tang, L.; Cheng, H.; Li, Z.; Li, C.; Zhang, X.; An, L. A five-component entropy-stabilized fluorite oxide. *J. Eur. Ceram. Soc.* **2018**, *38*, 4161–4164. [CrossRef]

13. Gild, J.; Samiee, M.; Braun, J.L.; Harrington, T.; Vega, H.; Hopkins, P.E.; Vecchio, K.; Luo, J. High-entropy fluorite oxides. *J. Eur. Ceram. Soc.* **2018**, *38*, 3578–3584. [CrossRef]
14. Jiang, S.; Hu, T.; Gild, J.; Zhou, N.; Nie, J.; Qin, M.; Harrington, T.; Vecchio, K.; Luo, J. A new class of high-entropy perovskite oxides. *Scr. Mater.* **2018**, *142*, 116–120. [CrossRef]
15. Sarkar, A.; Wang, Q.; Schiele, A.; Chellali, M.R.; Wang, D.; Brezesinski, T.; Hahn, H.; Velasco, L.; Breitung, B.; Sarkar, A.; et al. High-Entropy Oxides: Fundamental Aspects and Electrochemical Properties. *Adv. Mater.* **2019**, *31*, 1806236. [CrossRef]
16. Breitung, B.; Wang, Q.; Schiele, A.; Tripkovic, D.; Sarkar, A.; Velasco, L.; Wang, D.; Bhattacharya, S.S.; Hahn, H.; Brezesinski, T. Gassing Behavior of High-Entropy Oxide Anode and Oxyfluoride Cathode Probed Using Differential Electrochemical Mass Spectrometry. *Batter. Supercaps* **2020**, *3*, 361–369. [CrossRef]
17. Wang, Q.; Sarkar, A.; Li, Z.; Lu, Y.; Velasco, L.; Bhattacharya, S.S.; Brezesinski, T.; Hahn, H.; Breitung, B. High entropy oxides as anode material for Li-ion battery applications: A practical approach. *Electrochem. Commun.* **2019**, *100*, 121–125. [CrossRef]
18. Sarkar, A.; Velasco, L.; Wang, D.; Wang, Q.; Talasila, G.; de Biasi, L.; Kuebel, C.; Brezesinski, T.; Bhattacharya, S.S.; Hahn, H.; et al. High entropy oxides for reversible energy storage. *Nat. Commun.* **2018**, *9*, 1–9. [CrossRef]
19. Hume-Rothery, W.; Powell, H.M. The theory of superlattice structure in alloys. *Z. Krist.* **1935**, *91*, 23–47. [CrossRef]
20. Goldschmidt, V.M. Laws of crystal chemistry. *Naturwissenschaften* **1926**, *14*, 477–485. [CrossRef]
21. Krivovichev, V.G.; Charykova, M.V.; Krivovichev, S.V. The concept of mineral systems and its application to the study of mineral diversity and evolution. *Eur. J. Mineral.* **2018**, *30*, 219–230. [CrossRef]
22. Tomilin, I.A.; Kaloshkin, S.D. High entropy alloys—Semi-impossible regular solid solutions? *Mater. Sci. Technol.* **2015**, *31*, 1231–1234. [CrossRef]
23. Goldschmidt, V.M.; Ulrich, F.; Barth, T. Geochemical distribution laws of the elements. IV. The crystal structure of the oxides of the rare earth metals. In *Skrifter Utgit av det Norske Videnskap-Akademi i Oslo. (I) Matem.-Naturvid. Klasse*; Facsimile Publisher: London, UK, 1925; pp. 5–24.
24. Kennedy, B.J.; Avdeev, M. The structure of B-type Sm_2O_3. A powder neutron diffraction study using enriched 154Sm. *Solid State Sci.* **2011**, *13*, 1701–1703. [CrossRef]
25. Glushkova, V.B.; Boganov, A.G. Polymorphism of rare-earth sesquioxides. *Bull. Acad. Sci. USSR Div. Chem. Sci.* **1965**, *14*, 1101–1107. [CrossRef]
26. Rustad, J.R. Density functional calculations of the enthalpies of formation of rare-earth orthophosphates. *Am. Mineral.* **2012**, *97*, 791–799. [CrossRef]
27. Guo, B.; Harvey, A.S.; Neil, J.; Kennedy, I.M.; Navrotsky, A.; Risbud, S.H. Atmospheric pressure synthesis of heavy rare earth sesquioxides nanoparticles of the uncommon monoclinic phase. *J. Am. Ceram. Soc.* **2007**, *90*, 3683–3686. [CrossRef]
28. Gutowski, K.E.; Bridges, N.J.; Rogers, R.D. Actinide Structural Chemistry. In *The Chemistry of the Actinide and Transactinide Elements*; Morss, L.R., Edelstein, N.M., Fuger, J., Eds.; Springer: Dordrecht, The Netherlands, 2011.
29. Foex, M.; Pierre Traverse, J. Polymorphism of rare earth sesquioxides at high temperatures. *Bull. Soc. Fr. Miner. Cristal.* **1966**, *89*, 184–205.
30. Foex, M.; Traverse, J.P. Investigations about crystalline transformation in rare earths sesquioxides at high temperatures. *Rev. Int. Hautes Temp. Refract.* **1966**, *3*, 429–453.
31. Aldebert, P.; Traverse, J.P. Neutron diffraction study of the high temperature structures of lanthanum oxide and neodymium oxide. *Mater. Res. Bull.* **1979**, *14*, 303–323. [CrossRef]
32. Pavlik, A.; Ushakov, S.V.; Navrotsky, A.; Benmore, C.J.; Weber, R.J.K. Structure and thermal expansion of Lu_2O_3 and Yb_2O_3 up to the melting points. *J. Nucl. Mater.* **2017**, *495*, 385–391. [CrossRef]
33. Coutures, J.; Rouanet, A.; Verges, R.; Foex, M. High-temperature study of systems formed by lanthanum sesquioxide and lanthanide sesquioxides. I. Phase diagrams (1400°C < T < T liquid). *J. Solid State Chem.* **1976**, *17*, 171–182.
34. Tracy, C.L.; Lang, M.; Zhang, F.; Trautmatm, C.; Ewing, R.C. Phase transformations in Ln_2O_3 materials irradiated with swift heavy ions. *Phys. Rev. B Condens. Matter Mater. Phys.* **2015**, *92*, 174101. [CrossRef]
35. Schneider, S.J.; Roth, R.S. Phase equilibriums in system involving the rare earth oxides. II. Solid-state reactions in trivalent rare earth oxide systems. *J. Res. Natl. Bur. Stand. Sect. A* **1960**, *64*, 317–332. [CrossRef]

36. Chudinovych, O.V.; Korichev, S.F.; Andrievskaya, E.R. Interaction of Yttrium, Lanthanum, and Samarium Oxides at 1600 °C. *Powder Metall. Met. Ceram.* **2020**, *58*, 599–607. [CrossRef]
37. Qi, J.; Guo, X.; Mielewczyk-Gryn, A.; Navrotsky, A. Formation enthalpies of LaLn'O$_3$ (Ln'=Ho, Er, Tm and Yb) interlanthanide perovskites. *J. Solid State Chem.* **2015**, *227*, 150–154. [CrossRef]
38. Artini, C.; Pani, M.; Lausi, A.; Costa, G.A. Stability of interlanthanide perovskites ABO$_3$ (A≡La-Pr; B≡Y, Ho-Lu). *J. Phys. Chem. Solids* **2016**, *91*, 93–100. [CrossRef]
39. Artini, C. Crystal chemistry, stability and properties of interlanthanide perovskites: A review. *J. Eur. Ceram. Soc.* **2017**, *37*, 427–440. [CrossRef]
40. Pavunny, S.P.; Kumar, A.; Misra, P.; Scott, J.F.; Katiyar, R.S. Properties of the new electronic device material LaGdO$_3$. *Phys. Status Solidi B* **2014**, *251*, 131–139. [CrossRef]
41. Siai, A.; Haro-Gonzalez, P.; Horchani Naifer, K.; Ferid, M. Optical temperature sensing of Er^{3+}/Yb^{3+} doped LaGdO$_3$ based on fluorescence intensity ratio and lifetime thermometry. *Opt. Mater.* **2018**, *76*, 34–41. [CrossRef]
42. Gutierrez-Cano, V.; Rodriguez, F.; Gonzalez, J.A.; Valiente, R. Upconversion and Optical Nanothermometry in LaGdO$_3$: Er^{3+} Nanocrystals in the RT to 900 K Range. *J. Phys. Chem. C* **2019**, *123*, 29818–29828. [CrossRef]
43. Fagg, D.P.; Marozau, I.P.; Shaula, A.L.; Kharton, V.V.; Frade, J.R. Oxygen permeability, thermal expansion and mixed conductivity of Gd$_x$Ce0.8−xPr0.2O2−δ, x=0, 0.15, 0.2. *J. Solid State Chem.* **2006**, *179*, 3347–3356. [CrossRef]
44. Tseng, K.-P.; Yang, Q.; McCormack, S.J.; Kriven, W.M. High-entropy, phase-constrained, lanthanide sesquioxide. *J. Am. Ceram. Soc.* **2020**, *103*, 569–576. [CrossRef]
45. Aldebert, P.; Dianoux, A.J.; Traverse, J.P. Neutron scattering evidence for fast ionic oxygen diffusion in the high temperature phases of lanthanum oxide. *J. Phys.* **1979**, *40*, 1005–1012. [CrossRef]
46. Erukhimovitch, V.; Mordekoviz, Y.; Hayun, S. Spectroscopic study of ordering in non-stoichiometric magnesium aluminate spinel. *Am. Mineral.* **2015**, *100*, 1744. [CrossRef]
47. Ushakov, S.V.; Shvarev, A.; Alexeev, T.; Kapush, D.; Navrotsky, A. Drop-and-catch (DnC) calorimetry using aerodynamic levitation and laser heating. *J. Am. Ceram. Soc.* **2017**, *100*, 754–760. [CrossRef]
48. Toby, B.H.; Von Dreele, R.B. GSAS-II: The genesis of a modern open-source all purpose crystallography software package. *J. Appl. Cryst.* **2013**, *46*, 544–549. [CrossRef]
49. Rietveld, H.M. The Rietveld method. *Phys. Scr.* **2014**, *89*, 098002. [CrossRef]
50. Weber, J.K.R.; Tamalonis, A.; Benmore, C.J.; Alderman, O.L.G.; Sendelbach, S.; Hebden, A.; Williamson, M.A. Aerodynamic levitator for in situ x-ray structure measurements on high temperature and molten nuclear fuel materials. *Rev. Sci. Instrum.* **2016**, *87*, 073902. [CrossRef]
51. Guazzoni, G.E. High-Temperature Spectral Emittance of Oxides of Erbium, Samarium, Neodymium and Ytterbium. *Appl. Spectrosc.* **1972**, *26*, 60–65. [CrossRef]
52. Ushakov, S.V.; Navrotsky, A.; Weber, R.J.K.; Neuefeind, J.C. Structure and Thermal Expansion of YSZ and La$_2$Zr$_2$O$_7$ Above 1500°C from Neutron Diffraction on Levitated Samples. *J. Am. Ceram. Soc.* **2015**, *98*, 3381–3388. [CrossRef]
53. Ushakov, S.V.; Maram, P.S.; Kapush, D.; Pavlik, A.J., III; Fyhrie, M.; Gallington, L.C.; Benmore, C.J.; Weber, R.; Neuefeind, J.C.; McMurray, J.W.; et al. Phase transformations in oxides above 2000°C: Experimental technique development. *Adv. Appl. Ceram.* **2018**, *117*, s82–s89. [CrossRef]
54. McCormack, S.J.; Tamalonis, A.; Weber, R.J.K.; Kriven, W.M. Temperature gradients for thermophysical and thermochemical property measurements to 3000 °C for an aerodynamically levitated spheroid. *Rev. Sci. Instrum.* **2019**, *90*, 015109. [CrossRef] [PubMed]
55. Toby, B.H. Estimating observed structure factors without a structure. In *International Tables for Crystallography*; Gilmore, C.J., Kaduk, J.A., Schenk, H., Eds.; International Union of Crystallography Oxford University Press: Oxford, UK, 2019.
56. Boettinger, W.J.; Kattner, U.R.; Moon, K.-W.; Perepezko, J.J. *DTA and Heat-flux DSC Measurements of Alloy Melting and Freezing*; National Institute of Standards and Technology: Washington, DC, USA, 2006.
57. Ushakov, S.V.; Navrotsky, A. Direct measurements of fusion and phase transition enthalpies in lanthanum oxide. *J. Mater. Res.* **2011**, *26*, 845–847. [CrossRef]
58. Ushakov, S.V.; Navrotsky, A. Experimental approaches to the thermodynamics of ceramics above 1500 °C. *J. Am. Ceram. Soc.* **2012**, *95*, 1463–1482. [CrossRef]
59. Navrotsky, A.; Ushakov, S.V. Hot matters—Experimental methods for high-temperature property measurement. *Am. Ceram. Soc. Bull.* **2017**, *96*, 22–28.

60. Preston-Thomas, H. The International Temperature Scale of 1990 (ITS-90). *Metrologia* **1990**, *27*, 3–10. [CrossRef]
61. Schneider, S.J. Cooperative determination of the melting point of alumina. *Pure Appl. Chem.* **1970**. [CrossRef]
62. Hlavac, J. Melting temperatures of refractory oxides. Part I. *Pure Appl. Chem.* **1982**, *54*, 681–688. [CrossRef]
63. Kaufman, L.; Ågren, J. CALPHAD, first and second generation—Birth of the materials genome. *Scr. Mater.* **2014**, *70*, 3–6. [CrossRef]
64. Lukas, H.L.; Fries, S.G.; Sundman, B. *Computational Thermodynamics: The CALPHAD Method*; Cambridge University Press: Cambridge, UK, 2007.
65. Sundman, B.; Kattner, U.R.; Palumbo, M.; Fries, S.G. OpenCalphad—A free thermodynamic software. *Integr. Mater. Manuf. Innov.* **2015**, *4*, 1. [CrossRef]
66. Pelton, A.D.; Kang, Y.-B. Modeling short-range ordering in solutions. *Int. J. Mater. Res.* **2007**, *98*, 907–917. [CrossRef]
67. Zinkevich, M. Thermodynamics of rare earth sesquioxides. *Prog. Mater. Sci.* **2007**, *52*, 597–647. [CrossRef]
68. Zhang, Y.; Jung, I.-H. Critical evaluation of thermodynamic properties of rare earth sesquioxides (RE = La, Ce, Pr, Nd, Pm, Sm, Eu, Gd, Tb, Dy, Ho, Er, Tm, Yb, Lu, Sc and Y). *Calphad Comput. Coupling Phase Diagr.* **2017**, *58*, 169–203. [CrossRef]
69. Konings, R.J.M.; Beneš, O.; Kovács, A.; Manara, D.; Sedmidubský, D.; Gorokhov, L.; Iorish, V.S.; Yungman, V.; Shenyavskaya, E.; Osina, E. The Thermodynamic Properties of the f-Elements and their Compounds. Part 2. The Lanthanide and Actinide Oxides. *J. Phys. Chem. Ref. Data* **2014**, *43*, 013101. [CrossRef]
70. Zinkevich, M. Thermodynamic Database for Rare Earth Sesquioxides. Available online: https://materialsdata.nist.gov/handle/11256/965 (accessed on 16 April 2020).
71. Shannon, R. Revised effective ionic radii and systematic studies of interatomic distances in halides and chalcogenides. *Acta Crystallogr. Sect. A* **1976**, *32*, 751–767. [CrossRef]
72. Eyring, L. Chapter 27 The binary rare earth oxides. In *Handbook on the Physics and Chemistry of Rare Earths*; Elsevier: Amsterdam, The Netherlands, 1979; Volume 3, pp. 337–399.
73. Foex, M. Study on yttrium oxide melting point. *High Temp. High Press.* **1977**, *9*, 269–282.
74. Senkov, O.N.; Miracle, D.B.; Chaput, K.J.; Couzinie, J.-P. Development and exploration of refractory high entropy alloys—A review. *J. Mater. Res.* **2018**, *33*, 3092–3128. [CrossRef]
75. Zhang, C.; Zhang, F.; Chen, S.; Cao, W. Computational Thermodynamics Aided High-Entropy Alloy Design. *Jom* **2012**, *64*, 839–845. [CrossRef]
76. Zhong, Y.; Sabarou, H.; Yan, X.; Yang, M.; Gao, M.C.; Liu, X.; Sisson, R.D., Jr. Exploration of high entropy ceramics (HECs) with computational thermodynamics—A case study with LaMnO3±δ. *Mater. Des.* **2019**, *182*, 108060. [CrossRef]
77. Gong, W.; Liu, Y.; Xie, Y.; Zhao, Z.; Ushakov, S.V.; Navrotsky, A. Thermodynamic assessment of BaO–Ln$_2$O$_3$ (Ln = La, Pr, Eu, Gd, Er) systems. *J. Am. Ceram. Soc.* **2020**, *103*, 3896–3904. [CrossRef]
78. Gong, W.; Ushakov, S.V.; Agca, C.; Navrotsky, A. Thermochemistry of BaSm$_2$O$_4$ and thermodynamic assessment of the BaO–Sm$_2$O$_3$ system. *J. Am. Ceram. Soc.* **2018**, *101*, 5827–5835. [CrossRef]
79. Roth, R.S.S.; Schneider, S.J. Phase Equilibria in Systems Involving the Rare-Earth Oxides. Part I. Polymorphism of the Oxides of the Trivalent Rare-Earth Ions. *J. Res. Natl. Bur. Stand. A Phys. Chem.* **1960**, *64*, 309–316. [CrossRef] [PubMed]
80. Rouanet, A.; Coutures, J.; Foex, M. High-temperature phase diagram of the lanthanum(III) oxide-ytterbium(III) oxide system. *J. Solid State Chem.* **1972**, *4*, 219–222. [CrossRef]
81. Maister, I.M.; Lopato, L.M.; Shevchenko, A.V.; Nigmanov, B.S. Yttrium oxide-erbium oxide system. *Izv. Akad. Nauk SSSR Neorg. Mater.* **1984**, *20*, 446–448.
82. Norgren, S. Thermodynamic assessment of the Ho-Tb, Ho-Dy, Ho-Er, Er-Tb, and Er-Dy systems. *J. Phase Equilibria* **2000**, *21*, 148–156. [CrossRef]
83. Okamoto, H.; Massalski, T.B. Thermodynamically improbable phase diagrams. *J. Phase Equilibria* **1991**, *12*, 148–168. [CrossRef]
84. Spedding, F.H.; Sanden, B.; Beaudry, B.J. Erbium-yttrium, terbium-holmium, terbium-erbium, dysprosium-holmium, dysprosium-erbium, and holmium-erbium phase systems. *J. Less-Common Met.* **1973**, *31*, 1–14. [CrossRef]
85. Hoekstra, H.R. Phase relationships in the rare earth sesquioxides at high pressure. *Inorg. Chem.* **1966**, *5*, 754–757. [CrossRef]
86. Bai, L.; Liu, J.; Li, X.; Jiang, S.; Xiao, W.; Li, Y.; Tang, L.; Zhang, Y.; Zhang, D. Pressure-induced phase transformations in cubic Gd$_2$O$_3$. *J. Appl. Phys.* **2009**, *106*, 073507. [CrossRef]

87. Irshad, K.A.; Anees, P.; Sahoo, S.; Sanjay Kumar, N.R.; Srihari, V.; Kalavathi, S.; Chandra Shekar, N.V. Pressure induced structural phase transition in rare earth sesquioxide Tm_2O_3: Experiment and ab initio calculations. *J. Appl. Phys.* **2018**, *124*, 155901. [CrossRef]
88. Liu, D.; Lei, W.; Li, Y.; Ma, Y.; Hao, J.; Chen, X.; Jin, Y.; Liu, D.; Yu, S.; Cui, Q.; et al. High-Pressure Structural Transitions of Sc_2O_3 by X-ray Diffraction, Raman Spectra, and Ab Initio Calculations. *Inorg. Chem.* **2009**, *48*, 8251–8256. [CrossRef] [PubMed]
89. Taylor, D. Thermal expansion data: III. Sesquioxides, M_2O_3, with the corundum and the A-, B- and C-M_2O_3 structures. *Br. Ceram. Trans. J.* **1984**, *83*, 92–98.
90. Ploetz, G.L.; Krystyniak, C.W.; Dumas, H.E. Sintering characteristics of rare-earth oxides. *J. Am. Ceram. Soc.* **1958**, *41*, 551–554. [CrossRef]
91. Coutures, J.P.; Rand, M.H. Melting temperatures of refractory oxides: Part II. Lanthanoid sesquioxides. *Pure Appl. Chem.* **1989**, *61*, 1461–1482. [CrossRef]
92. Nakanishi, B.R.; Allanore, A. Electrochemical Investigation of Molten Lanthanum-Yttrium Oxide for Selective Liquid Rare-Earth Metal Extraction. *J. Electrochem. Soc.* **2019**, *166*, E420–E428. [CrossRef]
93. Rouanet, A. Zirconium dioxide—Lanthanide oxide systems close to the melting point. *Rev. Int. Hautes Temp. Refract.* **1971**, *8*, 161–180.
94. Fabrichnaya, O.; Savinykh, G.; Zienert, T.; Schreiber, G.; Seifert, H.J. Phase relations in the ZrO_2-Sm_2O_3-Y_2O_3-Al_2O_3 system: Experimental investigation and thermodynamic modelling. *Int. J. Mater. Res.* **2012**, *103*, 1469–1487. [CrossRef]
95. Shevchenko, A.V.; Nigmanov, B.S.; Zajtseva, Z.A.; Lopato, L.M. Interaction of samarium and gadolinium oxdes with yttrium oxides. *Izv. Akad. Nauk SSSR Neorg. Mater.* **1986**, *22*, 775–779.

© 2020 by the authors. Licensee MDPI, Basel, Switzerland. This article is an open access article distributed under the terms and conditions of the Creative Commons Attribution (CC BY) license (http://creativecommons.org/licenses/by/4.0/).

Article

Effect of Structure and Composition of Non-Stoichiometry Magnesium Aluminate Spinel on Water Adsorption

Yuval Mordekovitz [1], Yael Shoval [1], Natali Froumin [1,2] and Shmuel Hayun [1,2,*]

1. Department of Materials Engineering, Ben-Gurion University of the Negev, P.O. Box 653, Beer-Sheva 84105, Israel; yuvalmor@post.bgu.ac.il (Y.M.); yaelsh5@gmail.com (Y.S.); nfrum@bgu.ac.il (N.F.)
2. Ilsa katz Institute for Nanoscience and Technology, Ben-Gurion University of the Negev, P.O. Box 653, Beer-Sheva 84105, Israel
* Correspondence: hayuns@bgu.ac.il; Tel.: +972-8-6428742

Received: 11 June 2020; Accepted: 14 July 2020; Published: 17 July 2020

Abstract: $MgAl_2O_4$ is used in humidity sensing and measurement, and as a catalyst or catalyst support in a wide variety of applications. For such applications, a detailed understanding of the surface properties and defect structure of the spinel, and, in particular, of the gas interactions at the spinel surface is essential. However, to the best of our knowledge, very limited experimental data regarding this subject is currently available. In this work, four spinel samples with an Al_2O_3 to MgO ratio (n) between 0.95 and 2.45 were synthesized and analyzed using X-ray photoelectron spectroscopy and water adsorption micro-calorimetry. The results showed that the spinel composition and its consequent defect structure do indeed have a distinct effect on the spinel-water vapor surface interactions. The adsorption behavior at the spinel-water interface showed changes that resulted from alterations in types and energetic diversity of adsorption sites, affecting both H_2O uptake and overall energetics. Furthermore, changes in composition following appropriate thermal treatment were shown to have a major effect on the reducibility of the spinel which enabled increased water uptake at the surface. In addition to non-stoichiometry, the impact of intrinsic anti-site defects on the water-surface interaction was investigated. These defects were also shown to promote water uptake. Our results show that by composition modification and subsequent thermal treatments, the defect structure can be modified and controlled, allowing for the possibility of specifically designed spinels for water interactions.

Keywords: water adsorption; defect structure; reducibility; magnesium aluminate spinel

1. Introduction

Magnesium aluminate spinel ($MgAl_2O_4$, MAS) has been shown to be useful for humidity sensing and measurement applications, and as a catalyst or catalyst support for various organic reactions [1–7]. For all these applications, a detailed understanding of the surface properties of the spinel, and specifically the nature of spinel gas-surface interactions, is paramount. Nevertheless, to date, very limited data is available regarding water-surface interactions on MAS and their relation to spinel defect structure, and the information that is available is largely based on theoretical calculations [8]. MAS is the only stable intermediate phase in the Al_2O_3-MgO system, and at elevated temperatures (i.e., over 1300 °C), this system exhibits a large non-stoichiometric range [9], which can also exist at lower temperatures as a metastable nanomaterial [10–12].

The spinel structure has a general formula of AB_2O_4, with the lattice comprising an almost perfect, close-packed cubic arrangement containing 32 oxygen anions. In this arrangement, the A and B cations are situated inside the tetrahedral and octahedral interstitials, respectively. In MAS, eight Mg^{2+} cations are located in the tetrahedral sites, and sixteen Al^{3+} cations occupy the octahedral sites [10,13–15]. The defect structure of spinel is comprised of intrinsic (i.e., Frenkel, Schottky and anti-site) defects and extrinsic (i.e., non-stoichiometric, dopant or impurity) defects. It has been established that the dominating intrinsic defect in $MgAl_2O_4$ spinel is the anti-site defect (AKA inversion), in which two cations switch places. Specifically, an Mg^{2+} occupies the Al^{3+} octahedral site and vice versa [16–20].

The inversion level (i.e., the number of tetrahedral sites occupied by Al^{3+} cations) is controlled by three major factors. The first is related to the thermal history of the material, reflecting the *"intrinsic"* defect concentration. The other two are due to extrinsic parameters, the first of which is MAS stoichiometry (i.e., the. *"stoichiometric"* factor), and the second is residual disorder resulting from thermal effects and stresses in the material synthesis process (i.e., the *"residual"* inversion).

The *intrinsic* component can be calculated using a thermodynamic model, such as that developed by O'Neill and Navrotsky [21], using experimental parameters [10]. The *stoichiometric* component can be quantified from the total defect concentration [13]. The residual inversion, resulting from the synthesis process, cannot be assessed accurately, but this type of inversion defect can be manipulated and subsequently reordered using external fields [10,13].

For MAS, the shift away from idealized stoichiometry can be considered to be due to the dissolution of MgO or Al_2O_3 in the spinel matrix. Departing from the stoichiometric ratio in either the Al_2O_3- or MgO-rich direction, results in different structural defects. Spinel crystals with excess Al_2O_3 are characterized by $Al^{\bullet}_{Mg}$, which can be charge-compensated by V'''_{Mg}, V'''_{Al} or their combination [16,22–25], as demonstrated in the following equations:

$$4 \cdot Al_2O_3 \rightarrow 5 \cdot Al^X_{Al} + 12 \cdot O^X_O + 3 \cdot Al^{\bullet}_{Mg} + V'''_{Al} \quad (1)$$

$$4 \cdot Al_2O_3 \rightarrow 6 \cdot Al^X_{Al} + 12 \cdot O^X_O + 2 \cdot Al^{\bullet}_{Mg} + V''_{Mg} \quad (2)$$

$$12 \cdot Al_2O_3 \rightarrow 16 \cdot Al^X_{Al} + 36 \cdot O^X_O + 8 \cdot Al^{\bullet}_{Mg} + V''_{Mg} + 2 \cdot V'''_{Al} \quad (3)$$

Alternatively, spinel crystals with excess MgO incorporate Mg'_{Al} defects. In this case, the preferred charge compensation would be in the form of $V^{\bullet\bullet}_O$ [16,22–25], as described by:

$$3MgO \rightarrow 2Mg'_{Al} + Mg^X_{Mg} + 3O^X_O + V^{\bullet\bullet}_O \quad (4)$$

Using the Brouwer diagram, the defect types in MAS can be described in terms of the Al_2O_3 content [23] according to the following guidelines:

$$\frac{Mg'_{Al}/V^{\bullet\bullet}_O}{Low\ Al_2O_3} \leftarrow \frac{Al^{\bullet}_{Mg}/Mg'_{Al}}{MgAl_2O_4} \rightarrow \frac{Al^{\bullet}_{Mg}/V'''_{Al}}{Moderate\ Al_2O_3} \rightarrow \frac{Al^{\bullet}_{Mg}/V''_{Mg}}{High\ Al_2O_3} \quad (5)$$

The type and quantity of defects both in the bulk and on the surface of the material changes as a function of stoichiometry. Changes in defect structure can be used for tuning the properties of a material [14,26,27], including its surface properties [2]. Surface properties are also affected by the environmental state in which the material is maintained. For example, ambient, clean, reduced, oxidized or humid environments all have different effects on the surface state.

To the best of our knowledge, no experimental data regarding the effects of non-stoichiometry on surface-water interactions in the $MgO\bullet nAl_2O_3$ system have been published, although some theoretical work has been performed. This paper, therefore, aims to study the effect of the surface composition on the interactions between a non-stoichiometric $MgO\bullet nAl_2O_3$ spinel system and water vapor.

2. Materials and Methods

2.1. Materials

Nano-sized MgO•nAl$_2$O$_3$ powders with 0.95 < n < 2.45 were synthesized by the solution combustion method [28]. This entailed the mixing of appropriate amounts of magnesium and aluminum nitrate, Mg(NO$_3$)$_2$·6H$_2$O 96% metal basis, Al(NO$_3$)$_3$·9H$_2$O (96% metal basis, Fluka Analytical, Sigma Aldrich, St. Louis, MO, USA) in 200 mL of deionized water. To this solution, 30 g of citric acid (ACS reagent ≥99.5%) and 6 mL of ethylene glycol (anhydrous, 99.8%, Sigma Aldrich, St. Louis, MO, USA) were added. The solution was dried on a hot plate at 120 °C under agitation by magnetic stirring until high-viscosity foam-like colloids were formed. The foams were crushed using a mortar and pestle to a fine light brown powder, which served as precursors for the appropriate spinel. The precursors were calcined in air at 850 °C for 72 h to obtain fine white powders.

To study the effects of disorder on the adsorption process, some of the samples were heat-treated in the presence of an electric field, in order to reduce "residual inversion" defects. The samples were heated in air to 800 °C, and maintained at this temperature for 30 min. The heating rate was 10 °C/min, and the furnace was naturally cooled. An electric field of 200 V/cm was applied at all treatment stages. Figure 1 shows the crucible and capacitor setup used for the treatment. As can be seen, there was no contact between the powder and the electrodes, and there was no flow of current.

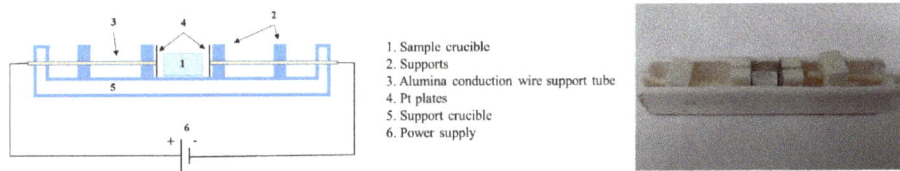

Figure 1. Crucible and cell used for the electric field heat treatments.

2.2. Characterization

X-ray diffraction (XRD) patterns of the samples were recorded using a Rigaku RINT 2100 (Tokyo, Japan) diffractometer with CuKα radiation. The operating parameters were 40 kV and 40 mA, with a 2θ step size of 0.02°. Si (NIST SRM 640c) served as internal standard for cell parameter determination. Crystallite sizes were refined from diffraction peak broadening using a whole profile fitting procedure, as implemented in the Jade software package (version 6.11, 2010, Materials Data Inc., Livermore, CA, USA).

Sample composition was determined by atomic absorption spectroscopy (AAS) using a Varian SpectrAA 240FS (currently Agilent Technologies, Santa-Clara, CA, USA).

Surface area was measured using the Brunauer–Emmett–Teller (BET) theory [29] using a Micrometrics ASAP 2020 (Micrometrics, Norcross, GA, USA) instrument. Fifteen-point adsorption isotherms of nitrogen were collected in the P/P$_0$ relative pressure range 0.05–0.30, where P$_0$ is the saturation pressure at −196 °C. Prior to analysis, each sample was degassed under vacuum at 700 °C for 4 h.

X-ray photoelectron spectroscopy (XPS) data was collected using an X-ray photoelectron spectrometer ESCALAB 250 (Thermo Fisher Scientific, Waltham, MA, USA) ultrahigh vacuum (10^{-9} bar) apparatus with an AlK$^\alpha$ X-ray source and a monochromator. The X-ray beam spot size was 500 µm, and survey spectra were recorded with pass energy (PE) of 150 eV. High-energy resolution spectra were recorded using 20 eV PE. To correct for charging effects, all spectra were calibrated relative to a carbon C1s peak positioned at 284.8 eV. Processing of the XPS results was carried out using the Thermo Scientific AVANTAGE program. For accurate surface characterization by XPS, a glove box was mounted on the XPS enter lock chamber to avoid adsorption of any species from the air on the samples. To ensure that all the samples were investigated under the same experimental conditions, all samples were equilibrated for 12 h in the entry lock chamber of the XPS prior to making the measurements.

IR spectra were recorded at room temperature using a Nicolet 6700 (Thermo Scientific, Madison, WI, USA) FT-IR spectrometer with a KBr-DTGS detector in the range spanning 400–4000 cm^{-1}. Mixtures containing 100 mg KBr and 1 mg spinel were compressed at 1 ton to generate thin plates. For each material, 64 scans of the spectrum were recorded and averaged. The spectrometer settings were at aperture of 150 and spectral resolution of 4 cm^{-1}. Peak positions and intensities were determined by OPUS software (Billerica, MA, USA) using the second derivative and standard methods. The averaged spectrum was used to calculate the inversion parameters of the samples, employing the method of Erukhimovitch et al. [10], which uses the intensity ratios of the γ_1 and γ_5 modes (FTIR peaks located at ~690 and ~830 cm^{-1}).

2.3. Water Adsorption Calorimetry

The heat of adsorption of the water–spinel surface interactions was measured using a custom-made apparatus, composed of a volumetric sorption system (ASAP2020, Micromeritics, Norcross, GA, USA) and a differential scanning calorimeter (Sensys Calvet, Setaram, Lion, France) [30]. The instrumental design and its operation have been discussed elsewhere. Here, approximately 100 mg of sample powder was put inside the sample tube, providing a total surface area of 1.0–5.6 m^2. The tube is than placed inside the calorimeter, it is also connected to the sorption system via a conceive tube. Prior to measurement, a degas procedure was performed. One was cooled to 25 °C at the end of the degassing process, whereas the second was exposed to an oxygen atmosphere prior to cooling down (i.e., oxidized/clean). All the samples were exposed to controlled, and incremental H$_2$O vapor doses until the partial pressure reached P/P$_0$ = 0.3. The incremental dose was set to provide 1 μmol of H$_2$O vapor per m^2 of sample surface. The heat of adsorption (ΔH_{ads}) for each dose was measured. The measurements were repeated 3–4 times for each sample to ensure reproducibility. A baseline run was performed to eliminate environmental and instrumental contributions to the signal.

3. Results and Discussion

This study focused on the effects of surface composition and state on water–surface interactions for four different MgO•nAl$_2$O$_3$ spinel powders. The samples investigated comprised a series of nano-sized (10–15 nm) metastable spinels, with composition (n) ranging between 0.95 and 2.45 (Table 1). Their lattice parameters, crystallite sizes and surface areas are summarized in Table 1. The lattice parameter increased inversely with the n ratio, in keeping with literature results [31,32]. The surface areas measured by the BET method differed significantly from those calculated theoretically from XRD crystallite size (assuming spherical approximation), indicating that the samples had undergone extensive sintering during the final stages of their synthesis.

Table 1. As synthesized characteristics: lattice parameter, crystallite size, surface and interface areas.

| n | (Mg$_x$Al$_y$O$_4$) | | Lattice Parameter, Å | Crystallite Size, nm | Surface Area, m^2/g | |
	(x) Mg	(y) Al			XRD	BET
0.95	1.04	1.97	8.089(2)	14.0 ± 0.2	117.7 ± 1.6	32.1 ± 0.2
1.07	0.95	2.03	8.078(2)	13.3 ± 0.2	123.9 ± 1.7	37.6 ± 0.2
1.15	0.72	2.18	8.065(6)	10.2 ± 0.3	161.6 ± 4.9	56.4 ± 0.2
2.45	0.48	2.35	7.989(4)	15.5 ± 0.8	106.3 ± 5.2	41.2 ± 0.2

Four types of surface states/conditions were addressed in this work: for simplicity, they will be referred to as "as synthesized" (AS), reduced (RD), clean (CL) and hydrated (HD). The AS condition refers to a sample after calcination. The RD condition refers to samples after the degassing procedure (Figure 2). It can be seen that the powders lost their original white color and became greyish-dark after the degassing procedure, with the samples that were richer in Al$_2$O$_3$ becoming markedly darker in the RD state, indicating that they were more easily reduced. Oxidation of the RD samples by exposure to 1 atm of oxygen at 700 °C and cooling to room temperature in a 1 atm oxygen environment resulted in the restoration of the original white color of the samples. It can be concluded that the dark color of

RD samples was a result of the formation of oxygen vacancies during the degassing stage, which was reversed in the oxidization step by "refilling" the oxygen vacancies formed by the initial degas, while simultaneously keeping the surface clean. The reoxidized samples were designated CL. Finally, either RD or CL samples were exposed to H₂O vapor in the water adsorption experiments to generate the HD states.

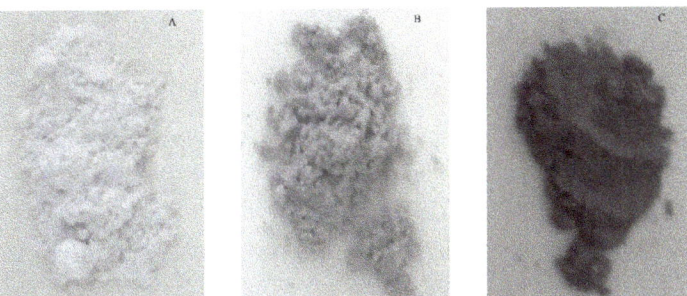

Figure 2. Different sample colors after degassing procedure showing different reduction level vs n: (**A**) n = 1.07; (**B**) n = 1.15; (**C**) n = 2.45.

3.1. Surface State Analysis

Typical XPS spectra in the Al2p and Mg2p regions for the four types of surface state (AS, RD, CL and HD) of MgO•1.07Al₂O₃ are shown Figures 3–6. Three types of bonds, specifically M–M, M–O and M–OH bonds (M = Al, Mg), were considered for each spectrum. In the Al2p spectra, these are assigned at ~73.0, 74.1 and 75.2 eV, respectively [33–35]. In the Mg2p spectra, their assignments are ~49.3, 50.9 and 51.8, respectively [36]. The dominant bond in AS samples (Figure 3) is the M–O bond, with less prominent, though not negligible, M–OH bonds. The RD samples (Figure 4) displayed different spectra due to the prolonged degassing procedure. Here, the amount of the hydroxyl surface species was reduced, and the presence of M–M bonds was detected. These M–M Bonds were formed due to oxygen deficiencies in the structure, as reflected in the color changes of the powders (Figure 2). In the CL samples (Figure 5), the M–M bonds disappeared and were replaced by M–O species, with the quantity of M–OH bonds being lower than in the AS samples. As expected, the HD samples exhibited the highest quantity of M–OH bonds, and no M–M bonds were identified (Figure 6). This behavior was found to be typical for all the samples in this study. The data for all samples are given in the Supplementary Materials in Tables S1 and S2.

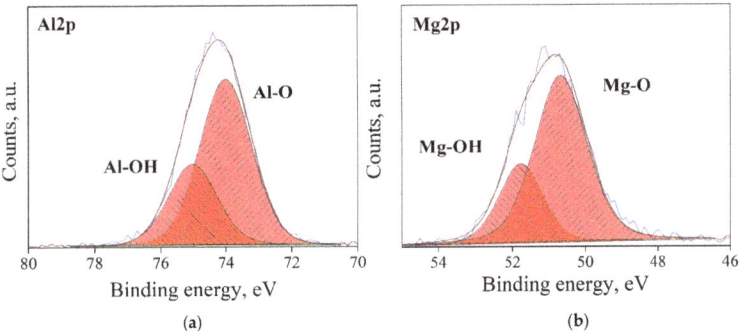

Figure 3. XPS spectra, of an AS, n = 1.07 sample: Al2p de-convoluted to Al–O and Al–OH peaks (**a**), and Mg2p, de-convoluted to Mg–O and Mg–OH peaks (**b**).

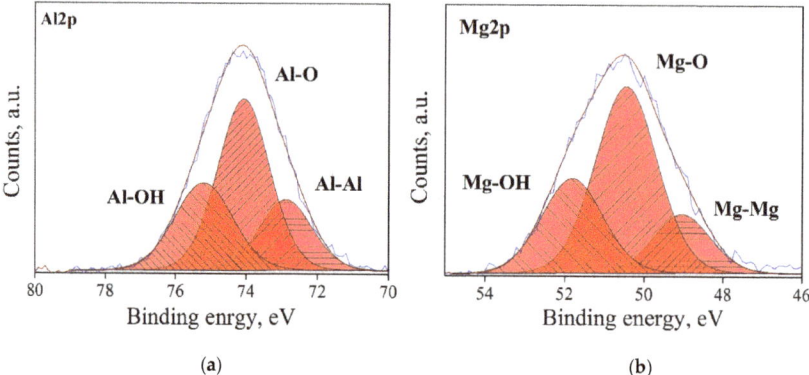

Figure 4. XPS spectra of an RD, n = 1.07 sample, after degassing and oxidation: Al2p de-convoluted to Al–O, Al–OH and Al–Al peaks (**a**), and Mg2p, de-convoluted to Mg–O, Mg–OH and Mg–Mg peaks (**b**).

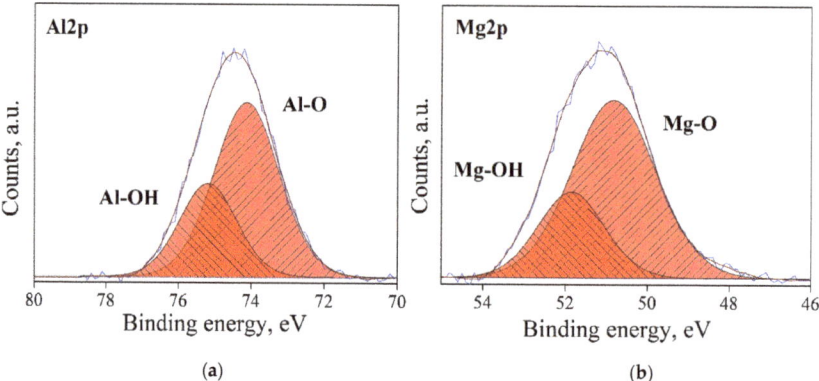

Figure 5. XPS spectra, of a CL, n = 1.07 sample, after degassing and oxidation: Al2p de-convoluted to Al–O and Al–OH peaks (**a**): and Mg2p deconvoluted to Mg–O and Mg–OH peaks (**b**).

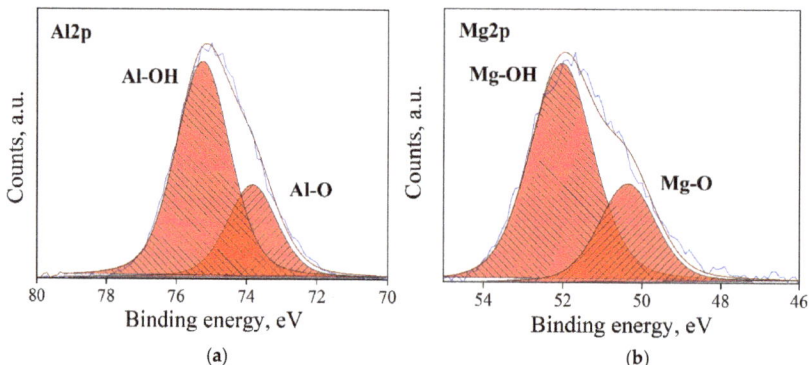

Figure 6. XPS spectra of an HD, n = 1.07 sample after degassing, oxidation, and exposure to water vapor: Al2p de-convoluted to Al–O and Al–OH peaks (**a**), and Mg2p, de-convoluted to Mg–O and Mg–OH peaks (**b**).

3.2. Water Adsorption Measurments

Heat of adsorption measurements were conducted for the reduced (RD) and fully oxidized (CL) samples. Typical heat of adsorption isotherms for MgO- and Al$_2$O$_3$-rich spinel samples are presented in Figure 7. It should be emphasized that in this calorimetric study, water molecules adsorbed with an enthalpy greater than −44 kJ/mol relative to vapor are referred to as "strongly bound", while those adsorbed with the enthalpy of condensation of liquid water (−44 kJ/mol) are considered "weakly bound". Furthermore, we should stress that such assignments are based solely on calorimetric data and do not reflect any structural studies of water adsorbed on the surface [37,38].

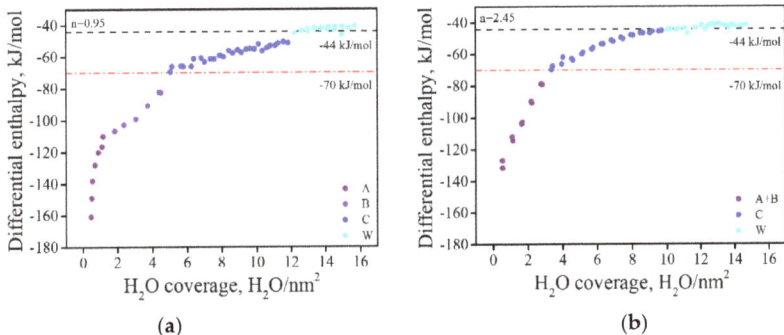

Figure 7. Differential enthalpies of adsorption as a function of water coverage on CL samples: an MgO-rich sample, n = 0.95 (**a**), and an Al$_2$O$_3$-rich sample, n = 2.45 (**b**). The blue line signifies the transition between strongly bonded and weakly bonded water, and the red line emphasizes enthalpy of −70 kJ/mol, where the site type of strongly bonded water is altered.

The heat of adsorption isotherms can be divided into segments, according to the change in their slope. Each segment of the isotherm may be considered as reflecting a different type of adsorption site, or site group [39,40]. The MgO-rich spinel displayed a four-step behavior (Figure 7a). In the first step, marked in purple (type A), a sharp decrease in the enthalpy of adsorption was seen, up to H$_2$O coverage of ~2 molecules/nm^2. From there, a second step was observed up to ~5 molecules/nm^2, marked in violet (type B), in which the slope changed direction, and the decrease in ΔH$_{ads}$ was moderated. At the transition point between the second and third steps ΔH$_{ads}$ reached a value of ≈−70 kJ/mol. From this point of the curve, the blue section (type C), the ΔH$_{ads}$ slowly decayed until the measured enthalpy corresponded to that of weakly bonded water. After reaching −44 kJ/mol (the cyan section of the curve, type D), the enthalpy fell to lower values (absolute) of about −39 kJ/mol. In general, the MgO-rich samples showed similar behavior to that of pure MgO, where surface hydroxides (i.e., Mg(OH)$_2$) are formed [41].

In the case of the Al$_2$O$_3$-rich samples, the first two steps observed for the MgO-rich samples were combined into a single step (up to ~4 molecules/nm^2, marked in purple, types A + B). After this point, a more moderate slope was identified, up to −44 kJ/mol (Figure 7b, blue colored, type C), with no further decreases.

The heat of the adsorption isotherms for all samples, in their CL and RD states, can be seen in Figure 8. These isotherms are a depiction of the gas adsorption amount (Figures S1 and S2) obtained in each dose with their corresponding energetic value. Table 2 summarizes the integral heat of adsorption of strongly bond water and the water coverage for all samples, as well as the amounts of Mg and Al hydroxides formed on the surface, deduced from the XPS analysis. The columns in Table 2 under the heading "Hydroxides" present differences in the surface hydroxide compositions for the initial RD and CL states and after their hydration. The measured values are in good agreement with the water coverage, except for the n = 0.95 sample. We believe that XPS measurements do not accurately account for the coverage in this sample, possibly due to the presence of surface contamination by adventitious species on the CL/RD samples. Such species are thought to readily adsorb due to the defective nature of MgO rich spinel [16,22–25].

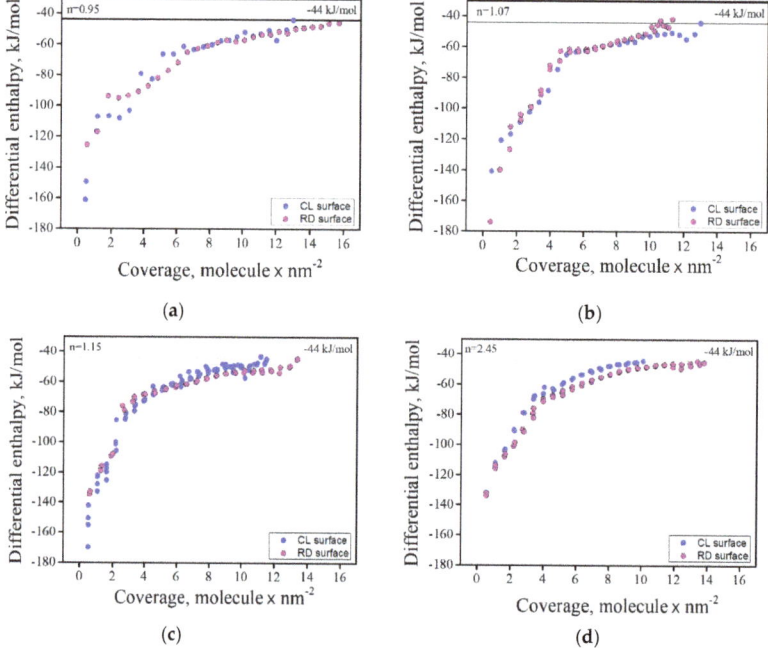

Figure 8. Heat of adsorption isotherms for all samples for RD and CL surfaces: (**a**) n = 0.95; (**b**) n = 1.07; (**c**) n = 1.15; (**d**) n = 2.45.

Table 2. Water adsorption data for Clean (CL) and Reduced (RD) spinel samples.

	n	Heat of Adsorption, kJ/mol	Hydroxides, mol. %			H_2O Coverage, Molecules/nm²
			Al–OH	Mg–OH	Total	
Clean	0.95	−75.1 ± 0.2	6.2 ± 0.3	13.7 ± 0.7	20.0 ± 1.0	12.2 ± 1.0
	1.07	−73.3 ± 0.4	34.0 ± 1.7	25.5 ± 1.3	59.4 ± 3.0	12.9 ± 0.1
	1.15	−71.0 ± 0.7	35.4 ± 1.8	12.4 ± 0.6	47.0 ± 2.4	11.5 ± 0.1
	2.45	−67.2 ± 0.2	32.1 ± 1.0	2.2 ± 0.1	34.3 ± 1.7	10.1 ± 0.1
Reduced	0.95	−71.0 ± 1.0	8.3 ± 0.4	16.9 ± 0.8	25.1 ± 1.2	15.3 ± 0.3
	1.07	−75.6 ± 0.9	30.3 ± 1.5	18.9 ± 0.9	48.2 ± 2.5	11.3 ± 0.2
	1.15	−68.5 ± 0.4	37.5 ± 1.9	11.5 ± 0.6	49.0 ± 2.4	13.4 ± 0.1
	2.45	−66.3 ± 0.3	41.3 ± 2.1	7.8 ± 0.4	49.0 ± 2.4	13.5 ± 0.2

The results showed a clear relation between the Al_2O_3 concentration and the water adsorption in the CL samples, in terms of both water uptake and energetics. In general, as the Al_2O_3 concentration persisted in the reduced samples (RD), with the exception of the n = 1.07 sample. The possible origins for this apparent anomaly will be discussed below. Notably, each sample in the RD state accumulated more adsorbed water than its CL analogue (excluding n = 1.07), as its surface defect structure was altered. It is possible that some of the adsorbed water may act as an oxidizing agent, but we believe that this role is limited due to the relatively low temperature of the adsorption process.

Water uptake in the RD state was dependent on Al_2O_3 concentration, but this dependency is not as simple as was observed for the CL samples, even if the n = 1.07 sample is excluded. This, and the apparently anomalous behavior of the n = 1.07 sample, are attributed to the differences in the level of reduction as was seen by the changes in sample coloring. Based on the color differences of the Al_2O_3-rich samples (Figure 2), we concluded that the level of reduction increased with Al_2O_3 concentration, but the extent of reduction was not quantitatively determined in this work. The effects of reduction in enhancing the extent of adsorption were evident in the increased (and similar) extent of water uptake for the n = 1.15 and n = 2.45 samples. However, the integral enthalpy of adsorption of the RD samples was lowered, relative to their CL counterparts, as the defect structure was progressively altered. These alterations, in turn, resulted in changes in the site population and its energetic diversity, the source of which lies in the newly induced material defect structure.

In spinel, water molecules are adsorbed in the vicinity of the metal ions, specifically Al_{Al}^x, Mg_{Mg}^x, $Al_{Mg}^{\bullet}$, and Mg'_{Al} [42]. As Al_2O_3 is added in excess, the Al^{3+} cations progressively occupy tetrahedral sites, substituting Mg^{2+} and disturbing the charge neutrality. To maintain charge neutrality, cation vacancies are formed, some of which are on the surface of the spinel structure [16,22–25]. Consequently, the quantity of surface cations is diminished and hence also the quantity of available adsorption sites which leads to lower quantities of adsorbed water.

An excess of Al_2O_3 also influences the proximity of the metal cations to the surface, as it affects which of the material planes have a higher tendency to be exposed. Cai et al. [43] calculated the surface stability of exposed spinel surfaces and concluded that in Al_2O_3-rich spinel 111_O_2(Al) plane tends to be exposed. In this plane, oxygen molecules are slightly elevated over the Al^{3+} cations, thus decreasing the energy of interaction at the surface. As the material becomes poorer in Al_2O_3, the 100_Al(O_2) plane is exposed [43]. This plane has a higher surface Gibbs free energy, thereby allowing for interactions at the surface that are more energetic.

It is important to note that the n = 0.95 samples are MgO-rich, and thus, the factors contributing to the defect structure are different. Here, oxygen vacancies compensate for excess Mg, which essentially allows for better exposure of the metal cations. Moreover, MgO is highly hygroscopic, forming a surface hydroxide phase that enables additional, more energetic water uptake [41]. In the RD state, oxygen vacancies were formed, their numbers growing with Al_2O_3 concentration, with consequent changes in the surface defect chemistry and electronic structure. In the RD samples, a decrease in the integral enthalpy of adsorption was registered, relative to their CL counterparts. To determine the origins of this decrease, a closer analysis of the energetic diversity of the adsorption sites is required.

Figure 9 shows normalized heat of adsorption isotherms for CL and RD samples. In these isotherms, the x-axis was normalized to the amount adsorbed at full coverage for each sample. This observation allows us to consider the energetic distribution of the sites. In samples n = 0.95 and 1.15, the isotherms for the RD and CL materials are almost overlapped, with the exception of the very first few sites at low relative coverage (type A), which were more energetic for the CL samples than for their RD counterparts. An inverse relationship was obtained for the n = 1.07 sample. These first few (low relative coverage) sites are very energetic and make a considerable contribution to the overall energetics. The new adsorption sites that were added after reduction are of type C, which suggests water was unable to re-oxidize the surface. Finally, the CL and RD isotherms for the n = 2.45 sample were in almost perfect alignment, suggesting that the energetic diversity of the adsorption sites was maintained regardless of the reduction process undergone by this material. Accordingly, the integral enthalpy of adsorption for this sample in the two states was similar.

At this point, it is important to address the anomalous behavior exhibited observed for the sample n = 1.07. This material has near stoichiometric composition, and, as is evident from Figure 2, was barely reduced by the degassing procedure. Nonetheless, some changes in the defect structure did occur. Jia et al. [42] showed computationally that for stoichiometric $ZnGa_2O_4$ spinel, such reduction-related defects do not always enhance water adsorption [42]. We assume that a similar explanation is appropriate here because of the proximity of the n = 1.07 sample to stoichiometry. This implies that in order for the reduction process to enhance the surface reactivity, the composition should not be near stoichiometric.

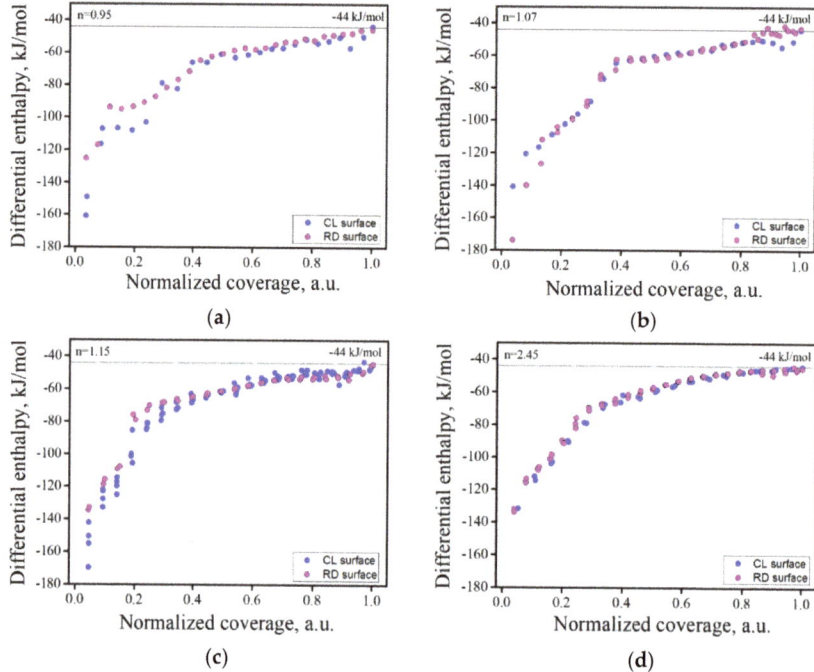

Figure 9. Heat of adsorption isotherms for all CL and RD samples. The coverage is normalized to the full coverage of strongly adsorbed water for each sample. (**a**) n = 0.95; (**b**) n = 1.07; (**c**) n = 1.15; (**d**) n = 2.45.

3.3. Effect of Anti-Site Defects

As discussed above, the spinel system is subject to extrinsic and intrinsic defects. The former, which can exert a considerable effect on water-surface interactions can, however, be influenced by controlling the composition of the material. The spinel system also presents intrinsic, anti-site defects that are not controllable or that are controllable only to a certain extent and require specific study and understanding. Thus, to assess the effects of anti-site defects on adsorption behavior, a MgO•2.45Al$_2$O$_3$ (n = 2.45) sample was heat-treated in the presence of a constant electric field (EF). FTIR spectra in the 400–1000 cm^{-1} range of the sample before and after the heat treatment are presented in Figure 10. The aim of the electric field treatment was essentially to rearrange the defects caused by the *residual* inversion without otherwise affecting the material. The thermal treatment was performed at 800 °C, a temperature lower than the calcination temperature of the sample (850 °C) so that any changes in the material as result of the heat treatment in the presence of the electric field can be attributed solely to the effect of the EF. The function of heating (to 800 °C) is to provide sufficient thermal energy to assist cation rearrangement, driven by the applied electric field. As a result of this treatment, highly disordered samples (I = 0.44) with n ratio of 2.45 underwent significant reordering (I = 0.33), as is qualitatively demonstrated in the FTIR spectrum by the decrease in the intensity of the γ_5 mode (~830 cm^{-1}) (Figure 10) [10].

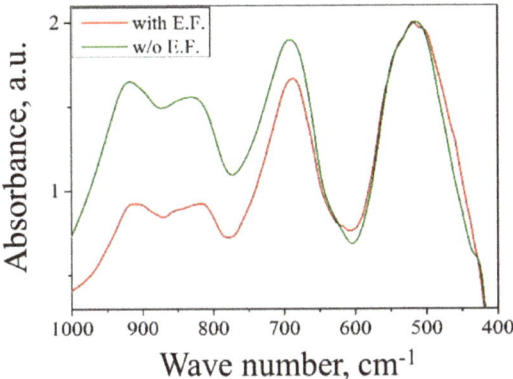

Figure 10. FTIR spectra of MgO•2.45Al$_2$O$_3$ before and after application of an electric field.

After the heat treatment and subsequent inversion parameter (i) decrease, water adsorption of this re-ordered sample was measured in the same way as for all the preceding materials. The adsorption enthalpy isotherms of the two spinel samples, before and after heat treatment with the electrical field are shown in Figure 11. From the data listed in Table 3, it is apparent that following reordering, the enthalpy of adsorption decreased, as did the coverage. The extent of formation of hydroxide bonds was determined using XPS (Table 3), and the results are in agreement with the results of the heat of adsorption experiments. Heat treatment together with the application of an electrical field led to significantly less hydroxides being formed, with the change in Al–OH bonding being more marked than that for Mg–OH. These experimental findings can be explained by the existence of excess charge when the Al^{+3} cations are located in the tetrahedral sites.

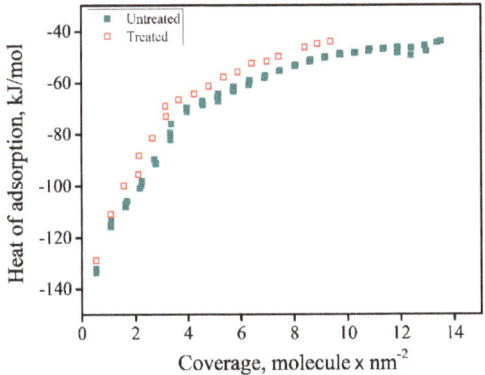

Figure 11. Differential enthalpy of adsorption as a function of water coverage of MgO•2.45Al$_2$O$_3$ samples before and after heat treatment in an electric field.

Table 3. Water adsorption by MgO•2.45Al$_2$O$_3$.

n		Inversion (i)	Heat of Adsorption, kJ/mol	Extent of Hydroxides Formed, mol. %			H$_2$O Coverage, Molecules/nm^2
				Al–OH	Mg–OH	Total	
2.45	Untreated	0.44	−67.2 ± 0.3	32.1 ± 1.6	2.2 ± 0.1	34.3 ± 1.7	10.1 ± 0.1
	Treated/(EF)	0.33	−62.5 ± 0.5	25.0 ± 1.3	1.8 ± 0.1	26.8 ± 1.4	8.8 ± 0.1

4. Conclusions

The relationship between the composition of spinels and their general properties is undeniable, and has been widely demonstrated in the literature. Surface behavior and water–surface interactions are no exception. The composition dictates, in part, the defect structure of a material. This, in turn, controls the surface behavior. Ultimately, these aspects govern the water–surface interaction.

We have shown that changes in stoichiometry alter the water adsorption behavior in our system. In general, an increase in the Al_2O_3 concentration lowers both water uptake and the energy of water adsorption. Changes in the Al_2O_3 concentration influence the defect structure of the material, thereby changing the adsorption site population and its energetic diversity. Furthermore, the material composition affects the reducibility of the material, and thus, its ability to host more defects. These defects promote water uptake while lowering the adsorption enthalpy.

In addition to the effects of non-stoichiometry, the effects of intrinsic defects in spinel should be considered when dealing with water–surface interactions. A spinel having a lower inversion parameter (i), i.e., a material with fewer anti-site defects, was shown to adsorb fewer strongly bonded water molecules and to present lower enthalpies of adsorption, indicating that the Al cation is more active when it occupies a tetrahedral site in the spinel structure.

Supplementary Materials: The following are available online at http://www.mdpi.com/1996-1944/13/14/3195/s1, Figure S1: Adsorption isotherms of the samples with clean surface, Figure S2: Adsorption isotherms of the samples with reduced surface, Table S1: Atomic percentage of Al and Mg oxides and hydroxides for adsorption processes with reduction, Table S2: Atomic percentage of Al and Mg oxides and hydroxides for adsorption processes with oxidation.

Author Contributions: Conceptualization, S.H.; samples synthesis and characterization, Y.M. and Y.S.; calorimetric measurements, Y.S.; data analysis, Y.M., Y.S., N.F. and S.H.; writing—original draft preparation, Y.M. and Y.S.; writing—review and editing, S.H.; visualization, Y.M. and Y.S. All authors have read and agreed to the published version of the manuscript.

Funding: This research was partially supported by the BSF United States-Israel Binational Science Foundation (grant 2010377) and the FP7-PEOPLE-2012-CIG (grant 321838-EEEF-GBE-CNS).

Acknowledgments: Yuval Mordekovitz gratefully acknowledges the Israeli Ministry of Science and Technology (MOST) for granting him the Ze'ev Jabotinsky scholarship.

Conflicts of Interest: The authors declare no conflict of interest.

References

1. Gusmano, G.; Montesperelli, G.; Traversa, E.; Mattogno, G. Microstructure and electrical properties of $MgAl_2O_4$ thin films for humidity sensing. *J. Am. Ceram. Soc.* **1993**, *76*, 743–750. [CrossRef]
2. Govindaraj, A.; Flahaut, E.; Laurent, C.; Peigney, A.; Rousset, A.; Rao, C.N.R. An investigation of carbon nanotubes obtained from the decomposition of methane over reduced $Mg_{1-x}M_xAl_2O_4$ spinel catalysts. *J. Mater. Res.* **1999**, *14*, 2567–2576. [CrossRef]
3. Mei, D.; Lebarbier Dagle, V.; Xing, R.; Albrecht, K.O.; Dagle, R.A. Steam reforming of ethylene glycol over $MgAl_2O_4$ supported Rh, Ni, and Co Catalysts. *ACS Catal.* **2016**, *6*, 315–325. [CrossRef]
4. Villa, A.; Gaiassi, A.; Rossetti, I.; Bianchi, C.L.; Van Benthem, K.; Veith, G.M.; Prati, L. Au on $MgAl_2O_4$ spinels: The effect of support surface properties in glycerol oxidation. *J. Catal.* **2010**, *275*, 108–116. [CrossRef]
5. Mei, D.; Glezakou, V.A.; Lebarbier, V.; Kovarik, L.; Wan, H.; Albrecht, K.O.; Gerber, M.; Rousseau, R.; Dagle, R.A. Highly active and stable $MgAl_2O_4$-supported Rh and Ir catalysts for methane steam reforming: A combined experimental and theoretical study. *J. Catal.* **2014**, *316*, 11–23. [CrossRef]
6. Mei, D.; Lebarbier, V.M.; Rousseau, R.; Glezakou, V.A.; Albrecht, K.O.; Kovarik, L.; Flake, M.; Dagle, R.A. Comparative investigation of benzene steam reforming over spinel supported Rh and Ir catalysts. *ACS Catal.* **2013**, *3*, 1133–1143. [CrossRef]
7. Li, W.Z.; Kovarik, L.; Mei, D.; Liu, J.; Wang, Y.; Peden, C.H.F. Stable platinum nanoparticles on specific $MgAl_2O_4$ spinel facets at high temperatures in oxidizing atmospheres. *Nat. Commun.* **2013**, *4*, 1–8. [CrossRef] [PubMed]
8. Fang, C.M.; de With, G. Computer simulation of dissociative adsorption of water on the surfaces of spinel $MgAl_2O_4$. *J. Am. Ceram. Soc.* **2001**, *84*, 1553–1558. [CrossRef]

9. Hallstedt, B. Thermodynamic assessment of the system MgO-Al$_2$O$_3$. *J. Am. Ceram. Soc.* **1992**, *75*, 1497–1507. [CrossRef]
10. Erukhimovitch, V.; Mordekoviz, Y.; Hayun, S. Spectroscopic study of ordering in non-stoichiometric magnesium aluminate spinel. *Am. Mineral.* **2015**, *100*, 1744–1751. [CrossRef]
11. Mordekovitz, Y.; Hayun, S. On the effect of lithium on the energetics and thermal stability of nano-sized nonstoichiometric magnesium aluminate spinel. *J. Am. Ceram. Soc.* **2016**, *99*, 1–9. [CrossRef]
12. Hilklin, T.R.; Laine, R.M. Synthesis of metastable phases in the magnesium spinel– alumina system. *Chem. Mater.* **2008**, *20*, 553–558. [CrossRef]
13. Halabi, M.; Ezersky, V.; Kohn, A.; Hayun, S. Charge distribution in nano-scale grains of magnesium aluminate spinel. *J. Am. Ceram. Soc.* **2017**, *100*, 800–811. [CrossRef]
14. Rubat du Merac, M.; Kleebe, H.-J.; Müller, M.M.; Reimanis, I.E. Fifty years of research and development coming to fruition; Unraveling the complex interactions during processing of transparent magnesium aluminate (MgAl$_2$O$_4$) spinel. *J. Am. Ceram. Soc.* **2013**, *96*, 3341–3365. [CrossRef]
15. Reimanis, I.; Kleebe, H.-J. A review on the sintering and microstructure development of transparent spinel (MgAl$_2$O$_4$). *J. Am. Ceram. Soc.* **2009**, *92*, 1472–1480. [CrossRef]
16. Ball, J.A.; Murphy, S.T.; Grimes, R.W.; Bacorisen, D.; Smith, R.; Uberuaga, B.P.; Sickafus, K.E. Defect processes in MgAl$_2$O$_4$ spinel. *Solid State Sci.* **2008**, *10*, 717–724. [CrossRef]
17. Rasmussen, M.K.; Foster, A.S.; Hinnemann, B.; Canova, F.F.; Helveg, S.; Meinander, K.; Martin, N.M.; Knudsen, J.; Vlad, A.; Lundgren, E.; et al. Stable cation inversion at the MgAl$_2$O$_4$(100) surface. *Phys. Rev. Lett.* **2011**, *107*, 2–5. [CrossRef]
18. Simeone, D.; Dondane-Thiriiet, C.; Gosset, D.; Daniel, P.; Beauvy, M. Comment—Disorder phase transition induced by swift ions in MgAl$_2$O$_4$ and ZnAl$_2$O$_4$ spinels. *J. Nucl. Mater.* **2002**, *300*, 151–160. [CrossRef]
19. Méducin, F.; Redfern, S.A.T.; Le Godec, Y.; Stone, H.J.; Tucker, M.G.; Dove, M.T.; Marshall, W.G. Study of cation order-disorder in MgAl$_2$O$_4$ spinel by in situ neutron diffraction up to 1600 K and 3.2 GPa. *Am. Mineral.* **2004**, *89*, 981–986. [CrossRef]
20. Wood, B.J.; Kirkpatrick, R.J.; Montez, B. Order-disorder phenomena in MgAl$_2$O$_4$ spinel. *Am. Mineral.* **1986**, *71*, 999–1006.
21. O'Neill, H.S.C.; Navrotsky, A. Simple spinels: Crystallographic parameters, cation radii, lattice energies, and cation distribution. *Am. Mineral.* **1983**, *68*, 181–194.
22. Sickafus, K.E.; Yu, N.; Nastasi, M. Radiation resistance of the oxide spinel: The role of stoichiometry on damage response. *Nucl. Instrum. Meth. B* **1996**, *116*, 85–91. [CrossRef]
23. Ting, C.J.; Lu, H.Y. Defect reactions and the controlling mechanism in the sintering of magnesium aluminate spinel. *J. Am. Ceram. Soc.* **1999**, *82*, 841–848. [CrossRef]
24. Chiang, Y.-M. *Grain Boundary Mobility and Segregation in Non-Stoichiometric Solid Solutions of Magnesium Aluminate Spinel*; Massachusetts Institute of Technology: Cambridge, MA, USA, 1980.
25. Chiang, Y.-M.; Kingery, W.D. Grain-boundary migration in nonstoichiometric solid solutions of magnesium aluminate spinel: I, grain growth studies. *J. Am. Ceram. Soc.* **1989**, *72*, 271–277. [CrossRef]
26. Sutorik, A.C.; Gilde, G.; Swab, J.J.; Cooper, C.; Gamble, R.; Shanholtz, E. Transparent solid solution magnesium aluminate spinel polycrystalline ceramic with the alumina-rich composition MgO·1.2 Al$_2$O$_3$. *J. Am. Ceram. Soc.* **2012**, *95*, 636–643. [CrossRef]
27. Barzilai, S.; Aizenshtein, M.; Mintz, M.H.; Hayun, S. Effect of adsorbed oxygen on the dissociation of water over gadolinium oxide surfaces: Density functional theory calculations and experimental results. *J. Phys. Chem. C* **2020**. [CrossRef]
28. Ianoş, R.; Lazău, I.; Păcurariu, C.; Barvinschi, P. Solution combustion synthesis of MgAl$_2$O$_4$ using fuel mixtures. *Mater. Res. Bull.* **2008**, *43*, 3408–3415. [CrossRef]
29. Brunauer, S.; Emmett, P.H.; Teller, E. Adsorption of gases in multimolecular layers. *J. Am. Chem. Soc.* **1938**, *60*, 309–319. [CrossRef]
30. Ushakov, S.V.; Navrotsky, A. Direct measurements of water adsorption enthalpy on hafnia and zirconia. *Appl. Phys. Lett.* **2005**, *87*, 1–3. [CrossRef]
31. Jing, S.-Y.; Lin, L.-B.; Houng, N.-K.; Zhang, J.; Lu, Y. Investigation on lattice constants of Mg-Al spinels. *J. Mater. Sci. Lett.* **2000**, *19*, 225–227. [CrossRef]
32. Navrotsky, A.; Wechsler, B.A.; Geisinger, K.; Seifert, F. Thermochemistry of MgAl$_2$O$_4$-Al$_{8/3}$O$_4$ defect spinels. *J. Am. Ceram. Soc* **1986**, *69*, 418–422. [CrossRef]

33. Corsi, J.S.; Fu, J.; Wang, Z.; Lee, T.; Ng, A.K.; Detsi, E. Hierarchical bulk nanoporous aluminum for on-site generation of hydrogen by hydrolysis in pure water and combustion of solid fuels. *ACS Sustain. Chem. Eng.* **2019**, *7*, 11194–11204. [CrossRef]
34. Hinnen, C.; Imbert, D.; Siffre, J.M.; Marcus, P. An in situ XPS study of sputter-deposited aluminium thin films on graphite. *Appl. Surf. Sci.* **1994**, *78*, 219–231. [CrossRef]
35. He, H.; Alberti, K.; Barr, T.L.; Klinowski, J. ESCA studies of aluminophosphate molecular sieves. *J. Phys. Chem.* **1993**, *97*, 13703–13707. [CrossRef]
36. Grigorova, E.; Khristov, M.; Peshev, P.; Nihtianova, D.; Velichkova, N.; Atanasova, G. Hydrogen sorption properties of a MgH_2–V_2O_5 composite prepared by ball milling. *Bulg. Chem. Commun.* **2013**, *45*, 280–287.
37. Shelly, L.; Schweke, D.; Zalkind, S.; Shamir, N.; Barzilai, S.; Gouder, T.; Hayun, S. Effect of U content on the activation of H2O on $Ce_{1-x}U_xO_{2+\delta}$ surfaces. *Chem. Mater.* **2018**, *30*, 8650–8660. [CrossRef]
38. Hayun, S.; Shvareva, T.Y.; Navrotsky, A. Nanoceria—Energetics of surfaces, interfaces and water adsorption. *J. Am. Ceram. Soc.* **2011**, *94*, 3992–3999. [CrossRef]
39. Uner, D.; Uner, M. Adsorption calorimetry in supported catalyst characterization: Adsorption structure sensitivity on Pt/γ-Al_2O_3. *Thermochim. Acta* **2005**, *434*, 107–112. [CrossRef]
40. Garcia-Cuello, V.; Moreno-Piraján, J.C.; Giraldo-Gutiérrez, L.; Sapag, K.; Zgrablich, G. Determination of differential enthalpy and isotherm by adsorption calorimetry. *Res. Lett. Phys. Chem.* **2008**, *2008*, 127328. [CrossRef]
41. Hayun, S.; Tran, T.; Ushakov, S.V.; Thron, A.M.; Van Benthem, K.; Navrotsky, A.; Castro, R.H.R. Experimental methodologies for assessing the surface energy of highly hygroscopic materials: The case of nanocrystalline magnesia. *J. Phys. Chem. C* **2011**, *115*, 23929–23935. [CrossRef]
42. Jia, C.; Fan, W.; Yang, F.; Zhao, X.; Sun, H.; Li, P.; Liu, L. A theoretical study of water adsorption and decomposition on low-index spinel $ZnGa_2O_4$ surfaces: Correlation between surface structure and photocatalytic properties. *Langmuir* **2013**, *29*, 7025–7037. [CrossRef] [PubMed]
43. Cai, Q.; Wang, J.G.; Wang, Y.; Mei, D. First-principles hermodynamics study of spinel $MgAl_2O_4$ surface stability. *J. Phys. Chem. C* **2016**, *120*, 19087–19096. [CrossRef]

© 2020 by the authors. Licensee MDPI, Basel, Switzerland. This article is an open access article distributed under the terms and conditions of the Creative Commons Attribution (CC BY) license (http://creativecommons.org/licenses/by/4.0/).

Article

High-Temperature Structural and Electrical Properties of BaLnCo$_2$O$_6$ Positrodes

Iga Szpunar [1,*], Ragnar Strandbakke [2,*], Magnus Helgerud Sørby [3], Sebastian Lech Wachowski [1], Maria Balaguer [4], Mateusz Tarach [4], José M. Serra [4], Agnieszka Witkowska [1], Ewa Dzik [1], Truls Norby [2], Maria Gazda [1] and Aleksandra Mielewczyk-Gryń [1,*]

1. Nanotechnology Centre A, Faculty of Applied Physics and Mathematics and Advanced Materials Centre, Gdańsk University of Technology, ul. Narutowicza 11/12, 80-233 Gdańsk, Poland; sebastian.wachowski@pg.edu.pl (S.L.W.); agnieszka.witkowska@pg.edu.pl (A.W.); ewa.dzik@pg.edu.pl (E.D.); maria.gazda@pg.edu.pl (M.G.)
2. Department of Chemistry, Centre for Materials Science and Nanotechnology, University of Oslo, FERMiO, Gaustadalléen 21, NO-0349 Oslo, Norway; truls.norby@kjemi.uio.no
3. Department for Neutron Materials Characterization, Institute for Energy Technology, Instituttveien 18, 2007 Kjeller, Norway; magnus.sorby@ife.no
4. Instituto de Tecnología Química, Universitat Politècnica de València, Consejo Superior de Investigaciones Científicas, Av. Naranjos s/n, E-46022 Valencia, Spain; mabara@upvnet.upv.es (M.B.); mata8@itq.upv.es (M.T.); jmserra@itq.upv.es (J.M.S.)
* Correspondence: iga.lewandowska@pg.edu.pl (I.S.); ragnar.strandbakke@kjemi.uio.no (R.S.); alegryn@pg.edu.pl (A.M.-G.)

Received: 7 July 2020; Accepted: 9 September 2020; Published: 11 September 2020

Abstract: The application of double perovskite cobaltites BaLnCo$_2$O$_{6-\delta}$ (Ln = lanthanide element) in electrochemical devices for energy conversion requires control of their properties at operating conditions. This work presents a study of a series of BaLnCo$_2$O$_{6-\delta}$ (Ln = La, Pr, Nd) with a focus on the evolution of structural and electrical properties with temperature. Symmetry, oxygen non-stoichiometry, and cobalt valence state have been examined by means of Synchrotron Radiation Powder X-ray Diffraction (SR-PXD), thermogravimetry (TG), and X-ray Absorption Spectroscopy (XAS). The results indicate that all three compositions maintain mainly orthorhombic structure from RT to 1000 °C. Chemical expansion from Co reduction and formation of oxygen vacancies is observed and characterized above 350 °C. Following XAS experiments, the high spin of Co was ascertained in the whole range of temperatures for BLC, BPC, and BNC.

Keywords: positrode; cobaltites; synchrotron powder diffraction; X-ray absorption spectroscopy; ceramics; thermal expansion; chemical expansion

1. Introduction

The double perovskite cobaltites BaLnCo$_2$O$_{6-\delta}$ (where Ln is a lanthanide) have been a subject of research attention in recent years due to their transport properties and possible application as electrodes in electrochemical devices for energy conversion. The substantial number of oxygen vacancies in oxygen-deficient layers of LnO$_\delta$ is believed to be beneficial for fast oxygen ion diffusion in these materials, while an overlap of Co3d and O2p orbitals in the Co-O slabs is favorable for enhanced electronic transport [1,2]. Furthermore, the flexibility of the double perovskite structure allows for substantial variations in oxygen non-stoichiometry and effective cobalt valence state, giving rise to high mixed ionic and electronic conductivity and fast surface kinetics [3].

In this study, we report the influence of temperature on structure and oxygen non-stoichiometry for a series of BaLnCo$_2$O$_{6-\delta}$ compositions. High-temperature structural Synchrotron Radiation

Powder X-ray Diffraction (SR-PXD) has been used to determine the temperature-dependent evolution of the materials' structure accompanied by the oxygen loss due to reduction (). In combination with thermogravimetry (TG), thermal and chemical unit cell expansion has been established for $BaLaCo_2O_{6-\delta}$ (BLC), $BaPrCo_2O_{6-\delta}$ (BPC), and $BaNdCo_2O_{6-\delta}$ (BNC). X-ray Absorption Spectroscopy (XAS) (room temperature) was undertaken for analysis of electronic states, with particular emphasis put on cobalt spin states analysis for BNC, BPC, and BLC.

2. Materials and Methods

The polycrystalline $BaLnCo_2O_{6-\delta}$ samples used in this study were prepared through a conventional solid-state reaction. Powders of La_2O_3 (99.99% Alfa Aesar, Ward Hill, MA, USA, preheated at 900 °C for 5 h), Pr_6O_{11} (99.99% Sigma-Aldrich, St. Louis, MO, USA), or Nd_2O_3 (99.9%, Chempur, preheated at 900 °C for 5 h) were used as lanthanide sources. Stoichiometric quantities of $BaCO_3$ (99.9% Sigma-Aldrich), Co_2O_3 (99.8% Alfa Aesar), and the respective lanthanide oxide were ground in an agate mortar and pressed into pellets at 1.5 MPa. Pellets were then annealed at 1100 °C in static air for 48 h in a tube furnace at a heating/cooling rate of 2 °C/min. $BaLaCo_2O_{6-\delta}$ was further annealed in Ar flow at 1050 °C for 24 h and in the airflow at 350 °C for 3 h to adopt a layered double perovskite structure.

Oxygen stoichiometry and oxidation state of cobalt were determined using iodometric titration at room temperature. Experimental procedure and details can be found elsewhere [4]. Approximately 15–20 mg of sample was used in this procedure.

Powder X-ray diffraction was used for the determination of sample purity, quality, and composition. X-ray diffractograms were in the 2θ angle range of 20 to 90° and were acquired using Philips X'Pert Pro diffractometer (Almelo, The Netherlands) with Cu Kα radiation. Samples were ground into powders and placed on the zero background slides for analysis.

The thermogravimetric oxidation studies were performed in synthetic air as purge gas (40 mL/min air purge gas, 20 mL/min protective N_2) atmosphere using a Netzsch Tarsus 401 thermal analyser (Selb, Germany) in the temperature range RT-900 °C, with a temperature step of 2 °C/min on both cooling and heating.

High-temperature synchrotron powder X-ray diffraction (SR-PXRD) patterns were collected at the Elettra-Synchrotron Trieste, Trieste, Italy at the Materials Characterization by X-ray Diffraction (MCX) beamline. Diffraction patterns were collected at 20 KeV energy and 2θ angle range of 1 to 35° at from RT to 1000 °C in air, the heating rate between each temperature step 3 °C/min. Experimental data were analyzed by the Rietveld method using GSAS-II software (Argonne National Laboratory, Argonne, IL, USA) [5,6].

The combination of thermogravimetric oxidation studies and high-temperature synchrotron powder X-ray diffraction was used to determine the volumetric chemical expansion coefficient. Continuous equilibrium during heating and cooling was ensured by additional thermo-gravimetric temperature ramps, where heating and cooling rates of 2°/min were compared to 3°/min, thus eliminating any uncertainties due to different heating rates applied in the two techniques (Figure S1).

UV–Vis spectra of the powders were recorded on a Varian 5000 UV–Vis–NIR spectrophotometer (Varian, Inc., Palo Alto, CA, USA) in the range of 200 to 800 nm using $BaSO_4$ as reference material and with a lamp change at 350 nm. The optical gap was then determined using the Kubelka–Munk theory.

X-ray Absorption Spectroscopy measurements were performed at the Solaris National Synchrotron Radiation Centre in Kraków, Poland. A dedicated PEEM/XAS bending magnet beamline was utilized to measure $Re-M_{4,5}$, $Co-L_{2,3}$, and O-K edges. Powders of samples were mounted on the carbon tape and placed on the Omicron plates for measurements.

3. Results and Discussion

The structure of double perovskite cobaltites has been the subject of numerous studies [7–12]. Those materials are reported to adopt both tetragonal (*P4/mmm*) [9,10,13] and orthorhombic (*Pmmm*) [9,12] structures. The difference in tetragonal and orthorhombic unit cells results from oxygen vacancies ordering along the b-axis. Figure 1 presents the differences between tetragonal and orthorhombic structures with different oxygen content. The obtained crystal structure of BaLnCo$_2$O$_{6-\delta}$ strongly depends on oxygen stoichiometry [4,7,8], as well as synthesis procedure [9]. As the difference between the two polymorphs, resulting from the oxygen vacancy ordering, is very subtle, the structural studies require very high-quality data such as synchrotron radiation X-ray powder diffraction collected with 2D detectors. Neutron diffraction, which is much more sensitive to oxygen, is a powerful complementary technique for determination of the oxygen vacancy concentrations at different sites [11,12,14–19].

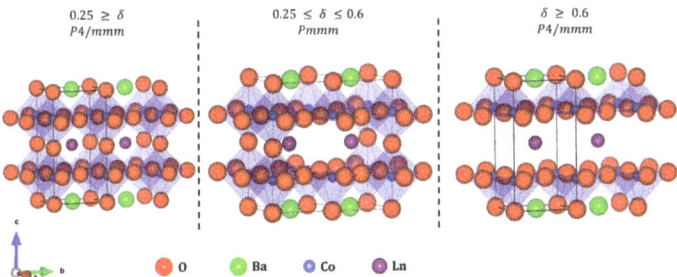

Figure 1. Unit cell of tetragonal (*P4/mmm*) and orthorhombic (*Pmmm*) BaLnCo$_2$O$_{6-\delta}$ with the different oxygen content.

Figure 2 depicts the SR-PXD patterns of BLC, BPC, and BNC as a function of temperature. No phase transitions are observed in the interval RT-1000 °C. The shift of peak positions towards lower angles is due to the chemical and thermal expansion. Rietveld refinements were performed with orthorhombic *Pmmm* structures (a$_p$ × 2a$_p$ × 2a$_p$) for all three compositions. The orthorhombic reflections are very subtle and hard to detect using methods basing on X-ray radiation. However, in our previous study [12], we showed by the combined use of SR-XRD and neutron diffraction that BLC, BPC, and BNC adopt orthorhombic symmetry at room temperature. Therefore, even though higher symmetry refinement (*P4/mmm*) is possible for BNC in this study, we follow the previous refinements of higher quality data and ascribe orthorhombic structure also to BNC. BLC and BNC showed additional minority phases of, respectively, tetragonal (*P4/mmm*) and cubic (*Pm$\bar{3}$m*) symmetry.

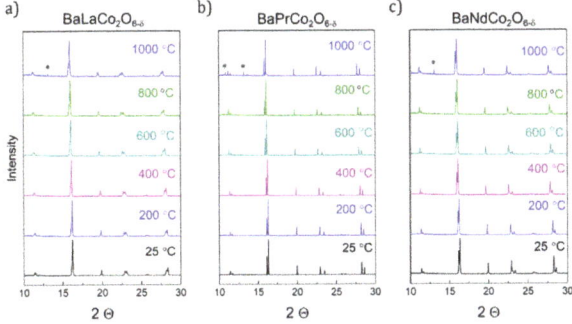

Figure 2. High-temperature Synchrotron Radiation Powder X-ray Diffraction (SR-PXD) diffractograms of BLC (**a**), BPC (**b**), and BNC (**c**). * denotes reflections from a reaction product between the sample and the silica capillary.

Figure 3 presents the room temperature SR-PXD diffraction patterns with Rietveld refinement profiles for the three compositions. The detailed results of Rietveld refinement are collected in Supplementary Information (Tables S1–S27). The lattice parameter doubling along the *c*-axis confirms A-site cation ordering, while the double *b*-parameter results from oxygen vacancy ordering [12].

BPC has been reported to form both tetragonal and orthorhombic structure, strongly depending on oxygen stoichiometry [4,6–8,20], as well as synthesis procedure [21], but is reported to be stabilized in orthorhombic structure when oxygen non-stoichiometry (δ) is between 0.25 and 0.6 [14].

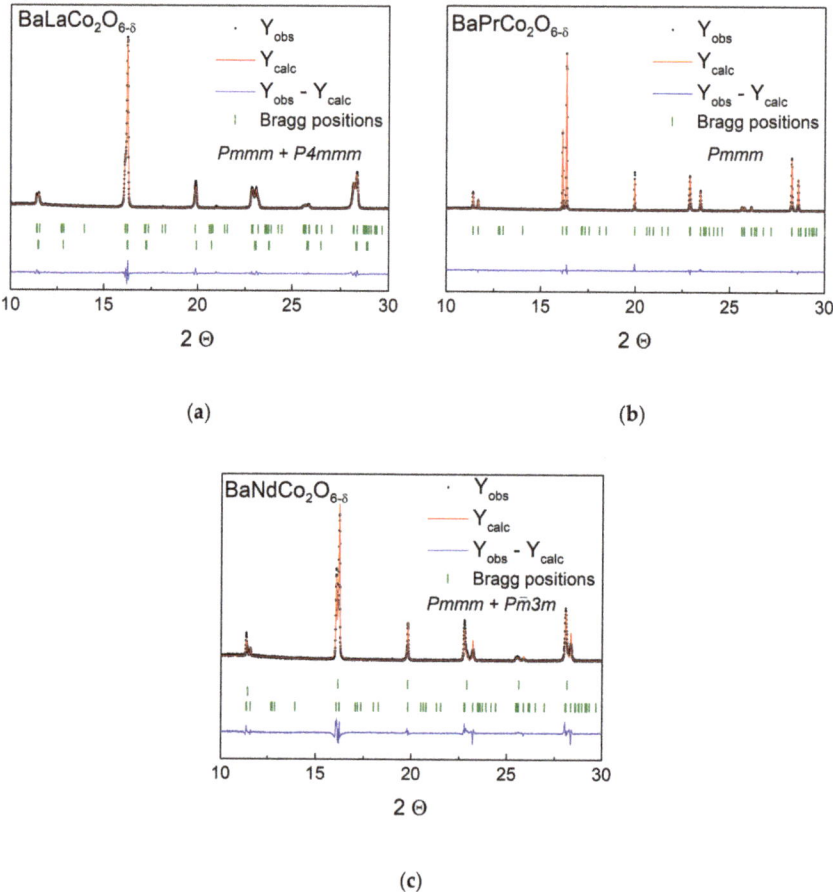

Figure 3. Observed (dotted lines), calculated (solid lines), and difference (bottom) SR-PXD for BLC (**a**), BPC (**b**), and BNC (**c**) at RT.

In this study, δ for the orthorhombic BPC structure was 0.36, according to iodometric titration results at room temperature, which is in agreement with the previous reports [20,22,23].

The temperature evolution of the lattice parameters for BNC and BPC is presented in Figure 4. We have kept the orthorhombic structure as refinement basis over the whole temperature range for all three compositions given our background data [12], and the fact that the orthorhombic phase for BNC and BPC is generally reported to be stabilized in $0.25 \leq \delta \leq 0.6$, while the tetragonal structure is adopted if the oxygen content is lower than 5.4, or higher than 5.75 [13,14,16,23]. Following this, and given that $a \neq b/2$ at all temperatures (Figure S2), our data do not support any phase transition between RT and 1000 °C for any of the compositions.

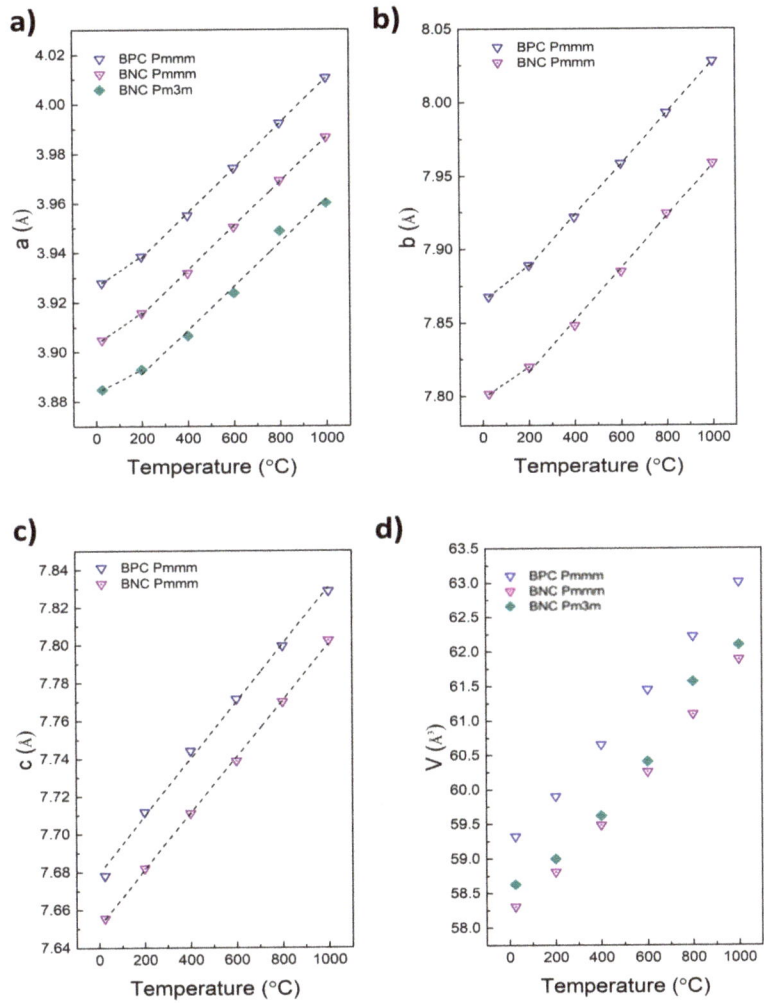

Figure 4. Temperature dependence of lattice parameters and unit cell volume of BPC and BNC. *a*-parameter (**a**) *b*-parameter (**b**) *c*-axis (**c**) unit cell volume (**d**).

The lattice parameters and the cell volumes increase linearly, but with different slopes below and above approximately 350 °C for the *a*- and *b*-parameter, where a combination of thermal and chemical expansion can be seen, as expected for reducible metal-based mixed conducting oxides [24]. The thermal expansion is related only to the inherent vibrational properties of the crystal lattice, while the chemical expansion of the *ab*-plane results from the increasing concentration of oxygen vacancies [24]. Oxidation studies of all three compositions show that δ changes very little between RT and approximately 300 °C, following the literature [16]. Table 1 reports the values of the linear expansion coefficient in two separate temperature ranges calculated as a slope of relation presented in Figure 4 for particular unit cell parameters. The lower temperature value (up to 200 °C) is a product of thermal expansion, while for higher temperature (above 200 °C) it is a combination of thermal and chemical expansion.

Table 1. Thermal expansion coefficients.

Material (Direction)	Expansion Coefficient (below 200 °C) (×10⁻⁶)	Expansion Coefficient (above 200 °C) (×10⁻⁶)
BPC (a)	15.3	23.0
BPC (b)	15.5	22.1
BPC (c)	19.7	
BNC *Pmmm* (a)	15.9	22.8
BNC *Pmmm* (b)	13.4	22.6
BNC *Pmmm* (c)	19.5	
BNC *Pm3m* (a)	11.9	22.6

The fractions of majority *Pmmm* and minority *P4/mmm* phases in BLC changes with temperature (Figure 5a). As the temperature increases, a reduction in the tetragonal phase fraction is observed, which can be correlated with oxygen loss. Oxygen vacancy ordering increases with increasing oxygen non-stoichiometry [16], gradually turning *P4/mmm* into *Pmmm* by ordering oxygen vacancies along the *b*-axis. As the orthorhombic phase for BNC and BPC is generally stabilized in $0.25 \leq \delta \leq 0.6$, the tetragonal structure is adapted if the oxygen content is lower than 5.4, or higher than 5.75 [13,14,16]. The scheme of the phase transition from tetragonal to the orthorhombic structure is presented in Figure 6. Oxygen vacancy formation in the tetragonal phase leads to the transformation of a particular unit cell to the orthorhombic structure, increasing the majority phase content. The oxygen content in the remaining tetragonal phase is thus constant, although oxygen is released simultaneously from both phases.

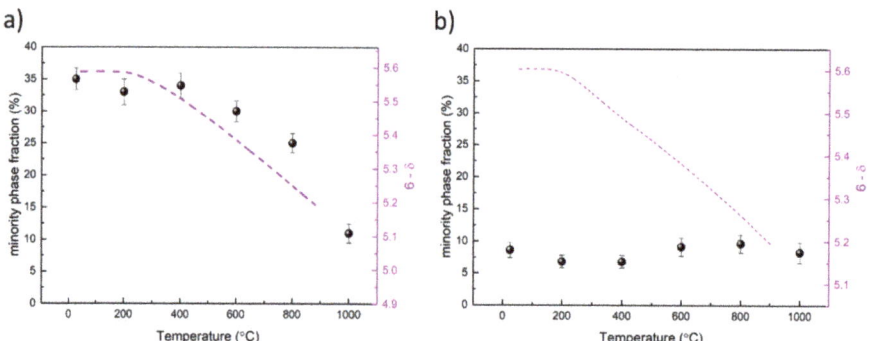

Figure 5. The fraction of the minority phases (dot) P4/mmm for BLC (**a**) and Pm3 m for BNC (**b**), and oxygen stoichiometry (dashed line) as a function of temperature. The oxygen stoichiometry changes were calculated based on thermogravimetry using titration results as a starting point.

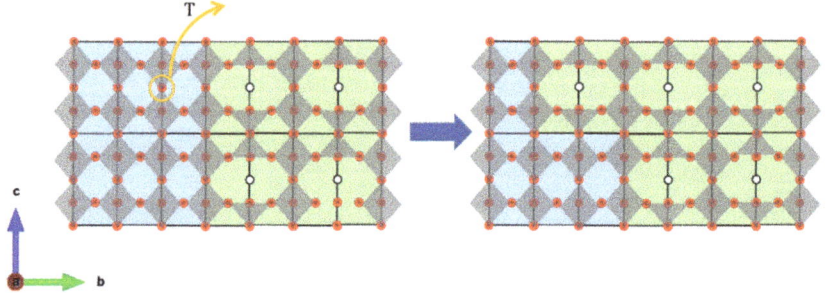

Figure 6. The scheme of the phase transition from tetragonal (blue unit cells) to orthorhombic (green unit cells) phase.

Such a relation between oxygen stoichiometry and phase composition was not observed for BNC, where a minority cubic phase fraction of ~7% was constant in the measured T-range (Figure 5b).

Oxygen loss in tetragonal BLC leads to a phase transition to the orthorhombic structure. The temperature evolution of lattice parameters (Figure 7) suggests that oxygen stoichiometry in the remaining tetragonal phase is constant in the whole temperature range. The *a*-parameter of tetragonal BLC increases linearly, while in the orthorhombic phase deviates upwards from linear relation, which is typical of chemical expansion upon reduction and confirmed by the dependence on the increasing concentration of oxygen vacancies (Figure 7b) [24]. The temperature dependence of the two-unit cell volumes shows a similar behavior as the *a*-parameters, being linear for the tetragonal phase and nonlinear for the orthorhombic one. Interestingly, both the *c*- and *b*-parameters of orthorhombic BLC change linearly with temperature, indicating anisotropic chemical expansion.

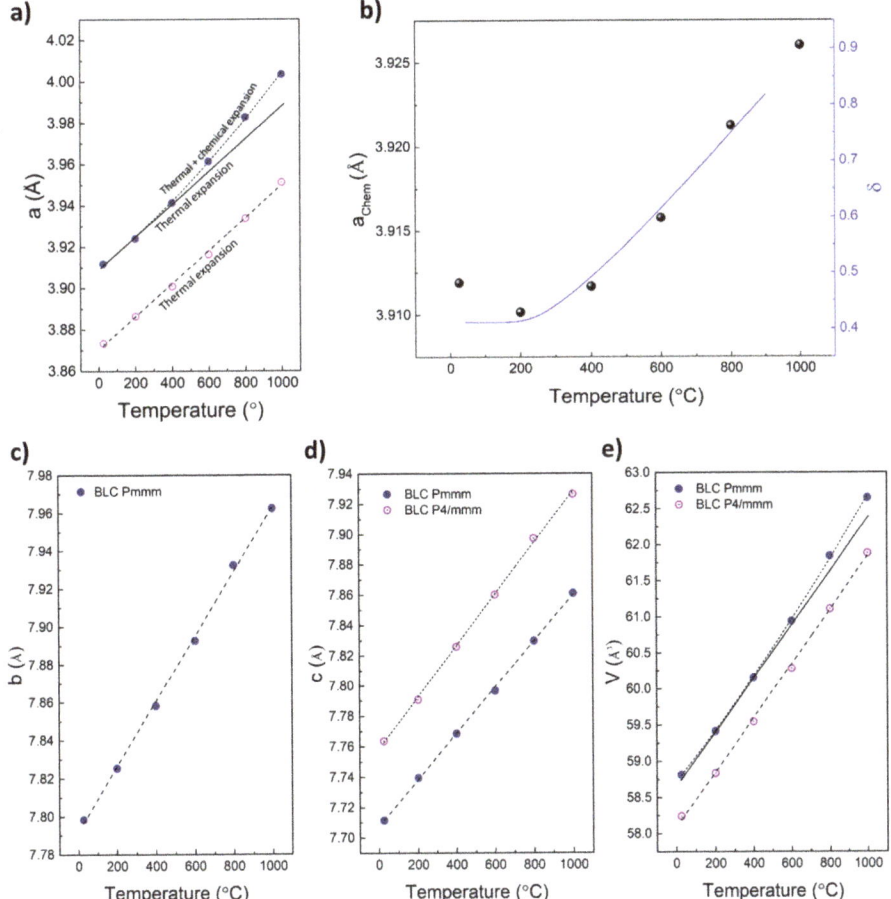

Figure 7. Temperature evolution of lattice parameters of orthorhombic and tetragonal BLC (**a,c–e**). Correlation between change in a parameter, resulting from chemical expansion, and the concentration of oxygen vacancies (**b**).

Figure 8 depicts the chemical expansion of BLC, BNC, and BPC as a relative change in orthorhombic unit cell volume vs. oxygen non-stoichiometry. The change due to the thermal expansion was subtracted from the total volume, thus the presented changes result only from the chemical expansion of the orthorhombic phase.

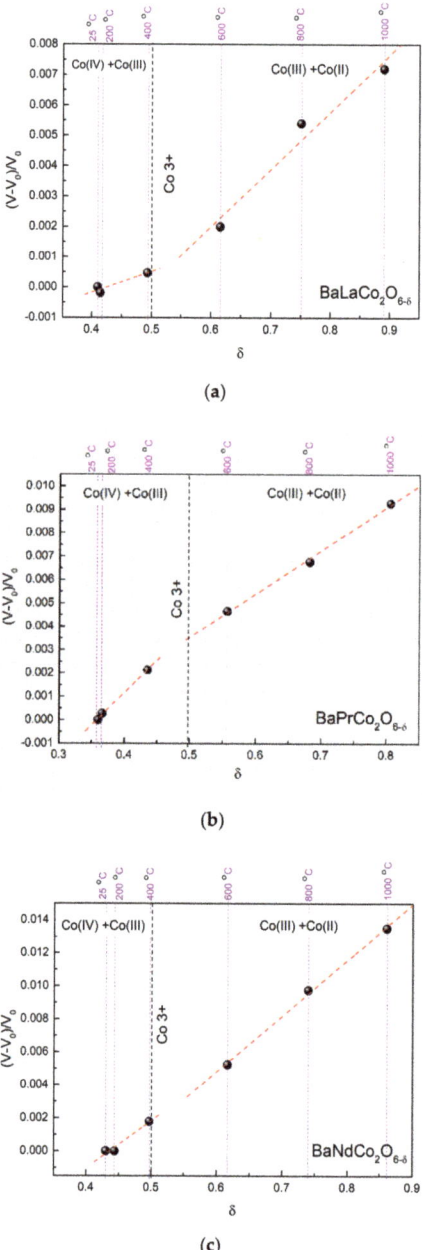

Figure 8. Relative change of unit cell volume from the chemical expansion of the orthorhombic phase as a function of oxygen non-stoichiometry for BLC (**a**), BPC (**b**), and BNC (**c**). Thermal expansion is subtracted.

As can be seen from Figures 4 and 7, the chemical expansion is anisotropic. The slope change upon oxygen loss is observed only along the *a*- and *b*-axes for BPC and BNC, and along the *a*-axis alone for BLC. The analysis of directional chemical expansion requires information of anisotropic shape and size of the cobalt ions, and localization of oxygen vacancies. In this study, the chemical expansion is analyzed non-directionally and given as total volumetric chemical expansion. The data was divided into two regions, where different chemical expansion models were applied. The chemical expansion results from both formation of oxygen vacancies and reduction of cobalt ions. Due to the different ionic sizes of Co^{2+}, Co^{3+}, and Co^{4+}, the chemical expansion in the two regimes should be described separately. However, in both cases, the chemical reaction driving the chemical expansion is oxygen exchange. Therefore, the chemical expansion coefficient (β) always consists of two parts: one originating from the formation of oxygen vacancies ($\beta_{V_O^{\bullet\bullet}}$) and the second from cobalt reduction (β_{Co}) (Equation (1)). The β_{Co} is doubled because there are two reduced cobalt ions per one oxygen vacancy. The total chemical expansion coefficient upon reduction can be determined experimentally, analyzing the temperature evolution of unit cell volume minus the effect of thermal expansion. Then, two components of chemical expansion can be considered separately.

$$\beta = \beta_{V_O^{\bullet\bullet}} + 2\beta_{Co} \tag{1}$$

The obtained values of the chemical expansion coefficients are given in Table 2. The derivations are given in the Appendix A.

Table 2. Chemical expansion coefficients for BLC, BPC, and BNC.

Material	Oxidation Range	β_{red}	β_{Co}	$\beta_{V_O^{\bullet\bullet}}$
BLC	$\delta < 0.5$	$\beta_{red} = 0.007$	$\beta_{Co\ 4\to 3} = 0.355$	$\beta_{V_O^{\bullet\bullet}} = -0.703$
	$\delta > 0.5$	$\beta_{red} = 0.019$	$\beta_{Co\ 3\to 2} = 0.822$	$\beta_{V_O^{\bullet\bullet}} = -1.624$
BNC	$\delta < 0.5$	$\beta_{red} = 0.029$	$\beta_{Co\ 4\to 3} = 0.353$	$\beta_{V_O^{\bullet\bullet}} = -0.677$
	$\delta > 0.5$	$\beta_{red} = 0.034$	$\beta_{Co\ 3\to 2} = 0.822$	$\beta_{V_O^{\bullet\bullet}} = -1.610$
BPC	$\delta < 0.5$	$\beta_{red} = 0.028$	$\beta_{Co\ 4\to 3} = 0.361$	$\beta_{V_O^{\bullet\bullet}} = -0.694$
	$\delta > 0.5$	$\beta_{red} = 0.019$	$\beta_{Co\ 3\to 2} = 0.822$	$\beta_{V_O^{\bullet\bullet}} = -1.625$

As shown, Co^{3+} exhibits a high spin (HS) state in BLC. This is in line with previous studies [23,25] stating that the HS of Co^{3+} and Co^{2+} is energetically more favorable. Thus, the ionic radii for cobalt at HS were used for calculations, with values of 0.53 Å, 0.61 Å, and 0.745 Å for Co^{4+}, Co^{3+}, and Co^{2+}, respectively [26,27]. In both regimes, the chemical expansion coefficient β_{red} is positive, meaning that the reduced oxygen content leads to unit cell volume increase. However, the total chemical expansion is a result of two separate effects. The reduction of cobalt ions gives a positive contribution to the chemical expansion. The negative value of the chemical expansion coefficient related to oxygen vacancy formation gives the information that the oxygen vacancy formation itself leads to the unit cell contraction.

Comparable values of the chemical expansion coefficient were reported in previous studies on perovskite oxides [19,28–30]. The most studied system is $La_{1-x}Sr_xCo_yFe_{1-y}O_{3-\delta}$ (LSCF) [24], where the total chemical expansion coefficient ranges from 0.022 for x = 0.4 and y = 0.8 to 0.059 for x = 0.5 and y = 0. The reported values refer to high temperatures, thus it should be compared to the regime $\delta > 0.5$.

The approach of separating the effect of cation reduction and oxygen vacancy formation on total expansion is still uncommon and based mostly on DFT studies, but the available studies confirm that the oxygen vacancies cause unit cell contraction [24]. The $\beta_{V_O^{\bullet\bullet}}$ significantly differs in the two investigated regimes. The determined values of $\beta_{V_O^{\bullet\bullet}}$ were used to calculate the volume of oxygen vacancy, according to Equation (A27) in the Appendix A. In this case, the oxygen vacancy size is a measure of lattice deformation.

Figures 9–11 present the results of the X-ray absorption studies (XAS) of the as-prepared samples at room temperature for the Co $L_{2,3}$-edges, Ba $M_{4,5}$-edges, O K-edge, and Pr $M_{4,5}$-edges spectra, respectively. Figure 7a shows the XAS data for Co $L_{2,3}$ and Ba $M_{4,5}$ edges collected for all studied compounds and reference samples. The intensity of the white lines (WL) attributed to both cobalt and barium orbitals varies between compositions. With the decrease in ionic radius from lanthanum (1.172 Å) to gadolinium (1.078 Å) (for six-fold coordination [25,26]), the WL of cobalt increases while barium WL intensity decreases, indicating that the decrease of the ionic radius is causing an increase in the density of the unoccupied electron 3d Co states and a decrease in the density of unoccupied electronic 4f Ba states. This relation is accompanied by the differences observed on the lower energy slope of Co L-edges (especially of L_3-edge).

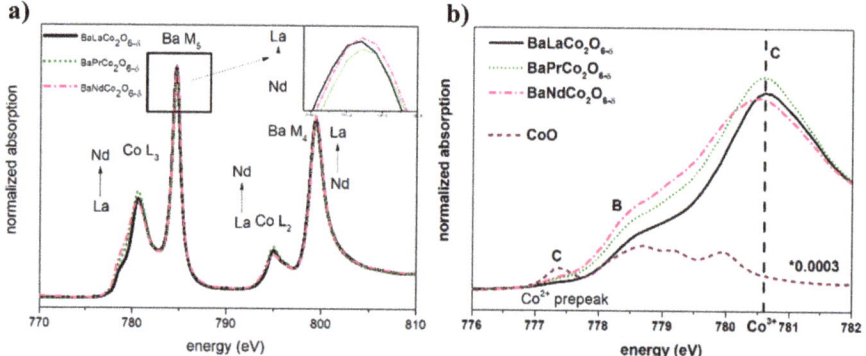

Figure 9. (a) X-ray Absorption Spectroscopy (XAS) spectrum of Co $L_{2,3}$ and Ba $M_{4,5}$ edges. (b) XANES (X-ray Absorption Near Edge Structure) spectra of Co L_3-edge for studied materials and reference sample (CoO).

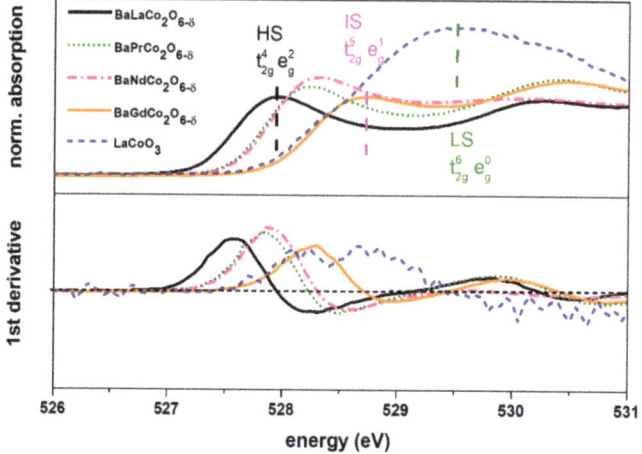

Figure 10. X-ray absorption spectra of oxygen K-edge in the Co 3d bands region. For comparison, intermediate spin (BaGdCo$_2$O$_6$) and low spin (LaCoO$_3$) reference samples are included.

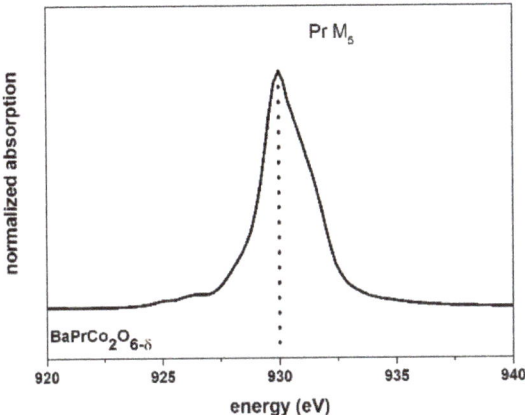

Figure 11. XANES praseodymium M_5 edge for $BaPrCo_2O_6$.

Figure 9 presents the Co L_3-edge (normalized XANES) for all investigated compositions along with the reference (CoO). The reference CoO spectrum has been scaled down to highlight the correlation between the pre-edge structure feature (A and B) and edge intensity (C). An almost undetectable pre-peak (A) in the spectra of the investigated samples suggests a negligible concentration of Co^{2+} species. The main line attributed to Co^{3+} (C) is present at the same energy for BLC, BPC, and BNC (780.6 eV). Note that there are no significant differences in the oxygen non-stoichiometry of the measured samples (Table 3).

Table 3. Co average oxidation state and oxygen non-stoichiometry obtained with iodometric titration BLC, BPC, and BNC for the samples before XAS studies.

Nominal Composition	Co Average Oxidation State	Oxygen Non-Stoichiometry
$BaLaCo_2O_6$	3.2880	0.2100
$BaPrCo_2O_6$	3.2540	0.2460
$BaNdCo_2O_6$	3.2332	0.2668

The shoulder on B—reflecting Co^{2+}—is visible for all samples; however, it is more extensive for BPC and BNC than for BLC. This may suggest that the more oxidized BLC—with more octahedrally coordinated Co—exhibits a more HS character. Such a relation has been previously reported for samples with fixed $\delta = 0.5$ [31]. The complementary to XAS cobalt L-edges data is oxygen K-edge analysis (Figure 10).

Oxygen K-edge is sensitive to the density of empty cobalt 3d t_{2g} and e_g states through hybridization with oxygen 2p orbitals. $LaCoO_3$ has been chosen as a reference for the analysis of O K-edge because it contains solely Co^{3+} in a low spin (LS) state ($t_{2g}^6 e_g^0$, S = 0) [32] along with $BaGdCo_2O_6$ which was previously reported to exhibit IS [31]. The comparison of recorded spectra supports the analysis of cobalt L-edges. Evaluation of the pre-peak position (between 526 and 531 eV) and comparison with the references reveals the highest density of 3d t_{2g} empty states (in the octahedral environment) for the BLC sample. On the other hand, the BGC spectra indicate a lower density of unoccupied t_{2g} state in comparison to the samples with larger lanthanides, which is partially caused by higher oxygen non-stoichiometry and the lower ionic radius of Gd with respect to La, Pr, and Nd. These results suggest that in the cobaltites with Pr and Nd, mixed HS/IS state of cobalt has been detected. This means that in the case of BLC exhibiting HS, the crystal field splitting is lower than in the case of BPC and BNC.

Figure 11 presents the Pr M_5 edge recorded for the BPC sample. The WL position for this spectrum (930 eV) suggests the dominance of Pr^{3+}. Herrero-Martin et al. reported the WL shift towards lower energies for higher Pr^{3+} content relative to Pr^{4+}. In their simulation, the WL peak for 85:15 Pr^{3+} to Pr^{4+} content in $Pr_{0.5}Ca_{0.5}CoO_3$ should be observed for ~934 eV, while the experimental data for Pr_2O_3 shows the WL at 931 eV [33].

UV–Vis (Figure 12) absorption experiments were performed to study the influence of the lanthanide dopant on the bandgap. The bandgap normally refers to the energy difference between valence and conduction bands. The optical band-gap energy is thus normally comparable to the thermal bandgap related to the formation of electron–hole pairs. Such an intrinsic formation of electrons and holes is the Co charge disproportionation reaction:

$$2Co^{3+} \rightarrow Co^{2+} + Co^{4+} \qquad (2)$$

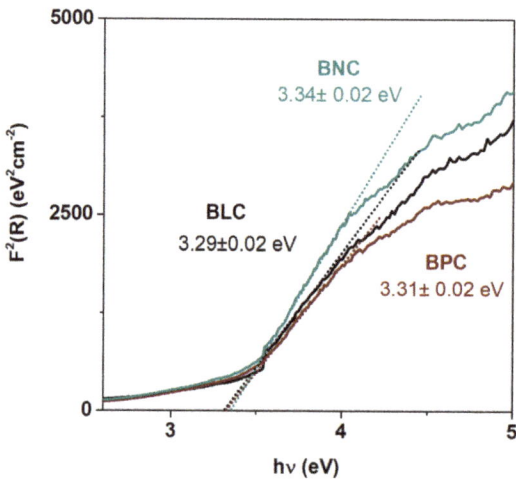

Figure 12. Plot of Kubelka–Munk function vs. energy for all measured compositions as well as values of energy gap at RT for BLC, BPC, and BNC.

The Kubelka–Munk function as given below was used to determine the optical band gap of investigated materials.

$$F(R) = \frac{(1-R)^2}{2R} \qquad (3)$$

where R is diffuse reflectance emanating from an infinitely thick sample [34].

This evaluation involves the plotting of the obtained $(h\nu F(RN)^2)$ as a function of $h\nu$. The bandgap E_g can be obtained by extrapolating a tangent line drawn in the point of inflexion of the curve to zero, i.e., the point of intersection with the $h\nu$ horizontal axis. Figure 8 shows the room temperature UV–Vis absorption spectra of the BLC, BNC, and BPC.

As observed, the lanthanide is not causing any shift of the absorption edge. The optical bandgap values for all investigated samples are similar and around 3.3 eV within the error range. The obtained values for the optical band gap are surprisingly high for the black powders. However, as the investigated materials may be considered as degenerate semiconductors, their color can be a result of either intra-band electronic transition or the transitions related to the in-gap states presented. It is reported that electrons and holes are transferred in a partially filled, degenerate O 2p Co 3d band, and that it is thus located at the top of the valence band. The measured optical bandgap may therefore not represent the electrical band-gap. The presence of band states within the bandgap of partially filled anti-bonding σ *-bands, wherein electronic conduction can occur, is previously reported for double perovskite cobaltites [35–37].

4. Conclusions

We have investigated the electrical and structural properties of chosen BaLnCo$_2$O$_{6-\delta}$ (Ln = La, Pr, and Nd) double perovskites. All measured compositions (BLC, BNC, and BPC) were refined to orthorhombic (Pmmm) structure up to 1000 °C. Moreover, BLC and BNC showed additional tetragonal (P4/mmm) and cubic (Pm$\overline{3}$m) minority phases, respectively. The thermal evolution of the unit cells shows that after subtracting expansion from thermal lattice vibrations, partial reduction of cobalt and formation of oxygen vacancies gives positive and negative contributions, respectively, to the chemical expansion. The spectroscopic studies show that cobalt is present only in the intermediate or high-spin state for all compositions at room temperature. The optical bandgap is characterized, showing values of ~3.3 eV, which is not consistent with high electronic conductivity. We ascribe this to partially filled antibonding states at the valence band maximum with high mobility for electrons and electron holes [36,37].

Supplementary Materials: The following are available online at http://www.mdpi.com/1996-1944/13/18/4044/s1, Figure S1: The temperature evolution of oxygen stoichiometry in BaGdCo$_2$O$_{6-\delta}$ with different heating rates, Figure S2: The temperature evolution of unit cell parameters a and b/2, Table S1: BaLaCo$_2$O$_{6-\delta}$ goodness of fit, Table S2: Atomic coordinates and U$_{iso}$ refined for orthorhombic phase of BaLaCo$_2$O$_{6-\delta}$. At room temperature, Table S3: Atomic coordinates and U$_{iso}$ refined for orthorhombic phase of BaLaCo$_2$O$_{6-\delta}$. At 200 °C, Table S4: Atomic coordinates and U$_{iso}$ refined for orthorhombic phase of BaLaCo$_2$O$_{6-\delta}$. At 400°C, Table S5: Atomic coordinates and U$_{iso}$ refined for orthorhombic phase of BaLaCo$_2$O$_{6-\delta}$. At 600 °C, Table S6: Atomic coordinates and U$_{iso}$ refined for orthorhombic phase of BaLaCo$_2$O$_{6-\delta}$. At 800°C, Table S7: Atomic coordinates and U$_{iso}$ refined for orthorhombic phase of BaLaCo$_2$O$_{6-\delta}$. At 1000 °C, Table S8: Atomic coordinates and U$_{iso}$ refined for tetragonal phase of BaLaCo$_2$O$_{6-\delta}$ at room temperature, Table S9: Atomic coordinates and U$_{iso}$ refined for tetragonal phase of BaLaCo$_2$O$_{6-\delta}$ at 200 °C, Table S10: Atomic coordinates and U$_{iso}$ refined for tetragonal phase of BaLaCo$_2$O$_{6-\delta}$ at 400 °C, Table S11: Atomic coordinates and U$_{iso}$ refined for tetragonal phase of BaLaCo$_2$O$_{6-\delta}$ at 600 °C, Table S12: Atomic coordinates and U$_{iso}$ refined for tetragonal phase of BaLaCo$_2$O$_{6-\delta}$ at 800 °C, Table S13: Atomic coordinates and U$_{iso}$ refined for tetragonal phase of BaLaCo$_2$O$_{6-\delta}$ at 1000 °C, Table S14: BaPrCo$_2$O$_{6-\delta}$ goodness of fit, Table S15: Atomic coordinates and U$_{iso}$ refined for orthorhombic phase of BaPrCo$_2$O$_{6-\delta}$ at room temperature, Table S16: Atomic coordinates and U$_{iso}$ refined for orthorhombic phase of BaPrCo$_2$O$_{6-\delta}$ at 200 °C, Table S17: Atomic coordinates and U$_{iso}$ refined for orthorhombic phase of BaPrCo$_2$O$_{6-\delta}$ at 400 °C, Table S18: Atomic coordinates and U$_{iso}$ refined for orthorhombic phase of BaPrCo$_2$O$_{6-\delta}$ at 600 °C, Table S19: Atomic coordinates and U$_{iso}$ refined for orthorhombic phase of BaPrCo$_2$O$_{6-\delta}$ at 800 °C, Table S20: Atomic coordinates and U$_{iso}$ refined for orthorhombic phase of BaPrCo$_2$O$_{6-\delta}$ at 1000 °C, Table S21: BaNdCo$_2$O$_{6-\delta}$ goodness of fit, Table S22: Atomic coordinates and U$_{iso}$ refined for orthorhombic phase of BaNdCo$_2$O$_{6-\delta}$ at room temperature, Table S23: Atomic coordinates and U$_{iso}$ refined for orthorhombic phase of BaNdCo$_2$O$_{6-\delta}$ at 200 °C, Table S24: Atomic coordinates and U$_{iso}$ refined for orthorhombic phase of BaNdCo$_2$O$_{6-\delta}$ at 400 °C, Table S25: Atomic coordinates and U$_{iso}$ refined for orthorhombic phase of BaNdCo$_2$O$_{6-\delta}$ at 600 °C, Table S26: Atomic coordinates and U$_{iso}$ refined for orthorhombic phase of BaNdCo$_2$O$_{6-\delta}$ at 800 °C, Table S27: Atomic coordinates and U$_{iso}$ refined for orthorhombic phase of BaNdCo$_2$O$_{6-\delta}$ at 1000 °C.

Author Contributions: Conceptualization, R.S., S.L.W., and A.M.-G.; Funding acquisition, R.S., M.H.S., J.M.S., A.M.-G., and T.N.; Investigation, I.S., S.L.W., M.B., A.W., M.G., E.D., and A.M.-G.; Methodology, R.S., M.H.S., S.L.W., and A.M.-G.; Supervision, J.M.S., M.G., E.D., A.M.-G., and T.N.; Writing—original draft, I.S., R.S., S.L.W., and A.M.-G.; Writing—review and editing, I.S., R.S., M.H.S., M.B., M.T., J.M.S., A.W., M.G., E.D., A.M.-G. and T.N. All authors have read and agreed to the published version of the manuscript.

Funding: The research has been supported by the National Science Centre Poland (2016/22/Z/ST5/00691), the Spanish Ministry of Science and Innovation (PCIN-2017-125, RTI2018-102161 and IJCI-2017-34110), and the Research Council of Norway (Grant n° 272797 "GoPHy MiCO") through the M-ERA.NET Joint Call 2016. We acknowledge the CERIC-ERIC Consortium for the access to MCX beamline at Elettra Sinchrotrone Trieste (proposal no 20187079). We also acknowledge Solaris National Radiation Centre Poland for access to the XAS/PEEM beamline (proposal no 181MS001).

Conflicts of Interest: The authors declare no conflict of interest.

Appendix A

Appendix A.1 Chemical Expansion of Cobalt Ions upon Reduction

Using a Kröger–Vink-compatible notation, the following description of Co ions can be introduced (Table A1) [35]:

Table A1. Cobalt ions notation.

Ion	Notation
$Co^{3.5+}$	Co_{Co}^{X}
Co^{4+}	$Co_{Co}^{\frac{1}{2}\bullet}$
Co^{3+}	$Co_{Co}^{\frac{1}{2}/}$
Co^{2+}	$Co_{Co}^{\frac{3}{2}/}$

Two reaction equations linking the reduction of cobalt and the formation of oxygen vacancies can be written. One would be the reduction of Co^{4+} to Co^{3+} associated with vacancy formation (Equation (A1)), and the second can be formulated similarly for the reduction of Co^{3+} to Co^{2+} (Equation (A2)).

$$2\,Co_{Co}^{\frac{1}{2}\bullet} + O_O^X \rightarrow 2\,Co_{Co}^{\frac{1}{2}/} + v_O^{\bullet\bullet} + \frac{1}{2}O_{2\,(g)} \tag{A1}$$

$$2\,Co_{Co}^{\frac{1}{2}/} + O_O^X \rightarrow 2\,Co_{Co}^{\frac{3}{2}/} + v_O^{\bullet\bullet} + \frac{1}{2}O_{2\,(g)} \tag{A2}$$

The regimes can be distinguished with regard to the presence of Co oxidation states in the material: oxidized state for $\delta < 0.5$ and reduced state for $\delta > 0.5$. In the former $\left[Co_{Co}^{\frac{1}{2}\bullet}\right] + \left[Co_{Co}^{\frac{1}{2}/}\right] \gg \left[Co_{Co}^{\frac{3}{2}/}\right]$ and in the latter $\left[Co_{Co}^{\frac{1}{2}/}\right] + \left[Co_{Co}^{\frac{3}{2}/}\right] \gg \left[Co_{Co}^{\frac{1}{2}\bullet}\right]$. The transition value of δ is 0.5, where the average cobalt oxidation state is 3.0.

The chemical expansion of cobalt results from the difference in size between cobalt oxidation states, and it is defined as

$$\beta_{Co} = \frac{1}{\Delta\delta} \cdot \frac{\Delta V_{Co}}{V_{Co_0}} \tag{A3}$$

where $\Delta\delta$ denotes the change of oxygen stoichiometry, which can be detailed as

$$\beta_{Co} = \frac{1}{\delta - \delta_0} \cdot \frac{V_{Co(\delta)} - V_{Co_0}}{V_{Co_0}} \tag{A4}$$

Here, $V_{Co(\delta)}$ is the average volume occupied by Co at any given δ and V_{Co_0} is the average volume occupied by Co at room temperature (δ_0).

Appendix A.2 Reduction from Co^{4+} to Co^{3+}

The average cobalt oxidation state (Co^{AVG}) can be calculated from

$$2Co^{AVG} = 2\left[Co_{Co}^{\frac{3}{2}/}\right] + 3\left[Co_{Co}^{\frac{1}{2}/}\right] + 4\left[Co_{Co}^{\frac{1}{2}\bullet}\right] \tag{A5}$$

In the oxidized regime, this gives

$$Co^{AVG} = \frac{3}{2}\left[Co_{Co}^{\frac{1}{2}/}\right] + 2\left[Co_{Co}^{\frac{1}{2}\bullet}\right] \tag{A6}$$

Co^{AVG} can be related to oxygen non-stoichiometry according to

$$Co^{AVG} = 3.5 - \delta \tag{A7}$$

The concentration of cobalt ions in the oxidized regime is given by

$$\left[Co_{Co}^{\frac{1}{2}\bullet}\right] + \left[Co_{Co}^{\frac{1}{2}/}\right] = 2 \tag{A8}$$

The combination of Equations (A6)–(A8) gives the relations between the concentration of Co^{3+} and Co^{4+} with oxygen non-stoichiometry (Equation (A9)) in the oxidized regime.

$$\left[Co_{Co}^{\frac{1}{2}\bullet}\right] = 1 - 2\delta \tag{A9}$$

$$\left[Co_{Co}^{\frac{1}{2}/}\right] = 2\delta + 1 \tag{A10}$$

The volume of cobalt can be described as the weighted arithmetic mean of Co^{4+} and Co^{3+} volume, where the concentration of each species is a weight (Equation (A11)).

$$V_{Co} = \frac{\left[Co_{Co}^{\frac{1}{2}\bullet}\right]\cdot V_{Co^{4+}} + \left[Co_{Co}^{\frac{1}{2}/}\right] V_{Co^{3+}}}{2} \tag{A11}$$

The average cobalt volume as a function of δ (Equation (A10)) can be obtained by including the Equations (A9) and (A10) to the Equation (A11).

$$V_{Co}(\delta) = \frac{(1-2\delta)\cdot V_{Co^{4+}} + (2\delta+1)\cdot V_{Co^{3+}}}{2} \tag{A12}$$

Similarly, the average volume occupied by Co at room temperature relates to δ_0:

$$V_{Co_0} = \frac{(1-2\delta_0)\cdot V_{Co^{4+}} + (2\delta_0+1)\cdot V_{Co^{3+}}}{2} \tag{A13}$$

Inserting Equations (A12) and (A13) into Equation (A5) leads to the expression on chemical expansion coefficient of Co^{4+} to Co^{3+} reduction (Equations (A14) and (A15)),

$$\beta_{Co\,4\to 3} = \frac{1}{\Delta\delta}\cdot\frac{2((\delta_0-\delta)\cdot V_{Co^{4+}} + (\delta-\delta_0)\cdot V_{Co^{3+}})}{(1-2\delta_0)\cdot V_{Co^{4+}} + (2\delta+1)\cdot V_{Co^{3+}}} \tag{A14}$$

$$\beta_{Co\,4\to 3} = \frac{2\cdot(V_{Co^{3+}} - V_{Co^{4+}})}{(1-2\delta_0)\cdot V_{Co^{4+}} + (2\delta_0+1)\cdot V_{Co^{3+}}} \tag{A15}$$

Cobalt volume can be calculated as sphere volume, leading to the Equation (A16).

$$\beta_{Co\,4\to 3} = \frac{2\cdot(r_{Co^{3+}}^3 - r_{Co^{4+}}^3)}{(1-2\delta_0)\cdot r_{Co^{4+}}^3 + (2\delta_0+1)\cdot r_{Co^{3+}}^3} \tag{A16}$$

Appendix A.3 Reduction of Co^{3+} to Co^{2+}

The analogous consideration can be made in the reduced regime. The average cobalt oxidation state is given with Equation (A17) and cobalt ion concentrations in the reduced regime are given by Equation (A18).

$$Co^{AVG} = \frac{3}{2}\cdot\left[Co_{Co}^{\frac{1}{2}/}\right] + \left[Co_{Co}^{\frac{3}{2}/}\right] \tag{A17}$$

$$\left[Co_{Co}^{\frac{1}{2}/}\right] + \left[Co_{Co}^{\frac{3}{2}/}\right] = 2 \tag{A18}$$

The relation between the concentration of cobalt species with δ is given by Equations (A19) and (A20).

$$\left[Co_{Co}^{\frac{1}{2}/}\right] = 3 - 2\delta \tag{A19}$$

$$\left[Co_{Co}^{\frac{3}{2}/}\right] = 2\delta - 1 \tag{A20}$$

In the reduced state, the cobalt volume is also a weighted mean of Co^{3+} and Co^{2+} volume (Equation (A21)).

$$V_{Co} = \frac{\left[Co_{Co}^{\frac{3}{2}/}\right] \cdot V_{Co^{3+}} + \left[Co_{Co}^{\frac{3}{2}/}\right] \cdot V_{Co^{2+}}}{2} \tag{A21}$$

Including Equations (A19) and (A20) to Equations (A21) and (A22) gives the relation between cobalt average volume and oxygen non-stoichiometry.

$$V_{Co}(\delta) = \frac{(3-2\delta) \cdot V_{Co^{3+}} + (2\delta-1) \cdot V_{Co^{2+}}}{2} \tag{A22}$$

With the Equation (A21) the chemical expansion coefficient of Co^{3+} to Co^{2+} reduction can be calculated:

$$\beta_{Co\,3\to 2} = \frac{1}{\delta - \delta_0} \cdot \frac{\Delta V_{Co}}{V_{Co_0}} \tag{A23}$$

In this regime, $\delta_0 = 0.5$ and $V_{Co_0} = V_{Co^{3+}}$.

$$\beta_{Co\,3\to 2} = \frac{1}{\delta - 0.5} \cdot \frac{(1-2\delta) \cdot V_{Co^{3+}} + (2\delta-1) \cdot V_{Co^{2+}}}{2 \cdot V_{Co^{3+}}} \tag{A24}$$

$$\beta_{Co\,3\to 2} = \frac{1}{\delta - 0.5} \cdot \frac{(0.5-\delta) \cdot V_{Co^{3+}} + (\delta-0.5) \cdot V_{Co^{2+}}}{V_{Co^{3+}}} \tag{A25}$$

$$\beta_{Co\,3\to 2} = \frac{V_{Co^{2+}} - V_{Co^{3+}}}{V_{Co^{3+}}} \tag{A26}$$

Equation (A24) is equivalent to Equation (A4) with $\delta = 1.5$, and where all Co is in oxidation state 2+. The volume of cobalt can be related to the cobalt ionic radius, which leads to Equation (A26).

$$\beta_{Co\,3\to 2} = \frac{r_{Co^{2+}}^3 - r_{Co^{3+}}^3}{r_{Co^{3+}}^3} \tag{A27}$$

Appendix A.4 Chemical Expansion of Oxygen Vacancies Formation

The expression of the chemical expansion coefficient of oxygen vacancies is the same in both δ ranges and may be defined with the Equation (A28).

$$\beta_{v_O^{\bullet\bullet}} = \frac{1}{\Delta\delta} \cdot \frac{\Delta V_O}{V_{O_0}} \tag{A28}$$

As a V_O is the average volume of oxygen site volume, which can also be calculated as weighted arithmetic means of oxygen ions and oxygen vacancies volume (Equation (A29)).

$$V_O = \frac{\left[O_O^x\right] \cdot V_{O_O^x} + \left[v_O^{\bullet\bullet}\right] \cdot V_{v_O^{\bullet\bullet}}}{6} \tag{A29}$$

The molar concentration of oxygen vacancies is by definition equal to δ, thus the relation between the volume of oxygen site and δ can be written (Equation (A30)).

$$V_O = \frac{(6-\delta) \cdot V_{O_O^x} + \delta \cdot V_{v_O^{\bullet\bullet}}}{6} \tag{A30}$$

Substituting Equation (A27) to Equation (A25), the expression on the chemical expansion coefficient of oxygen vacancies is obtained (Equations (A31) and (A32)). The value of δ_0 is now equivalent to δ at RT.

$$\beta_{v_O^{\bullet\bullet}} = \frac{1}{\delta - \delta_0} \cdot \frac{(\delta_0 - \delta) \cdot V_{O_O^x} + (\delta - \delta_0) \cdot V_{V_O^{\bullet\bullet}}}{(6 - \delta_0) \cdot V_{O_O^x} + \delta_0 \cdot V_{V_O^{\bullet\bullet}}} \quad (A31)$$

$$\beta_{v_O^{\bullet\bullet}} = \frac{V_{V_O^{\bullet\bullet}} - V_{O_O^x}}{(6 - \delta_0) \cdot V_{O_O^x} + \delta_0 \cdot V_{V_O^{\bullet\bullet}}} \quad (A32)$$

Subtracting the calculated values of chemical expansion coefficient upon cobalt reduction from the total value, the chemical expansion coefficient of oxygen vacancies formation may be obtained; knowing the ionic radii of the oxygen ion, the size of oxygen vacancy can be determined.

References

1. Kim, J.-H.; Manthiram, A. LnBaCo$_2$O$_{5+\delta}$ Oxides as cathodes for intermediate-temperature solid oxide fuel cells. *J. Electrochem. Soc.* **2008**, *155*, B385. [CrossRef]
2. Tarancón, A.; Marrero-López, D.; Peña-Martínez, J.; Ruiz-Morales, J.C.; Núñez, P. Effect of phase transition on high-temperature electrical properties of GdBaCo$_2$O$_{5+x}$ layered perovskite. *Solid State Ion.* **2008**, *179*, 611–618. [CrossRef]
3. Kim, G.; Wang, S.; Jacobson, A.J.; Reimus, L.; Brodersen, P.; Mims, C.A. Rapid oxygen ion diffusion and surface exchange kinetics in PrBaCo$_2$O$_{5+x}$ with a perovskite related structure and ordered a cations. *J. Mater. Chem.* **2007**, *17*, 2500–2505. [CrossRef]
4. Szpunar, I.; Wachowski, S.; Miruszewski, T.; Dzierzgowski, K.; Górnicka, K.; Klimczuk, T.; Sørby, M.H.; Balaguer, M.; Serra, J.M.; Strandbakke, R.; et al. Electric and magnetic properties of lanthanum barium cobaltite. *J. Am. Ceram. Soc.* **2020**, *103*, 1809–1818. [CrossRef]
5. Larson, A.C.; Von, R.B.; Lansce, D. *General Structure Analysis System*; Los Alamos National Laboratory: Los Alamos, NM, USA, 2000.
6. Toby, B.H. EXPGUI, a graphical user interface for GSAS. *J. Appl. Crystallogr.* **2001**, *34*, 210–213. [CrossRef]
7. Conder, K.; Podlesnyak, A.; Pomjakushina, E.; Stingaciu, M. Layered cobaltites: Synthesis, oxygen nonstoichiometry, transport and magnetic properties. *Acta Phys. Pol. A* **2007**, *111*, 7–14. [CrossRef]
8. Fauth, F.; Suard, E.; Caignaert, V.; Domengès, B.; Mirebeau, I.; Keller, L. Interplay of structural, magnetic and transport properties in thelayered Co-based perovskite LnBaCo$_2$O$_5$ (Ln = Tb, Dy, Ho). *Eur. Phys. J. B* **2001**, *21*, 163–174. [CrossRef]
9. Streule, S.; Podlesnyak, A.; Mesot, J.; Medarde, M.; Conder, K.; Pomjakushina, E.; Mitberg, E.; Kozhevnikov, V. Effect of oxygen ordering on the structural and magnetic properties of the layered perovskites PrBaCo$_2$O$_{5+\delta}$. *J. Phys. Condens. Matter* **2005**, *17*, 3317–3324. [CrossRef]
10. Bernuy-Lopez, C.; Høydalsvik, K.; Einarsrud, M.-A.; Grande, T. Effect of A-Site Cation Ordering on Chemical Stability, Oxygen Stoichiometry and Electrical Conductivity in Layered LaBaCo$_2$O$_{5+\delta}$ Double Perovskite. *Materials* **2016**, *9*, 154. [CrossRef]
11. Garcés, D.; Setevich, C.F.; Caneiro, A.; Cuello, G.J.; Mogni, L. Effect of cationic order-disorder on the transport properties of LaBaCo$_2$O$_{6-\delta}$ and La$_{0.5}$Ba$_{0.5}$CoO$_{3-\delta}$ perovskites. *J. Appl. Crystallogr.* **2014**, *17*, 325–334. [CrossRef]
12. Wachowski, S.L.; Szpunar, I.; Sørby, M.H.; Mielewczyk–Gryń, A.; Balaguer, M.; Ghica, C.; Istrate, M.C.; Gazda, M.; Gunnæs, A.E.; Serra, J.M.; et al. Structure and water uptake in BaLnCo$_2$O$_{6-\delta}$ (Ln = La, Pr, Nd, Sm, Gd, Tb and Dy). *Acta Mater.* **2020**. [CrossRef]
13. Pralong, V.; Caignaert, V.; Hebert, S.; Maignan, A.; Raveau, B. Soft chemistry synthesis and characterizations of fully oxidized and reduced NdBaCo$_2$O$_{5+\delta}$ phases δ = 0, 1. *Solid State Ion.* **2006**, *177*, 1879–1881. [CrossRef]
14. Frontera, C.; Caneiro, A.; Carrillo, A.E.; Oró-Solé, J.; García-Muñoz, J.L. Tailoring oxygen content on PrBaCo$_2$O$_{5+\delta}$ layered cobaltites. *Chem. Mater.* **2005**, *17*, 5439–5445. [CrossRef]
15. Burley, J.C.; Mitchell, J.F.; Short, S.; Miller, D.; Tang, Y. Structural and Magnetic Chemistry of NdBaCo$_2$O$_{5+\delta}$. *J. Solid State Chem.* **2003**, *170*, 339–350. [CrossRef]

16. Aksenova, T.V.; Gavrilova, L.Y.; TsvetkoV, D.S.; Voronin, V.I.; CherepanoV, V.A. Crystal structure and physicochemical properties of layered perovskite-like phases LnBaCo$_2$O$_{5+\delta}$. *Russ. J. Phys. Chem. A* **2011**, *85*, 427–432. [CrossRef]
17. García-Muñoz, J.L.; Frontera, C.; Llobet, A.; Carrillo, A.E.; Caneiro, A.; Aranda, M.A.G.; Ritter, C.; Dooryee, E. Study of the oxygen-deficient double perovskite PrBaCo$_2$O$_{5.75}$. *Phys. B Condens. Matter* **2004**, *350*, E277–E279. [CrossRef]
18. Mitchell, J.F.; Burley, J.; Short, S. Crystal and magnetic structure of NdBaCo$_2$O$_{5+\delta}$: Spin states in a perovskite-derived, mixed-valent cobaltite. *J. Appl. Phys.* **2003**, *93*, 7364–7366. [CrossRef]
19. McIntosh, S.; Vente, J.F.; Haije, W.G.; Blank, D.H.A.; Bouwmeester, H.J.M. Oxygen stoichiometry and chemical expansion of Ba$_{0.5}$Sr$_{0.5}$Co$_{0.8}$Fe$_4$O$_{3-\delta}$-measured by in situ neutron diffraction. *Chem. Mater.* **2006**, *18*, 2187–2193. [CrossRef]
20. Zhang, X.; Wang, X.-M.; Wei, H.-W.; Lin, X.-H.; Wang, C.-H.; Zhang, Y.; Chen, C.; Jing, X.-P. Effect of oxygen content on transport and magnetic properties of PrBaCo$_2$O$_{5.50+\delta}$. *Mater. Res. Bull.* **2015**, *65*, 80–88. [CrossRef]
21. Conder, K.; Pomjakushina, E.; Soldatov, A.; Mitberg, E. Oxygen content determination in perovskite-type cobaltates. *Mater. Res. Bull.* **2005**, *40*, 257–263. [CrossRef]
22. Choi, S.; Kucharczyk, C.J.; Liang, Y.; Zhang, X.; Takeuchi, I.; Ji, H., II; Haile, S.M. Exceptional power density and stability at intermediate temperatures in protonic ceramic fuel cells. *Nat. Energy* **2018**, *3*, 202–210. [CrossRef]
23. Politov, B.V.; Suntsov, A.Y.; Kellerman, D.G.; Leonidov, I.A.; Kozhevnikov, V.L. High temperature magnetic and transport properties of PrBaCo$_2$O$_{6-\delta}$ cobaltite: Spin blockade evidence. *J. Magn. Magn. Mater.* **2019**, *469*, 259–263. [CrossRef]
24. Løken, A.; Ricote, S.; Wachowski, S. Thermal and Chemical Expansion in Proton Ceramic Electrolytes and Compatible Electrodes. *Crystals* **2018**, *8*, 365. [CrossRef]
25. Korotin, M.A.; Ezhov, S.Y.; Solovyev, I.V.; Anisimov, V.I.; Khomskii, D.I.; Sawatzky, G.A. Intermediate-spin state and properties of LaCoO$_3$. *Phys. Rev. B* **1996**, *54*, 5309–5316. [CrossRef] [PubMed]
26. Shannon, R.D.; Prewitt, C.T. Effective ionic radii in oxides and fluorides. *Acta Crystallogr. Sect. B Struct. Crystallogr. Cryst. Chem.* **1969**, *25*, 925–946. [CrossRef]
27. Shannon, R.D. Revised effective ionic radii and systematic studies of interatomic distances in halides and chalcogenides. *Acta Crystallogr. Sect. A* **1976**, *32*, 751–767. [CrossRef]
28. Kriegel, R.; Kircheisen, R.; Töpfer, J. Oxygen stoichiometry and expansion behavior of Ba$_{0.5}$Sr$_{0.5}$Co$_{0.8}$Fe$_{0.2}$O$_{3-\delta}$. *Solid State Ion.* **2010**, *181*, 64–70. [CrossRef]
29. Grande, T.; Tolchard, J.R.; Selbach, S.M. Anisotropic thermal and chemical expansion in Sr-substituted LaMnO$_{3+\delta}$: Implications for chemical strain relaxation. *Chem. Mater.* **2012**, *24*, 338–345. [CrossRef]
30. Kharton, V.V.; Yaremchenko, A.A.; Patrakeev, M.V.; Naumovich, E.N.; Marques, F.M.B. Thermal and chemical induced expansion of La$_{0.3}$Sr$_{0.7}$(Fe,Ga)O$_{3-\delta}$ ceramics. *J. Eur. Ceram. Soc.* **2003**, *23*, 1417–1426. [CrossRef]
31. Padilla-Pantoja, J. Spin-Lattice Coupling in Strongly Correlated Cobalt Oxides Investigated by Sychrotron and Neutron Techniques. Ph.D. Thesis, Universitat Autonoma de Barcelona, Barcelona, Spain, 2016.
32. Toulemonde, O.; N'Guyen, N.; Studer, F.; Traverse, A. Spin state transition in LaCoO$_3$ with temperature or strontium doping as seen by XAS. *J. Solid State Chem.* **2001**, *158*, 208–217. [CrossRef]
33. Herrero-Martín, J.; García-Muñoz, J.L.; Valencia, S.; Frontera, C.; Blasco, J.; Barón-González, A.J.; Subías, G.; Abrudan, R.; Radu, F.; Dudzik, E.; et al. Valence change of praseodymium in Pr$_{0.5}$Ca$_{0.5}$CoO$_3$ investigated by x-ray absorption spectroscopy. *Phys. Rev. B Condens. Matter Mater. Phys.* **2011**, *84*, 1–6. [CrossRef]
34. Kubelka, P. New contributions to the optics of intensely light-scattering materials part I. *J. Opt. Soc. Am.* **1948**, *38*, 448. [CrossRef] [PubMed]
35. Vøllestad, E.; Schrade, M.; Segalini, J.; Strandbakke, R.; Norby, T. Relating defect chemistry and electronic transport in the double perovskite Ba1-xGd$_{0.8}$La$_{0.2+x}$Co$_2$O$_{6-\delta}$ (BGLC). *J. Mater. Chem. A* **2017**, *5*, 15743–15751. [CrossRef]
36. Taskin, A.A.; Ando, Y. Electron-hole asymmetry in GdBaCo$_2$O$_{5+x}$: Evidence for spin blockade of electron transport in a correlated electron system. *Phys. Rev. Lett.* **2005**, *95*, 176603. [CrossRef] [PubMed]
37. Vøllestad, E.; Strandbakke, R.; Tarach, M.; Catalán-Martínez, D.; Fontaine, M.L.; Beeaff, D.; Clark, D.R.; Serra, J.M.; Norby, T. Mixed proton and electron conducting double perovskite anodes for stable and efficient tubular proton ceramic electrolysers. *Nat. Mater.* **2019**, *18*, 752–759. [CrossRef] [PubMed]

© 2020 by the authors. Licensee MDPI, Basel, Switzerland. This article is an open access article distributed under the terms and conditions of the Creative Commons Attribution (CC BY) license (http://creativecommons.org/licenses/by/4.0/).

Review

Lifetime Prediction Methods for Degradable Polymeric Materials—A Short Review

Angelika Plota and Anna Masek *

Institute of Polymer and Dye Technology, Faculty of Chemistry, Lodz University of Technology, Stefanowskiego 12/16, 90-924 Lodz, Poland; angelika.plota@gmail.com
* Correspondence: anna.masek@p.lodz.pl

Received: 20 September 2020; Accepted: 8 October 2020; Published: 12 October 2020

Abstract: The determination of the secure working life of polymeric materials is essential for their successful application in the packaging, medicine, engineering and consumer goods industries. An understanding of the chemical and physical changes in the structure of different polymers when exposed to long-term external factors (e.g., heat, ozone, oxygen, UV radiation, light radiation, chemical substances, water vapour) has provided a model for examining their ultimate lifetime by not only stabilization of the polymer, but also accelerating the degradation reactions. This paper presents an overview of the latest accounts on the impact of the most common environmental factors on the degradation processes of polymeric materials, and some examples of shelf life of rubber products are given. Additionally, the methods of lifetime prediction of degradable polymers using accelerated ageing tests and methods for extrapolation of data from induced thermal degradation are described: the Arrhenius model, time–temperature superposition (TTSP), the Williams–Landel–Ferry (WLF) model and 5 isoconversional approaches: Friedman's, Ozawa-Flynn-Wall (OFW), the OFW method corrected by N. Sbirrazzuoli et al., the Kissinger–Akahira–Sunose (KAS) algorithm, and the advanced isoconversional method by S. Vyazovkin. Examples of applications in recent years are given.

Keywords: lifetime; degradation; accelerated aging; polymer; kinetic models; thermal analysis

1. Introduction

Nowadays, polymeric materials play a significant role in the development of modern civilization and are increasingly replacing traditional materials in almost every field, e.g., packaging, construction and in many other industries. Over the past several decades, the advance of plastics has made for economic development and brought huge benefits to our lives. The average annual production of these materials was only 1.5 million tons in the 1950s and had precipitously increased to nearly 400 million tons in 2018. Plastics are characterized by good durability, easy processability, light weight and low production costs for large, ready-made products [1].

The basic groups of application of polymeric materials in technology result from their specific functional properties. Most often they are used in the following areas: construction [2,3], medicine [4], agriculture [5], food [6], household [7], automotive [8,9] and chemical [10]. The application of plastic as packaging items is especially important, with around 40 million tonnes of plastic film produced from polyethylene alone [11].

Over many years of use, polymeric materials should exhibit good mechanical, physical and chemical properties and maintain their aesthetic qualities. However, during operation over time, they often change their original performance characteristics. The physical properties of elastomer products undergo various changes, as a result of which these products may become useless due to excessive hardening [12], softening [13], cracking [14] or other surface damage [15]. The reason is the susceptibility of polymers to oxidation and degradation processes. In outdoor applications, all polymeric materials degrade eventually [16].

Lifetime prediction of polymeric materials offers the benefit of isolating and identifying material failures against the fallouts of damage resulting in catastrophic harm. Therefore, accurate prediction of the service life of a material is very important in terms of safety, especially considering elastomeric construction materials (e.g., aircraft components, defense applications, nuclear reactor safety construction components) [15,17,18]. Additionally, this is a crucial factor for reliable use of polymers in medicine, engineering or consumer goods production [11].

This manuscript presents an overview of the latest accounts on the impact of the most common environmental factors on the degradation processes of polymeric materials. First, some general ways in which polymers environmentally degrade are set out, describing the factors, mechanisms and various changes caused by these processes. Then, some examples of shelf life for different polymeric materials and some generalities on accelerated aging are presented. The last part concentrates on different methods for lifetime prediction of degradable polymers using kinetic models from the literature for extrapolation of data from induced thermal degradation with examples of applications reported in recent years. The main goal of this work is to increase people's awareness of the importance of lifetime prediction for different materials due to their safety, influence on the environment and implications for long-term use. Based on polymer ageing research experience and a literature review, this paper provides a new perspective on this subject.

2. Degradation of Polymers

Three basic processes of polymer chain scission leading to a reduction in the molecular weight of polymers are known: depolymerization, destruction and degradation [19–21]. The depolymerization reaction consists of the thermal decomposition of the polymer to the monomer [19–21]. The destruction process represents the decomposition of polymer chains with separation of low molecular weight compounds other than the monomer [22]. The degradation process is the partial decomposition of the polymer, not into low molecular weight products, but into fragments with large but smaller molecular weights than the original polymer, and this is the most common process during the lifetime of a polymeric material [23–26].

2.1. Changes Caused by Degradation

Degradation is a process of structural changes that may result from physical or chemical transformations taking place in polymeric materials under the influence of long-term external factors (e.g., heat, ozone, oxygen, UV radiation, light radiation, chemical substances, water vapour, high energy radiation, dynamic stresses) which cause deterioration of the primary use properties [27–29]. As a result, primary properties are lost, and the first visible undesired signs are: a change in colour (e.g., yellowing), gloss (tarnishing) and texture [30,31]. In practice, there are much more complex systems of factors causing material degradation, which often makes it very difficult to distinguish which of these factors has a dominant influence because they act simultaneously [29].

The degradation process most often causes irreversible changes in the polymer, which are the result of such reactions as: crosslinking, chain cutting, thermal oxidation and even destruction [32]. The scheme of changes caused by the degradation process is shown in Figure 1.

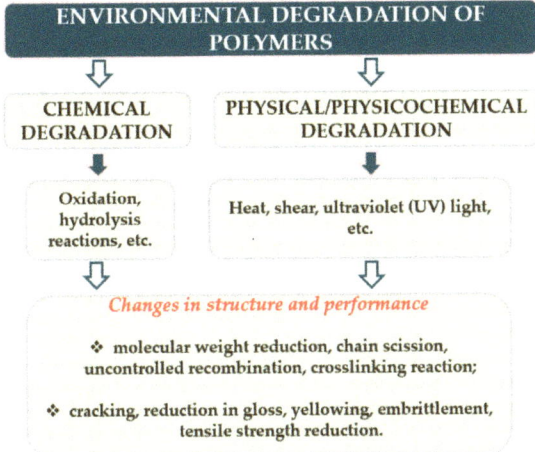

Figure 1. Changes in polymers caused by the chemical (oxidation, hydrolysis) and physicochemical (heat, shear, UV light) degradation factors.

Generally, this process can be carried out in an intentional and controlled way, thus gaining practical significance (improving processing operations or recovering mers from polymers) or it is uncontrolled and limits their practical application [29]. There are cases in which, in the first phase of material degradation, the factor causing the process has a positive effect on improving some properties (e.g., mechanical strength) by additional crosslinking of the elastomer under the influence of heat. However, in later stages of the degradation, the progress of ongoing processes, such as excessive cross linking or molecular weight reduction, begin to adversely affect the properties [33].

2.2. Types of Degradation

There are different types of polymer degradation such as thermal, photo, hydrolitic, bio and mechanical degradation [24]. Polymeric materials are susceptible to degradation to varying degrees. Degradation initiators and corresponding types of degradation are shown in Figure 2.

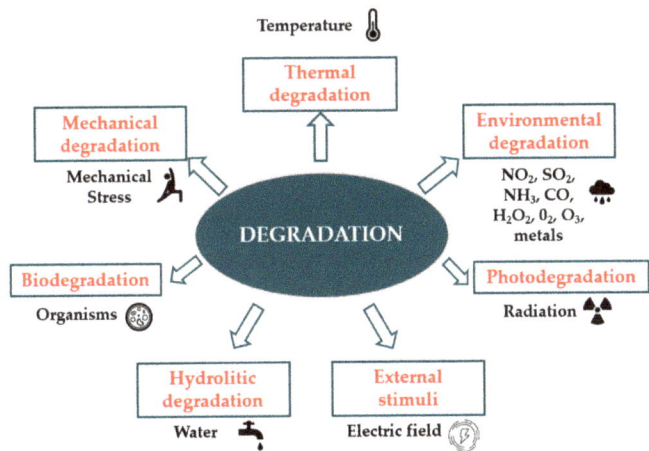

Figure 2. Chemical, physical and biological agents causing various types of degradation.

Of all the factors that affect the deterioration of elastomers, ozone, oxygen and elevated temperature are the most important. This is called ozone and thermo-oxidative aging [15]. There are strong links between the types of degradation presented above and, as mentioned, there are often several types of degradation at the same time [34]. An example is the simultaneous action of oxygen, light and other atmospheric factors [32].

The effect of the above factors also depends on the duration of their action (exposure time) and the type of polymer tested (including its molecular structure and its defects), the type and content of impurities, as well as the thickness and shape of the product [31].

Studies conducted so far on the process of polymer and plastic degradation show that the susceptibility of polymers depends to a large extent on the degree of crystallinity. Crystalline polymers are more resistant than amorphous ones. Additionally, polymers with a linear structure are more rapidly degraded than branched ones [35]. The polymer's decomposition also depends on its molecular weight, the larger the molecular weight, the slower the decomposition [36]. The last factor speeding up the degradation process is specific chemical groups. Amide, ester and urea groups are responsible for accelerating the degradation of a polymer because they easily hydrolyze [37].

2.3. General Mechanism of Thermal Degradation

Each type of polymer degradation is characterized by a specific molecular mechanism. Additionally, different mechanisms may be involved in the degradation of a material simultaneously [34]. The initial reaction during this process is always cracking of a bond in the polymer chain or other molecule that initiates degradation. A number of secondary reactions can occur as a result of bond breakage, which initiate further bond breakage, recombination or substitution [38].

The decomposition mechanism of polymeric materials is started by an initiation stage, during which free radicals are produced and hydrogen atoms are disconnected by energy from any source: radiation, light, heat or by the presence of an initiator [39], Equation (1):

$$R-H \xrightarrow{heat,\ light} R\cdot\ +\ H\cdot \tag{1}$$

The next stage is propagation, which is a series of reactions. Initially, a free radical reaction takes place with an oxygen molecule to form a peroxide radical, which then extracts a hydrogen atom from another polymer chain to form an unstable hydroperoxide group, which is divided into two new free peroxide or hydroxyl radicals. Because each initiating radical can generate 2 new free radicals, this process can be accelerated depending on how easy it is to remove the hydrogen from other polymeric chains, and how quickly the free radicals submit to termination by recombination and disproportionation. The reactions during the propagation stage are as follows [40,41]:

$$R\cdot\ +\ O_2 \rightarrow ROO\cdot$$
$$ROO\cdot\ +\ RH \rightarrow R\cdot\ +\ ROOH$$
$$ROOH \rightarrow RO\cdot\ +\ \cdot OH$$
$$RO\cdot\ +\ RH \rightarrow R\cdot\ +\ ROH$$
$$OH\ +\ RH \rightarrow R\cdot\ +\ H_2O$$

The final stage is termination, whereby two compounds with an unpaired electron form one inactive product. This can be a reaction between 2 peroxide radicals, alkyl radicals or 2 different radicals in the system [41]:

$$R\cdot + R\cdot \rightarrow R-R$$
$$2ROO\cdot \rightarrow ROOR + O_2$$
$$R\cdot + ROO\cdot \rightarrow ROOR$$
$$R\cdot + RO\cdot \rightarrow ROR$$
$$HO\cdot + ROO\cdot \rightarrow ROH + O_2$$

Focusing on degradable polymers, for example, polyolefins (polyethylene—PE, polypropylene—PP), two of the primary modes of degradation in industrial practice are thermal and photodegradation [42]. The initiation stage of polymer degradation via UV light is mainly associated with the presence of UV chromophores blended in with the polymer. Since saturated polyolefins do not themselves assimilate much UV light directly, the greatest impact of harmful UV is from absorption by chromophores in such compounds as pigments, flame retardants, catalyst residues, and any organic molecules that contain double bonds. As a result of the release of the absorbed part of the UV energy, the bond is cracked and free radicals are released, which initiate the degradation process. In the case of polypropylene, this process leads to chain scission, while for polyethylene, the cross-linking reaction is predominant [43]. Additionally, bonds of polyolefins themselves can be to some extent degraded by UV wavelength radiation—around 300 nm for PE and around 370 nm for PP [42]. An example of the loss of polyolefin properties due to the action of UV light is high density polyethylene (HDPE) in the form of a 1.5 mm wide plate, which can lose 80% of its strength after 2000 h of exposure to UV [43].

In the case of elastomers, they undergo various changes over time when exposed to ultraviolet light, heat, oxygen or ozone. Currently, there are a few standards that assess elastomers under elevated temperature or in chemical environment, such as ISO 188:2011 and ISO 1817:2015. When rubber is heated in the presence of air or oxygen, it loses strength, especially its tensile strength decreases. The determination of tensile strength change during material aging is not so simple because there are two opposing reactions that can take place simultaneously [44,45]. In order to explain the mechanical behaviour of crosslinked elastomer under exposure to various temperatures, Tobolsky et al. [46] offered a two-network theory. On the one hand, at higher temperatures softening can be observed, which is caused by the degradation of the molecular chains and crosslinks, and on the other hand, hardening can be observed as a result of additional cross linking. Depending on the type of vulcanization system, antidegradants used and type of filler, either softening or hardening reactions may prevail under any given aging conditions [44,45].

For example, crosslinking predominates in polybutadiene (BR) and its copolymers, such as nitrile-butadiene (NBR), styrene-butadiene-styrene (SBS), and in other diene rubbers with less active double bonds, whereas elastomers with electron donating side groups (–CH$_3$) attached to a carbon atom vicinal to the double bonds are susceptible to chain scission. This includes natural rubber (NR), isobutylene isoprene rubber (IIR), polyisoprene (IR), and any other unsaturated elastomer with electron donating groups [47,48].

More detailed description of the general mechanism of thermal degradation will be presented in a future publication.

During the thermoxidation, mechanochemical or photochemical processes that occur under the influence of radiation, the radical chain process plays a key role. Counteracting these unfavorable processes can be achieved by improving stability, i.e., the material's resistance to aging. There are various ways of modifying polymers by chemical and physical methods to increase resistance. The most common is modification with various additives such as stabilizers, antioxidants and UV absorbers [28,41,49].

3. Lifetime of Polymers

The durability of synthetic polymers is important for both manufacturers and users of plastic products, and above all for waste management. Unfortunately, not every type of polymer and plastic can be reprocessed by a recycling process. For this reason, the degradation processes of polymeric materials are constantly in the spotlight [50].

Lifetime (i.e., service life, shelf life, storage life) of polymeric materials at ambient or elevated temperature is a pivotal property that constitutes the acceptable lifetime (manufacturer's warranty). After this time, the material reaches the threshold (usually 50% of the initial value) of the measured value at service temperature [51,52]. The storage time depends on the conditions under which the product is stored and on the type of elastomer and type of product (e.g., gaskets, tyres, etc.), as well as application (e.g., space, automotive, etc.) [15].

Rubbers belong to the group of materials characterized by long shelf life when stored in appropriate conditions, such as: temperature below 25 °C, lack of light, moisture, oxygen, ozone and chemicals, when the durability of the material can be for 3–25 years, depending on the polymer used [15]. The storage life of products made of different rubbers is given in Table 1 [53].

Table 1. Shelf life of products made out of different types of rubber [51].

Rubber	Abbreviation	Recommended Storage Life Without Inspection (Years)	Storage Life Extension After Visual Inspection (Years)
Natural Rubber	NR	5	2
Butadiene-styrene	SBR	5	2
Nitrile	N	7	3
Nitrile-butadiene	HNBR	7	3
Acrylic	ACM	7	3
Chloroprene	C	7	3
Ethylene-Propylene	E	10	5
Viton™/FKM	V	10	5
Kalrez™/FFKM	KLZ	10	5
Silicone	S	10	5

Accelerated ageing tests are carried out by simulation of natural conditions in laboratory equipment using intensification of factors influencing the polymer and accelerating the ageing process. The accurate prediction of the material lifetime under the conditions of use is very important in terms of safety (especially considering elastomeric construction materials), environment (replacement of traditional polymers with new biopolymers that are more eco-friendly) and in many others fields [18,30].

4. Accelerated Aging of Polymeric Materials

Estimating the life of polymeric materials through accelerated aging is an essential tool that provides a quantitative or qualitative comparison depending on the methodology applied [54–56].

The ageing of material under operating conditions may take a very long time before changes are visible, so degradation processes are accelerated [57–59]. It was assumed that, in natural conditions, the time needed to assess changes in material properties for soft plastics is about 3 years, while for hard plastics it is not less than 5 years. Therefore, application of accelerated aging tests in laboratory conditions significantly simplifies the process and its analysis. Based on the accelerated aging test, an approximate assessment of the aging resistance of the material is obtained [28].

Due to the importance of predicting the lifetime of polymeric materials in areas such as defense applications, nuclear reactor safety components and aircraft components, there is strong emphasis on developing increasingly better methods of accelerated ageing [17]. The durability of synthetic polymers is important both for manufacturers of plastic products and their users, and above all from the point of view of waste management. Unfortunately, not every type of polymer can be recycled by re-processing. Therefore, it is necessary to determine the lifetime of polymeric materials and to create materials capable of long-term use [50]. In order to research the accelerated ageing of the polymers, the conditions under which the product will be operated should be determined and several elevated temperature values are then selected for the ageing process. Usually, the rate of a chemical reaction increases with temperature. As a result of subjecting polymeric samples to increased temperature, it is possible to determine the relationship between temperature and the rate of the degradation reaction [15,51]. By extrapolation, the degree of degradation of the material after a certain time or the time required to achieve a certain degree of degradation can be estimated for a specific temperature [60].

Accelerated ageing tests are carried out under more aggressive conditions than potential operating conditions (higher concentration of oxygen and/or ozone in the atmosphere, higher temperature). Accelerated (short term) ageing methods are based on appropriate selection of the set of external factors (e.g., temperature, light, heat, moisture) and the set of measured properties (e.g., changes in mechanical or dielectric properties). Apart from the selection of external factors and their intensity, the speed of these changes is also important. Results obtained from accelerated aging give an approximate assessment of resistance to aging [56,60–63].

5. Methods for Predicting the Lifetime of Polymeric Materials

5.1. Analysis of Thermal Degradation Kinetics Using Thermogravimetric Analysis

The study of polymer degradation processes using thermal analysis methods has been the subject of interest of many researchers for several decades. This is also evidenced by the large number of research papers devoted to this issue [64–69]. Recently, Seifi et al. (2020) [70] presented the applications of thermal analysis techniques in research in the past decade (2010–2020). In past years, this was mainly related to the search for new thermally resistant polymers, nowadays, it concerns problems related to environmental protection, i.e., the search for polymeric materials with a specific lifetime [11,50,67,69,71].

Thermal analysis methods, including most often thermogravimetry (TG) and differential scanning calorimetry (DSC), are also used to record the course of polymer degradation [64,72,73]. These methods give the opportunity to determine the changes in the state of the test sample when the temperature changes under different measuring conditions [64]. The methods of thermal analysis are used to study phase changes and chemical reactions that occur when heating or cooling the substance. They also enable the determination of thermodynamic and kinetic parameters of the reaction [74,75]. As a result of subjecting polymer samples to elevated temperatures, it is possible to determine the relationship between temperature and degradation reaction rate. By extrapolation, it is possible to estimate, for a given temperature, the degree of material degradation after a specified time, or the time required to achieve a certain degree of degradation [76]. Many authors adopt different kinetic models and their corresponding conversion functions [77,78]. Methods using a combination of thermal analysis with other techniques, such as infrared analysis and mass spectroscopy of degradation products, have proved particularly useful [79,80].

An important aspect of the degradation process model is the kinetic description, which consists of determining the activation energy (E_a), Arrhenius preexponential factor (A) and reaction order (n) [69,81–83]. These parameters are of theoretical and also practical importance. However, they must be supported by an understanding of the mechanism of the chemical reactions that occur in the process of polymer degradation. For non-isothermal tests, it should be assumed that the constant k is temperature dependent, e.g., according to the Arrhenius equation [64,84]. The differential equation representing the basic kinetic dependence of the degradation process has the form [78,85], Equation (2):

$$\frac{d\alpha}{dt} = k(T) \cdot f(\alpha) \qquad (2)$$

where k(T) is temperature-dependent reaction rate constant and f(α) is the model of reaction.

5.2. Arrhenius Model

The degree of degradation of polymers is measured by the deterioration of their properties [15,23,86]. Depending on the temperature, the rate of deterioration of these properties, i.e., the rate of degradation reaction, varies [86]. An increase in temperature usually results in an increase in the reaction rate, which is associated with an increase in the reaction rate constant (k) [87,88]. The relationship between the rate constant and temperature is described by the Arrhenius equation [78,89,90], Equation (3):

$$k = k(T) = A \cdot e^{\frac{-E_a}{RT}} \tag{3}$$

where A is the frequency factor, E_a is the activation energy, R is the gas constant and T is the absolute temperature.

The above equation is usually presented in a logarithmic form [15,91], Equation (4):

$$\ln k(T) = -\frac{E_a}{RT} + \ln A \tag{4}$$

The reaction rate is obtained from the change at any temperature in the particular property with exposure time at this temperature [92]. Due to different reaction rates k_i at different temperatures T_i, the same property level x_a is reached after different reaction times t_i (Figure 3) [15,93], Equation (5):

$$F(x_a) = k_i(T_i) t_i \tag{5}$$

where $F(x_a)$ is the reaction state function.

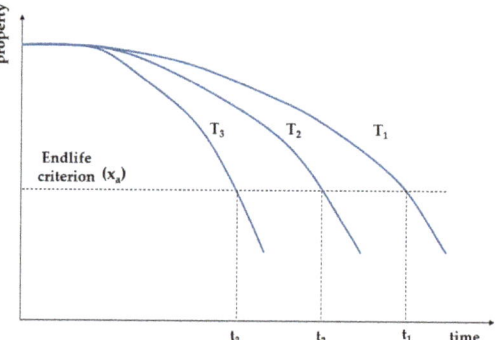

Figure 3. Change in property with time at three different temperatures $T_1 < T_2 < T_3$ [59].

By combining Equations (4) and (5), a logarithmic relationship is obtained [15,94], Equation (6):

$$\ln t_i = \frac{E_a}{RT} + B \tag{6}$$

where B is a constant.

Plotted against the inverse of temperature, the graph ln(t) forms a straight line and is known as the graphical form of the Arrhenius equation [95]. The lifetime of a polymeric material is read from the Arrhenius equation graph, which is extrapolated for a particular temperature and calculated from the following dependence [15], Equation (7):

$$e^{\ln t_i} = t_i \tag{7}$$

To eliminate the time dependence in Equation (2), which depends on temperature (T) and conversion rate (α), when heating at constant speed, the equation should be converted by dividing the differential equation by the heating rate [85], Equation (8):

$$\frac{d\alpha}{dT} = \frac{A}{\beta} \cdot e^{\frac{-E_a}{RT}} \cdot f(\alpha) \tag{8}$$

where β—$\frac{dT}{dt}$ is the heating rate.

By applying the time-dependent velocity equation and the linear transformation, kinetic parameters are obtained: activation energy (E) and response factor (A) [85], Equation (9):

$$\ln\left(\frac{\frac{d\alpha}{dT}}{f(\alpha)}\right) = \ln\left(\frac{A}{\beta}\right) - \frac{E}{RT} \tag{9}$$

Monitoring the ageing process with one exposure time is insufficient. It is, therefore, necessary to determine trend curves for material properties for several experimental conditions in order to extrapolate to the lifetime. Figure 4 presents an Arrhenius plot relating the time extrapolated as a function of inverse temperature.

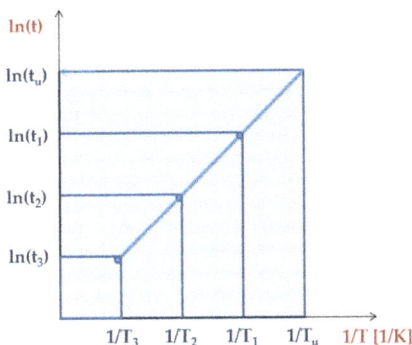

Figure 4. Arrhenius diagram showing time extrapolation as a function of inverse temperature [59].

Ln(t_i) is proportional to E_a/RT, E is the activation energy and R is the gas constant. This linear dependence enables prediction of the lifetime t_u extrapolated at T_u (temperature of use) [59].

The Arrhenius method assumes that the mechanism of degradation at elevated temperature is identical to that of degradation under the operating conditions. However, this assumption is not always true. The mechanism of this process at low temperature may be quite different, so that in such a situation it is not possible to extrapolate the value of material's lifetime using a linear time–temperature relationship [88].

Lewandowski et al. (2016) [15] used the Arrhenius and time–temperature superposition methods to predict the lifetime of different elastomers (e.g., nitrile-butadiene rubber—NBR, chloroprene rubber—CR, ethylene propylene diene monomer—EPDM) and, in their opinion, predictions using the Arrhenius method can be subject to large error if the degradation is of a complex nature, e.g., physico-chemical, or the degradation mechanism changes depending on the temperature. Also, Xiong et al. (2013) [96] applied this method to lifetime prediction of NBR composite sheet in aviation kerosene by using nonlinear curve fitting of ATR-FTIR spectra. The lifetime of the NBR composite sheet in the aviation kerosene was 11,113 days at 20 °C, 6467 days at 25 °C, 3831 days at 30 °C, 2309 days at 35 °C and 1414 days at 40 °C. In the author's opinion, the method based on the Arrhenius equation is valuable and the NBR composite sheet can still be safely used to store aviation kerosene after it has been used in aviation kerosene at room temperature for 8 years.

Madej-Kiełbik et al. (2019) [97] adopted the Arrhenius methodology for the accelerated aging of personal protectors for motorcyclists. The main constituents of the protectors were polyurethane and ethylene-vinyl acetate copolymer (EVA). They determined the endlife criterion (x_a) using the kinetics of changes in the selected property/parameter of the material, at temperature intervals of 10 °C. Finally, they were able to calculate the real aging time of the material.

Koga et al. (2019) [98] analyzed the degradation behavior of PVC resin under high temperature conditions using Fourier transform infrared spectroscopy (FTIR), tensile testing and small punch (SP) testing. Based on the results from these tests, they compared the activation energies and estimated lifetime using the Arrhenius method. In this case, it turned out that SP testing is the most accurate and minimally destructive lifetime prediction method that can estimate early deterioration.

Wang et al. (2019) [99] proposed a lifetime prediction model of aging natural gas polyethylene (PE) pipeline with various internal pressures by thermal-oxidative aging (TOA and oxidative induction time (OIT) tests. The Arrhenius relationship was interpreted as a linear correlation between the OIT (logarithmic scale) at different test temperatures and the inverse of temperature (1/T). Lifetime prediction over the range 0–0.4 MPa at 20 °C proved to be in excess of the 50 years' lifetime requirement. The authors confirmed that this method is very suitable for pressured urban gas PE pipes, and also very suitable for other plastic pipes in similar environments.

Based on these examples, it can be said that the Arrhenius model is useful for many different polymeric materials, e.g., NBR, EPDM, EVA, PVC, PE, etc. The results that are obtained with this method are valuable but only in the case when degradation is not limited by diffusion. The approach may be used for lifetime prediction within a temperature range where the degradation mechanism remains the same. Most authors choose this way of lifetime prediction because the method is reliable and time saving, and can calculate change of aging performance at any time.

5.3. Time–Temperature Superposition

The principle of time–temperature superposition (TTSP) is one of the bases of accelerated experimental methodology for polymeric materials [100]. The accuracy of predicting the lifetime of polymeric materials and activation energy can be improved by using the time–temperature superposition method [15].

This method is determined from an empirical assumption that the influence of temperature and time on the characteristics are equivalent. For a short-period of accelerated aging at higher temperature, the changes of characteristic variable are the same as those measured for longer times but at normal (lower) temperatures. The TTSP consists of determining the a_T factor (shift factor), which results in a single curve describing the correlation between the tested material property and temperature T or time t [91], Equation (10):

$$\ln a_T = \frac{E_a}{R}\left(\frac{1}{T_0} - \frac{1}{T}\right) \qquad (10)$$

where a_T is a shift factor, T is a temperature and T_0 is a reference temperature (usually the lowest value of the ageing temperature).

A simple scheme for this method is given in Figure 5.

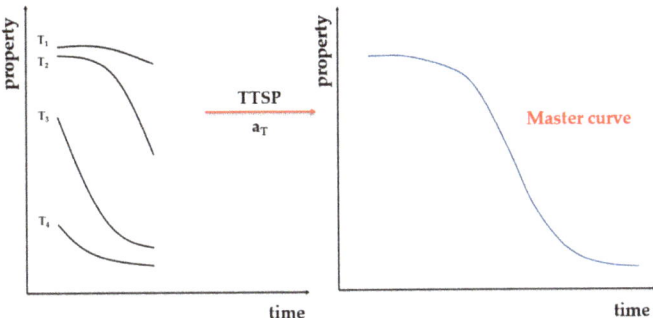

Figure 5. Schematic illustration of creating a master curve through the time–temperature superposition (TTSP) method [92].

For polymeric materials, it is possible to determine temperature functions that enable displacing respective isothermal segments of the selected response function along the logarithmic time scale and create a master curve, which is recorded at a reference temperature (T_0) [100–102].

By creating a straight line in the logarithmic diagram of empirically determined values of the shift coefficient in relation to the absolute temperature, the shift coefficient can be calculated for each desired temperature. Once the displacement coefficients are determined, they are used to extrapolate the changes in behaviour at ambient temperature [91]. Such a curve is called a complex curve and, unlike the Arrhenius method, it uses all experimental points for all ageing temperature values. From the complex curve it is possible to determine the time needed to obtain a threshold value at T_0 [103].

The lifetime of a polymeric material at operating temperature can be determined by extrapolating the log a_T value for operating temperature (a_{Te}) [15], Equation (11):

$$t_e = \frac{a_{T0}}{a_{Te}} \cdot t_0 \qquad (11)$$

where a_{T0} is a shift factor for the reference temperature (a_{Te} is an extrapolated shift factor for operating temperature (determined from the dependence of the logarithm (a_T) on the temperature) and t_0 is the lifetime of the material at the reference temperature.

Nakada et al. (2011) [104] used the time–temperature superposition principle to predict the long-term viscoelastic behavior of amorphous resin. The master curves of creep compliances could be constructed from measured data by shifting the a_T factors vertically as well as horizontally. Therefore, the long-term viscoelastic behavior at a temperature below T_g can be predicted accurately based on the short-term viscoelastic behavior at elevated temperatures using the TTSP with vertical shift as well as horizontal shift. The creep compliance was also tested by Fukushima et al. (2009) [105] using the TTSP method for long-term lifetime prediction of polymer composites. In the authors' opinion, it can be considered that the time–temperature shift factor is obtained accurately and easily from dynamic viscoelastic tests.

This has been confirmed by Krauklis et al. (2019) [101]. They applied the TTSP method to predict creep of a plasticized epoxy, which is a matrix for fiber reinforced polymer (FRP) composite materials. This methodology has turned out to be useful to predict the long-term viscoelastic behavior of plasticized polymers at temperatures below the T_g temperatures based on short-term creep experimental data.

Yin et al. (2019) [106] investigated the aging behavior of PMMA in a liquid scintillator at different temperatures under static tensile stress with dynamic mechanical analysis (DMA) and differential scanning calorimetry (DSC). Then, the service life of the PMMA was predicted based on the time–temperature superposition approach. The tensile strength of PMMA at different aging temperatures showed a downward trend with time of aging. For tests at 30 °C, 40 °C, 50 °C, linear behavior of the plot of the shift factors versus the inverse aging temperature was observed, indicating that the aging mechanism at these temperatures was consistent with the Arrhenius rule. The result at 55 °C deviated significantly from the straight line because the Arrhenius equation is valid only when the assumption of constant acceleration for chosen temperatures is fulfilled. However, when the material deviates from this assumption, the Arrhenius equation ceases to be useful, which could be observed in the case of the temperature of 55 °C. According to the Arrhenius model and extrapolation, the lifetime of PMMA was determined to be 25 years at 20 °C. A discussion about why the PMMA at 55 °C aged differently and what it means would be welcomed, although this is not needed for predicting lifetime.

Most papers that apply the time–temperature superposition principle concentrated on prediction of the long-term viscoelastic behavior of material at a temperature below T_g. It should be emphasised that this method is valid only in accordance with the assumption of constant acceleration. In other cases, it can be subject to measurement error and the value of temperature will deviate from the straight line.

5.4. Williams–Landel–Ferry (WLF) Model

The Williams–Landel–Ferry (WLF) method is also used to predict the lifetime of polymeric materials based on any physical property [107]. In the case of complicated dependence of a given material property on time or when the degradation process is limited by diffusion, better results can be obtained by using the WLF method [15,108].

The WLF equation is an essential tool to predict temperature induced physical/chemical changes as a function of processing and storage conditions. This equation allows for the estimation of the shift factor (a_T) for temperatures other than those for which the polymeric material was tested. Compared to the Arrhenius equation that describes the behaviour above the glass transition temperature (T_g), the WLF equation is applicable close to this temperature [109].

The WLF model uses a time–temperature superposition without any additional assumptions regarding the dependence of material properties on time at any temperature. In this method, the a_T factor is described by the formula [110–112], Equation (12):

$$\log a_T = \frac{C_1(T - T_0)}{C_2 + (T - T_0)} \quad (12)$$

where a_T is a reduced variables shift factor, C_1 and C_2 are the experimental constants, which depend on temperature and the material being tested, and T_0 is a reference temperature (usually $T_g \leq T_0 \leq T_g + 100$).

The WLF method can be used when the dependence of a given property on time at different temperature values is similar. The curves corresponding to the different temperature values are shifted parallel to each other by the shift factor a_T [112].

By comparing Equations (10) and (11) and converting the logarithms from natural to decimal, the following values of C_1 and C_2 are obtained [113], Equations (13) and (14):

$$C_1 = -\frac{1}{2.303} \cdot \frac{E_a}{RT_0} \quad (13)$$

$$C_2 = T_0 \quad (14)$$

This equation has been successfully applied to polymeric materials using the "universal values". However, the literature has shown that these constants can be different and should be measured experimentally [114].

Hu et al. (2013) [113] applied the WLF and Arrhenius equations to research on temperature and frequency dependent rheological behaviour of carbon black filled natural rubber. They confirmed that the temperature dependence of the shift factor is modelled well by both the Williams–Landel–Ferry equation and the Arrhenius equation.

Ljubic et al. (2014) [115] considered the possibilities of using time–temperature superposition with the WLF equation. This method is well established for bulk linear and homopolymers but for more complex polymeric materials (individual, blend or composite) with different structures and morphologies, it can sometimes be a challenge. The combination of TTSP with the WLF equation can be successfully applied to crosslinked polymers (polyurethanes and epoxy), polyolefins for biomedical application, Kevlar 49, polymer blends, biopolymers and polymer composites. In these cases, the dynamic mechanical and viscoelastic properties were tested, and modeling of the properties was performed by using TTSP and the WLF equation. In the authors' opinion, this combination of TTSP with WLF has versatile applications and can be a useful tool in the study of a broad range of polymeric materials, their properties and lifetime prediction for final products.

Chang et al. (2013) [116] used the TTSP method with the WLF equation to apply to a series of short-term creep tests of a wood–plastic composite (WPC). The success of the method implied that the WPC product studied is a thermo-rheologically simple material and only horizontal shifting is needed for the time–temperature superposition. Additionally, the temperature shift factor used to construct the master curve was fitted well by the WLF equation. Dan-asabe B. (2016) [117] used a long term TTSP (time–temperature superposition) performance prediction with WLF assumption to characterization of a banana (stem) particulate reinforced PVC composite as a piping material. The composite turned out to be rheologically simple as it satisfied the WLF condition. Application of long-term prediction showed that the composites have better long-term performance than PVC pipe over a period of 126 years of use.

As has been observed, the WLF equation is always used in combination with the TTSP principle. By applying this method, it is possible to obtain better lifetime prediction results than with the Arrhenius method when the time dependence of a property is complicated or when degradation is limited by diffusion.

5.5. Isoconversional Methods

Isoconversional methods are one of the more reliable kinetic methods for processing thermoanalytical data [118]. They are based on the isoconversional principle, with the main presumption of temperature independence of the pre-exponential factor and the activation energy. However, both the pre-exponential factor and the activation energy are still interrelated conversion functions [119]. These methods have their roots in the Ozawa–Flynn–Wall [120,121] and Friedman [122] methods developed in the 1960s.

The accuracy of the method has been improved by the Kissinger–Akahira–Sunose (KAS) [123] approach to calculating kinetic parameters. Starink [124] and then Gao et al. [125] proposed subsequent modifications to the calculation procedure and coefficients in the KAS equation to predict more accurate values of the pre-exposure factor and the activation energy. Vyazovkin has continuously developed an alternative calculation method of activation energy [126,127], giving access to the well-known advanced isoconversion method (AICM), now widely accepted as one of the most precise methods of estimating the activation energy from TGA experiments [128]. All of the abovementioned methods, except the Friedman method, are integral, i.e., the parameters are calculated on the basis of integral analysis of the measured TG signals [119].

The main benefits of isoconversional methods are that they allow the evaluation of the effective activation energy, E_a, without presuming any specific form of the reaction model, $f(\alpha)$ or $g(\alpha)$, and that a change in the variation of E_a, may generally involve a change in the reaction mechanism or the degree of limitation of the overall rate of reaction, as measured by thermoanalytical techniques [118].

Isoconversion methods require a series of experiments with different temperature programs and obtaining the effective value of activation energy E_a as a function of conversion degree α. A significant variability of E_a with α means that the process is kinetically complex, and the activation energy dependences assessed by an isoconversional method enable significant mechanistic and kinetic analyses and the understanding of multi-step processes, as well as reliable kinetic prognoses.

Model-free isoconversional methods are a powerful tool to obtain information about the complexity of the reaction by determining E_a [118,129,130].

5.5.1. Friedman's Method

The determination of activation energy without the need to adopt a specific kinetic model is possible using this method. The E_a value is based on thermal measurements at various heating rates. The Friedman's isoconversion method is a widely used differential method which, unlike conventional integral isoconversional methods, provides accurate values of the activation energy [119,131]. The modification of the general reaction rate equation results in the following expression [118], Equation (15):

$$\ln \frac{d\alpha}{dt} = \ln[Af(\alpha)] - \frac{E_a}{RT} \tag{15}$$

where α is a specific degree of degradation.

The term $[Af(\alpha)]$ constitutes the product between the mathematical function $f(\alpha)$ and the pre-exponential factor A that characterizes the reaction mechanism. After evaluating E_a and $[Af(\alpha)]$, the response rate $(d\alpha/dt)$ for each α value can be calculated [118].

By using the following linear regression, the activation energy at each level of isoconversion can be determined [132], Equation (16):

$$\ln \frac{d\alpha}{dt} = f \frac{1}{T_{jk}} \tag{16}$$

where T_{jk} are the temperatures at which the degree of conversion is achieved (α_k) at the heating rate (β_j).

Das et al. (2017) [133] made a comparison of different kinetic models in order to research the thermal decomposition behaviour of high and low-density polyethylene (LDPE and HDPE), polypropylene (PP) and poly(lactic acid) (PLA). One of them was the Friedman method. According to the opinion of these authors, it is difficult to choose one model that can extract the correct kinetic parameters from the complex reactions occurring, and it is better to use several models and make a comparison. Cui et al. (2014) [134] used the Friedman method to study the thermal degradation kinetics of photonically cured electrically conductive adhesives. To demonstrate the applicability of the Friedman method, the results calculated from this kinetic method were compared with those from experiments and they turned out to be in good agreement. In the authors' opinion, with the Friedman method, a comprehensive and in-depth understanding can be obtained of the thermal degradation kinetics of photonically cured electrically conductive adhesives.

Mittal et al. (2020) [135] studied the thermal decomposition kinetics and properties of grafted barley husk reinforced PVA/starch composite films for packaging applications, based on TGA measurements. They determined the activation energy for composite films by using Friedman (FR), Flynn–Wall–Ozawa (FWO), Kissinger–Akahira–Sunose (KAS), and modified Coasts Redfern (CR) methods. It was observed that the activation energy obtained using these all methods showed the same trend for various conversion factors (0.1–0.9) for all the films, which indicated the reliability of the values of activation energy obtained.

5.5.2. Ozawa–Flynn–Wall (OFW) Method

The Ozawa–Flynn–Wall (OFW) method is an integral isoconversional technique, in which the activation energy is related to the heating rate and temperature at a constant conversion rate [136]. The main advantage of this method is that it does not require any assumptions about the form of the kinetic equation except a temperature dependence of the Arrhenius type. This method is a model-free method that measures temperatures corresponding to constant values of α from experiments at various heating rates (β), and plotting $\ln(\alpha)$ in relation to 1/T. The slopes of such graphs give $-E_a/R$ [137].

In this method, by using Doyle's approximation of the temperature integral, the following equation is obtained [130,136], Equation (17):

$$\ln \beta = \ln \frac{AE_a}{Rg(\alpha)} - 5.331 - 1.052 \frac{E_a}{RT_p} \tag{17}$$

where β is a heating rate, E_a is the activation energy, R is the gas constant and T_p is the peak temperature.

According to the above equation, from the experimental thermogravimetric curves, which are recorded for several heating rates, for α = const, the linear regression of the relation $\ln(\beta) = f(1/T)$ is obtained and its slope can be used to determine the activation energy [136].

The OFW method is potentially suitable for use in systems with multiple reactions such that the activation energy changes over time. However, it is predicted that this method will fail if reactions of very different kinds occur simultaneously. Competitive reactions that have various products also make this method inapplicable. Moreover, the OFW method is less accurate than the Friedman's method. It has been shown that, if E_a depends on the degree of conversion, its values obtained by integral and isoconversional methods are different [137].

Ali et al. (2019) [138] calculated the activation energy E_a by applying Coats–Redfern, Ozawa–Flynn–Wall, Kissinger–Akahira–Sunose and Friedman models in order to research the thermo-catalytic decomposition of polystyrene waste. From the results obtained, it was observed that activation energy investigated using the OFW is the lowest, therefore this model is most suitable for explaining catalytic decomposition of expanded waste polystyrene.

Vassiliou et al. (2010) [139] used the OFW method to study the thermal degradation kinetics of organically modified PET with montmorillonite and fumed silica nanoparticles. It enabled them to determine a kinetic parameter in terms of the apparent activation energy.

Benhacine et al. (2014) [140] applied the Flynn–Wall–Ozawa method to determine the activation energy E_a of the degradation process of isotactic polypropylene/Algerian bentonite nanocomposites prepared via melt blending. Determination of activation energy was based on weight loss versus temperature data obtained at several heating rates, and this was necessary in order to deeply analyze the effect of incorporation of pure bentonite or organically modified bentonite into iPP matrix on the degradation mechanism of iPP.

5.5.3. Ozawa–Flynn–Wall Corrected Method by N. Sbirrazzuoli et al.

N. Sbirrazzuoli et al. [141] proposed another way to obtain corrected values of activation energy by applying a numerical integration of p(x). The first part of this method involves calculating Equation (18) in order to obtain an approximate value of activation energy for a given conversion (Ozawa–Flynn–Wall method). Next, the average temperature ($\overline{T_\alpha}$) for different heating rates is assessed and p(x) is extrapolated by numerical integration of the following equation, Equation (18):

$$g(\alpha) = \frac{AE}{R\beta}\left[\frac{\exp(-x)}{x} - \int_x^\infty \left(\frac{\exp(-x)}{x}\right) dx\right] = \frac{AE}{R\beta} p(x) \tag{18}$$

where x = E/RT is minimized activation energy at the temperature T. This equation assumes that the value of E is constant.

Finally, ln p(x) values are matched by a first order polynomial in the range $x = E_\alpha (1 \pm 0,2)/(R\overline{T_\alpha})$. [141].

5.5.4. Kissinger–Akahira–Sunose (KAS) Method

The KAS method is considered to be one of the best isoconversional methods because of the Coats–Redfern approximation of the temperature integral. It does not require knowledge of the exact thermal degradation mechanism [136,142]. Using this method, the activation energy is determined from the following equation [143], Equation (19):

$$\ln\left(\frac{\beta}{T_p^2}\right) = -\frac{E_a}{RT_p} + \text{const} \tag{19}$$

where T_p is a peak temperature.

Therefore, for α = const. the plot of $\ln(\beta/T_p^2)$ against $(1/T_p)$ obtained from thermogravimetric curves, recorded for different heating rates, is a straight line. The slope and intercept of this line can yield the pre-exponential factor and activation energy, respectively [136,143].

Wang et al. (2005) [144] described the possibility of using the KAS algorithm for modeling the cure kinetics of commercial phenol-formaldehyde (PF) resins. Additionally, this algorithm could also be used to predict the isothermal cure of PF resins from dynamic tests. It has also been confirmed by Gabilondo et al. (2007) [145] that the d KAS method seems useful for the dynamic cure prediction of that type of thermoset.

Additionally, Sun et al. (2014) [146] applied the Kissinger–Akahira–Sunose (KAS) method to determine activation energy (E_a), and investigated it as the change of conversion (α) in order to describe the curing behavior of epoxy resins by using differential scanning calorimetry.

5.5.5. Advanced Isoconversional Method by S. Vyazovkin

The advanced isoconversional method developed by Vyazovkin S. [126] is a non-linear integral method and is free of the approximations applied in OFW and KAS approaches. It is based on a direct numeric integration of the following equation [141], Equation (20):

$$g(\alpha) = \frac{A}{\beta}\int_0^T \exp(-E/RT)dT \tag{20}$$

where $\beta = dT/dt$ is a heating rate.

For a set of n experiments performed under various temperature programs $T_i(t)$, the activation energy is assessed at any specific value of α by finding the E_a value which minimises the function, Equation (21) [147]:

$$\phi(E_\alpha) = \sum_{i=1}^{n}\sum_{j\neq i}^{n}\frac{J[E_\alpha, T_i(t_\alpha)]}{J[E_\alpha, T_j(t_\alpha)]} \tag{21}$$

Recently, this method has been modified by integration, which is performed over small time segments that enables elimination of the errors related to integral methods when activation energy varies significantly with α, Equation (22) [147]:

$$J[E_\alpha, T_i(t_\alpha)] = \int_{t_\alpha - \Delta\alpha}^{t_\alpha} \exp\left[\frac{-E_\alpha}{RT_i(t)}\right]dt \tag{22}$$

In Equation (22), α varies from $\Delta\alpha$ to $(1 - \Delta\alpha)$ with a step $\Delta\alpha = m^{-1}$, where m is the number of intervals that are chosen for analysis [141].

The first experimental deployments of the isoconversional predictive procedure were made by Vyazovkin [148] and Vyazovkin and Sbirrazzuoli [149] concerning the application of nonisothermal data to predict epoxy curing kinetics under isothermal conditions. For both methods, the isoconversional predictions agreed well with the real measurements and were better that those obtained by other methods.

Dunne et al. (2000) [150] predicted isothermal and nonisothermal cure kinetics of an epoxy-based photo-dielectric dry film (ViaLux™ 81). They reported perfect agreement with actual nonisothermal measurements. However, the predictions for isothermal conditions deviated markedly from the actual data. The reason for this was that they were done to a temperature below the limiting glass transition temperature.

Li et al. (2001) [151] described several examples of successful kinetic predictions obtained by applying the isoconversional methodology to the curing process of polyurethane. He et al. (2005) [152] applied isoconversional predictions to demonstrate the accelerating effect of moisture in wood on curing of polymeric diphenylmethane diisocyanate.

Polli et al. (2005) [153] used the isoconversional analysis in order to predict the kinetics of thermal degradation of branched and linear polycarbonates in a wide temperature range. Vyazovkin et al. (2004) [154] certified that the isoconversional predictions of degradation of polymeric materials to the flash ignition temperature can be very usable in assessing potential fire resistance of polystyrene (PS) and PS–clay nanocomposite.

Over recent years, the use of isoconversional kinetic analysis has provided new opportunities in traditional areas of application such as polymer degradation or curing, and also efficiently delving into areas such as glass transition. The effectiveness of these methods comes from their ability to handle the complexity of different processes. The resulting activation energies can be used to perform reliable kinetic predictions, to obtain information about complex mechanisms, and also to access inner kinetic parameters.

6. Conclusions

Degradation of polymeric materials can be caused by many different factors: both chemical (working environment) and physical (temperature, radiation) and mechanical (stress). The mechanism of degradation can, therefore, be very complicated. There are many different kinetic models that are used to predict the lifetime of these materials. Those described in this paper are those used most often.

Predictions are among the most important practical features of kinetic analysis. They are widely used to evaluate the kinetic behavior of polymeric materials beyond the temperature regions of experimental measurements.

For the Arrhenius model, it assumes that the mechanism of degradation at elevated temperatures is identical to the mechanism of degradation under operating conditions. However, this assumption is not always true. For this reason, a time–temperature superposition is used, which increases the accuracy of predicting the lifetime of polymeric materials and activation energy. It should be remembered that degradable polymers degrade during thermal processing, which can of course affect the lifetime predictions. The lifetime of polymers can also be determined by the Williams–Landel–Ferry method (WLF), in which there are no assumptions concerning the dependence of the tested property on temperature and, in the case of a degradation process that is limited by diffusion, better results can be obtained by using this method.

There are also isoconversional methods, which are considered to be one of the more reliable kinetic methods of processing thermoanalytical data, and their main benefit is that they allow the evaluation of the effective activation energy, E_a, without presuming any specific form of the reaction model. The activation energy calculated from different isoconversional methods used mainly for prediction of lifetime is a feature to follow physico-chemical processes occurring inside the material. In fact, that the ageing process can result in scission, crystallization, oxidation, thermal decomposition, etc., and those factors are closely related. Moreover, in the great majority of publications, activation energies are determined using techniques like DTA and DSC, where endo- and exothermal transformations can be observed. Those changes can also be found during the ageing process. Therefore, activation energy, which can also be thought of as the magnitude of the potential barrier of molecules at the surface of the material, can bring necessary information on how fast the ageing process will occur. For instance, thermal stability can be estimated as the time to reach a certain extent of conversion at a given temperature. Kinetic predictions of this type can be easily achieved by using the activation energy dependence measured by an isoconversional method.

Scientists are constantly working to develop new methods or refine existing ones because of the great interest and need for such research on accelerated ageing and prediction of the lifetime of polymeric materials that are used in almost every area of our lives.

Author Contributions: Conceptualization, formal analysis, review and editing, methodology A.M.; conceptualization, data analysis, investigation, methodology and writing, A.P. All authors have read and agreed to the published version of the manuscript.

Funding: This research received no external funding.

Conflicts of Interest: The authors declare no conflict of interest.

References

1. Jim Jem, K.; Tan, B. The Development and Challenges of Poly (lactic acid) and Poly (glycolic acid). *Adv. Ind. Eng. Polym. Res.* **2020**, *3*, 60–70. [CrossRef]
2. Küçük, V.A.; Çınar, E.; Korucu, H.; Simsek, B.; Bilge Güvenç, A.; Uygunoglu, T.; Kocakerim, M. Thermal, electrical and mechanical properties of filler-doped polymer concrete. *Constr. Build. Mater.* **2019**, *226*, 188–199. [CrossRef]
3. Rod, K.A.; Nguyen, M.; Elbakhshwan, M.; Gills, S.; Kutchko, B.; Varga, T.; Mckinney, A.M.; Roosendaal, T.J.; Childers, M.I.; Zhao, C.; et al. Insights into the physical and chemical properties of a cement-polymer composite developed for geothermal wellbore applications. *Cem. Concr. Compos.* **2019**, *97*, 279–287. [CrossRef]
4. Zarrintaj, P.; Jouyandeh, M.; Ganjali, M.R.; Shirkavand, B.; Mozafari, M.; Sheiko, S.S. Thermo-sensitive polymers in medicine: A review. *Eur. Polym. J.* **2019**, *117*, 402–423. [CrossRef]
5. Chauhan, D.; Afreen, S.; Talreja, N.; Ashfaq, M. Multifunctional copper polymer-based nanocomposite for environmental and agricultural applications. In *Multifunctional Hybrid Nanomaterials for Sustainable Agri-Food and Ecosystems*; Abd-Elsalam, K.A., Ed.; Elsevier: Rotterdam, The Netherland, 2020; pp. 189–211. ISBN 9780128213544.
6. Zhong, Y.; Godwin, P.; Jin, Y.; Xiao, H. Biodegradable Polymers and Green-based Antimicrobial Packaging Materials: A mini-review. *Adv. Ind. Eng. Polym. Res.* **2020**, *3*, 27–35. [CrossRef]
7. Mishra, M. Household Goods: Polymers in Insulators: Polymers for High-Voltage Outdoor Use. In *Encyclopedia of Polymer Applications, 3 Volume Set*; CRC Press: Boca Raton, FL, USA, 2018; pp. 1549–1562. ISBN 9781351019422.
8. Patil, A.; Patel, A.; Purohit, R. An overview of Polymeric Materials for Automotive Applications. *Mater. Today Proc.* **2017**, *4*, 3807–3815. [CrossRef]
9. Jasso-Gastinel, C.F.; Soltero-Martínez, J.F.A.; Mendizábal, E. Introduction: Modifiable Characteristics and Applications. In *Modification of Polymer Properties*; William Andrew: Norwich, NY, USA, 2017; pp. 1–21. ISBN 9780323443982.
10. Kaur, R.; Marwaha, A.; Chhabra, V.A.; Kaushal, K.; Kim, K.; Tripathi, S.K. Facile synthesis of a Cu-based metal-organic framework from plastic waste and its application as a sensor for acetone. *J. Clean. Prod.* **2020**, *263*, 121492. [CrossRef]
11. Laycock, B.; Nikolić, M.; Colwell, J.M.; Gauthier, E.; Halley, P.; Bottle, S.; George, G. Lifetime prediction of biodegradable polymers. *Prog. Polym. Sci.* **2017**, *71*, 144–189. [CrossRef]
12. Barba, D.; Arias, A.; Garcia-gonzalez, D. Temperature and strain rate dependences on hardening and softening behaviours in semi-crystalline polymers: Application to PEEK. *Int. J. Solids Struct.* **2020**, *182–183*, 205–217. [CrossRef]
13. Yan, C.; Huang, W.; Lin, P.; Zhang, Y.; Lv, Q. Chemical and rheological evaluation of aging properties of high content SBS polymer modified asphalt. *Fuel* **2019**, *252*, 417–426. [CrossRef]
14. Awaja, F.; Zhang, S.; Tripathi, M.; Nikiforov, A.; Pugno, N. Cracks, microcracks and fracture in polymer structures: Formation, detection, autonomic repair. *Prog. Mater. Sci.* **2016**, *83*, 536–573. [CrossRef]
15. Lewandowski, M.; Pawłowska, U. Part I. Degradation of elastomers and prediction of lifetime. *Elastomery* **2016**, *20*, 24–30.

16. Gijsman, P.; Meijers, G.; Vitarelli, G. Comparison of the UV-degradation chemistry of polypropylene, polyethylene, polyamide 6 and polybutylene terephthalate. *Polym. Degrad. Stab.* **1999**, *65*, 433–441. [CrossRef]
17. Gillen, K.T.; Celina, M. The wear-out approach for predicting the remaining lifetime of materials. *Polym. Degrad. Stab.* **2000**, *71*, 15–30. [CrossRef]
18. Hondred, P.R. Polymer Damage Mitigation-Predictive Lifetime Models of Polymer Insulation Degradation and Biorenewable Thermosets through Cationic Polymerization for Self-Healing Applications. Ph.D. Thesis, Iowa State University, Ames, IA, USA, 2013.
19. Ashter, S.A. Mechanisms of Polymer Degradation. In *Introduction to Bioplastics Engineering*; William Andrew: Norwich, NY, USA, 2016; pp. 31–59. ISBN 978-0-323-39396-6.
20. Gogate, P.R.; Prajapat, A.L. Depolymerization using sonochemical reactors: A critical review. *Ultrason. Sonochem.* **2015**, *27*, 480–494. [CrossRef]
21. Godiya, C.B.; Gabrielli, S.; Materazzi, S.; Pianesi, M.S.; Stefanini, N.; Marcantoni, E. Depolymerization of waste poly(methyl methacrylate) scraps and purification of depolymerized products. *J. Environ. Manag.* **2019**, *231*, 1012–1020. [CrossRef]
22. Wiles, D.M.; Scott, G. Polyolefins with controlled environmental degradability. *Polym. Degrad. Stab.* **2006**, *91*, 1581–1592. [CrossRef]
23. Bhuvaneswari, G.H. Degradability of Polymers. In *Recycling of Polyurethane Foams*; Thomas, S., Rane, A.V., Kanny, K., Abitha, V.K., Thomas, G.M., Eds.; William Andrew: Norwich, NY, USA, 2018; pp. 29–44. ISBN 9780323511339.
24. Anju, S.; Prajitha, N.; Sukanya, V.S.; Mohanan, P.V. Complicity of degradable polymers in health-care applications. *Mater. Today Chem.* **2020**, *16*, 100236. [CrossRef]
25. Melnikov, M.; Seropegina, E.N. Photoradical ageing of polymers. *Int. J. Polym. Mater.* **1996**, *31*, 41–93. [CrossRef]
26. La Mantia, F.P.; Morreale, M.; Botta, L.; Mistretta, M.C.; Ceraulo, M.; Scaffaro, R. Degradation of polymer blends: A brief review. *Polym. Degrad. Stab.* **2017**, *145*, 79–92. [CrossRef]
27. White, J.R. Polymer ageing: Physics, chemistry or engineering? Time to reflect. *C. R. Chim.* **2006**, *9*, 1396–1408. [CrossRef]
28. Chmielnicki, B. Niektóre aspekty starzenia wytworów z poliamidów wzmocnionych. Cz. 1 Podatność poliamidów na procesy starzenia. *Przetwórstwo Tworzyw* **2009**, *15*, 116–122.
29. Rojek, M. *Metodologia Badań Diagnostycznych Warstwowych Materiałów Kompozytowych o Osnowie Polimerowej*; Dobrzański, L.A., Ed.; International OCSCO World Press: Gliwice, Poland, 2011; ISBN 83-89728-89-3.
30. Sobków, D.; Czaja, K. Influence of accelerated aging conditions on the process of polyolefines degradation. *Polimery* **2003**, *48*, 627–632. [CrossRef]
31. Gijsman, P. Review on the thermo-oxidative degradation of polymers during processing and in service. *E-Polymers* **2008**, *8*, 1–34. [CrossRef]
32. Moraczewski, K.; Stepczyńska, A.; Malinowski, R.; Karasiewicz, T.; Jagodziński, B.; Rytlewski, P. The Effect of Accelerated Aging on Polylactide Containing Plant Extracts. *Polymers* **2019**, *11*, 575. [CrossRef] [PubMed]
33. Masłowski, M.; Zaborski, M. Effect of thermooxidative and photooxidative aging processes on mechanical properties of magnetorheological elastomer composites. *Polimery* **2015**, *60*, 264–271. [CrossRef]
34. Król-Morkisz, K.; Pielichowska, K. Thermal Decomposition of Polymer Nanocomposites With Functionalized Nanoparticles. In *Polymer Composites with Functionalized Nanoparticles*; Elsevier: Rotterdam, The Netherland, 2019; pp. 405–435. ISBN 978-0-12-814064-2.
35. Mierzwa-Hersztek, M.; Gondek, K.; Kopeć, M. Degradation of Polyethylene and Biocomponent-Derived Polymer Materials: An Overview. *J. Polym. Environ.* **2019**, *27*, 600–611. [CrossRef]
36. Muthukumar, A.; Veerappapillai, S. Biodegradation of Plastics—A Brief Review. *Int. J. Pharm. Sci. Rev. Res.* **2015**, *31*, 204–209.
37. Lyu, S.; Untereker, D. Degradability of Polymers for Implantable Biomedical Devices. *Int. J. Mol. Sci.* **2009**, *10*, 4033–4065. [CrossRef]
38. Wojtala, A. The effect of properties of polyolefines and outdoor factors on the course of polyolefines degradation. *Polimery* **2001**, *46*, 120–124. [CrossRef]
39. Laurence, W.M. Introduction to the Effect of Heat Aging on Plastics. In *The Effect of Long Term Thermal Exposure on Plastics and Elastomers*; William Andrew: Norwich, NY, USA, 2014; pp. 17–42. ISBN 978-0-323-22108-5.

40. Reis, A.; Spickett, C.M. Chemistry of phospholipid oxidation. *Biochim. Biophys. Acta* **2012**, *1818*, 2374–2387. [CrossRef]
41. Kröhnke, C. Polymer Additives. In *Polymer Science: A Comprehensive Reference*; Elsevier, B.V.: Amsterdam, The Netherlands, 2012; Volume 8, pp. 349–375. ISBN 9780444533494.
42. Liu, X.; Gao, C.; Sangwan, P.; Yu, L.; Tong, Z. Accelerating the degradation of polyolefins through additives and blending. *J. Appl. Polym. Sci.* **2014**, *131*, 9001–9015. [CrossRef]
43. Tolinski, M. Ultraviolet Light Protection and Stabilization. In *Additives for Polyolefins*; William Andrew: Norwich, NY, USA, 2015; pp. 32–43. ISBN 9780323358842.
44. Lu, M.; Zhang, H.; Sun, L. Quantitative prediction of elastomer degradation and mechanical behavior based on diffusion–reaction process. *J. Appl. Polym. Sci.* **2020**, 1–9. [CrossRef]
45. Bin Samsuri, A.; Abdullahi, A.A. Degradation of Natural Rubber and Synthetic Elastomers. In *Reference Module in Materials Science and Materials Engineering*; Elsevier Ltd.: Amsterdam, The Netherlands, 2017; pp. 1–32. ISBN 9780128035818.
46. Tobolsky, A.V. *Properties and Structure of Polymers*; Wiley: New York, NY, USA, 1960.
47. Rybiński, P.; Kucharska-Jastrząbek, A.; Janowska, G. Thermal Properties of Diene Elastomers. *Modif. Polym.* **2014**, *56*, 477–486. [CrossRef]
48. Coquillat, M.; Verdu, J.; Colin, X.; Audouin, L.; Nevie, R. Thermal oxidation of polybutadiene. Part 1: Effect of temperature, oxygen pressure and sample thickness on the thermal oxidation of hydroxyl-terminated polybutadiene. *Polym. Degrad. Stab.* **2007**, *92*, 1326–1333. [CrossRef]
49. Masek, A. Flavonoids as Natural Stabilizers and Color Indicators of Ageing for Polymeric Materials. *Polymers* **2015**, *7*, 1125–1144. [CrossRef]
50. Thompson, R.C.; Moore, C.J.; Saal, F.S.; Swan, S.H. Plastics, the environment and human health: Current consensus and future trends. *Philos. Trans. R. Soc. B Biol. Sci.* **2009**, *364*, 2153–2166. [CrossRef]
51. ISO, the International Organization for Standardization. *Rubber, Vulcanized or Thermoplastic—Estimation of Life-Time and Maximum Temperature of Use (Standard No. ISO 11346:2014)*; International Organization for Standardization: Geneva, Switzerland; pp. 1–10.
52. Dobkowski, Z. Lifetime prediction for polymer materials using OIT measurements by the DSC method. *Polimery* **2005**, *50*, 213–215. [CrossRef]
53. ISO, the International Organization for Standardization. *Rubber Products—Guidelines for Storage (Standard No. ISO 2230:2002)*; International Organization for Standardization: Geneva, Switzerland; pp. 1–11.
54. Shah, C.S.; Patni, M.J. Accelerated Aging and Life Time Prediction Analysis of Polymer Composites: A New Approach for a Realistic Prediction Using Cumulative Damage Theory. *Polym. Test.* **1994**, *13*, 295–322. [CrossRef]
55. Kiliaris, P.; Papaspyrides, C.D.; Pfaendner, R. Influence of accelerated aging on clay-reinforced polyamide 6. *Polym. Degrad. Stab.* **2009**, *94*, 389–396. [CrossRef]
56. Calixto, E. Accelerated Test and Reliability Growth Analysis Models. In *Gas and Oil Reliability Engineering*; Gulf Professional Publishing: Houston, TX, USA, 2013; pp. 63–118. ISBN 978-0-12-391914-4.
57. Jachowicz, T.; Sikora, R. Methods of forecasting of the changes of polymeric products properties. *Polimery* **2006**, *51*, 177–185. [CrossRef]
58. Singh, H.K. Lifetime Prediction and Durability of Elastomeric Seals for Fuel Cell Applications. Ph.D. Thesis, Virginia Polytechnic Institute, Blacksburg, VA, USA, 2009.
59. Le Huy, M.; Evrard, G. Methodologies for lifetime predictions of rubber using Arrhenius and WLF models. *Die Angew. Makromol. Chem.* **1998**, *261–262*, 135–142. [CrossRef]
60. Gillen, K.T.; Bernstein, R.; Celina, M. Challenges of accelerated aging techniques for elastomer lifetime predictions. *Rubber Chem. Technol.* **2015**, *88*, 1–27. [CrossRef]
61. Käser, F.; Roduit, B. Lifetime prediction of rubber using the chemiluminescence approach and isoconversional kinetics. *J. Therm. Anal. Calorim.* **2008**, *93*, 231–237. [CrossRef]
62. Denis, L.; Grzeskowiak, H.; Trias, D.; Delaux, D. Accelerated Life Testing. In *Reliability of High-Power Mechatronic Systems 2*; ISTE Press—Elsevier: London, UK, 2017; pp. 1–56. ISBN 9781785482618.
63. Li, J.; Tian, Y.; Wang, D. Change-point detection of failure mechanism for electronic devices based on Arrhenius model. *Appl. Math. Model.* **2020**, *83*, 46–58. [CrossRef]
64. Budrugeac, P. Theory and practice in the thermoanalytical kinetics of complex processes: Application for the isothermal and non-isothermal thermal degradation of HDPE. *Thermochim. Acta* **2010**, *500*, 30–37. [CrossRef]

65. Pielichowski, J.; Pielichowski, K. Application of thermal analysis for the investigation of polymer degradation processes. *J. Therm. Anal.* **1995**, *43*, 505–508. [CrossRef]
66. Kutz, M. Thermal Degradation of Polymer and Polymer Composites. In *Handbook of Environmental Degradation of Materials*; William Andrew: Norwich, NY, USA, 2018; pp. 185–206. ISBN 9780323524728.
67. Ajitha, A.R.; Sabu, T. Applications of compatibilized polymer blends in automobile industry. In *Compatibilization of Polymer Blends. Micro and Nano Scale Phase Morphologies, Interphase Characterization and Properties*; Elsevier: Rotterdam, The Netherland, 2020; pp. 563–593. ISBN 9780128160060.
68. Lühr, C.; Pecenka, R. Development of a model for the fast analysis of polymer mixtures based on cellulose, hemicellulose (xylan), lignin using thermogravimetric analysis and application of the model to poplar wood. *Fuel* **2020**, *277*. [CrossRef]
69. Dai, L.; Wang, L.Y.; Yuan, T.Q.; He, J. Study on thermal degradation kinetics of cellulose-graft-poly(l-lactic acid) by thermogravimetric analysis. *Polym. Degrad. Stab.* **2014**, *99*, 233–239. [CrossRef]
70. Seifi, H.; Gholami, T.; Seifi, S.; Ghoreishi, S.M.; Salavati-Niasari, M. A review on current trends in thermal analysis and hyphenated techniques in the investigation of physical, mechanical and chemical properties of nanomaterials. *J. Anal. Appl. Pyrolysis* **2020**, *149*, 104840. [CrossRef]
71. Lau, K.S.Y. High-Performance Polyimides and High Temperature Resistant Polymers. In *Handbook of Thermoset Plastics*; Dodiuk, H., Goodman, S.H., Eds.; William Andrew: Norwich, NY, USA, 2014; pp. 297–424. ISBN 978-1-4557-3107-7.
72. Niu, S.; Yu, H.; Zhao, S.; Zhang, X.; Li, X.; Han, K.; Lu, C.; Wang, Y. Apparent kinetic and thermodynamic calculation for thermal degradation of stearic acid and its esterification derivants through thermogravimetric analysis. *Renew. Energy* **2019**, *133*, 373–381. [CrossRef]
73. Artiaga, R.; Cao, R.; Naya, S.; Garcia, A. Polymer Degradation from the Thermal Analysis Point of View. *Mater. Res. Soc.* **2004**, *851*, 499–510. [CrossRef]
74. Blanco, I. Lifetime prediction of polymers: To bet, or not to bet-is this the question? *Materials* **2018**, *11*, 1383. [CrossRef] [PubMed]
75. Nyombi, A.; Williams, M.; Wessling, R. Determination of kinetic parameters and thermodynamic properties for ash (Fraxinus) wood sawdust slow pyrolysis by thermogravimetric analysis. *Energy Sources Part A Recover. Util. Environ. Eff.* **2018**, *40*, 2660–2670. [CrossRef]
76. Woo, C.S.; Park, H.S.; Kwang, M.C. Design and applications: Evaluation of characteristics for chevron rubber spring. In *Constitutive Models for Rubber VIII*; Gil-Negrete, N., Alonso, A., Eds.; CRC Press: Boca Raton, FL, USA, 2013; pp. 621–626. ISBN 9781138000728.
77. Kaczmarek, H.; Kosmalska, D.; Malinowski, R.; Bajer, K. Advances in studies of thermal degradation of polymeric materials. Part I. Literature studies. *Polimery* **2019**, *64*, 239–314. [CrossRef]
78. Capart, R.; Khezami, L.; Burnham, A.K. Assessment of various kinetic models for the pyrolysis of a microgranular cellulose. *Thermochim. Acta* **2004**, *417*, 79–89. [CrossRef]
79. Materazzi, S.; Vecchio, S. Evolved Gas Analysis by Mass Spectrometry. *Appl. Spectrosc. Rev.* **2011**, *46*, 261–340. [CrossRef]
80. Qin, L.; Han, J.; Zhao, B.; Wang, Y.; Chen, W.; Xing, F. Thermal degradation of medical plastic waste by in-situ FTIR, TG-MS and TG-GC/MS coupled analyses. *J. Anal. Appl. Pyrolysis* **2018**, *136*, 132–145. [CrossRef]
81. Li, L.Q.; Guan, C.X.; Zhang, A.Q.; Chen, D.H.; Qing, Z.B. Thermal stabilities and thermal degradation kinetics of polyimides. *Polym. Degrad. Stab.* **2004**, *84*, 369–373. [CrossRef]
82. Jin, W.P.; Sea, C.O.; Hac, P.L.; Hee, T.K.; Kyong, O.Y. A kinetic analysis of thermal degradation of polymers using a dynamic method. *Polym. Degrad. Stab.* **2000**, *67*, 535–540. [CrossRef]
83. Yang, K.K.; Wang, X.L.; Wang, Y.Z.; Wu, B.; Yin, Y.D.; Yang, B. Kinetics of thermal degradation and thermal oxidative degradation of poly(p-dioxanone). *Eur. Polym. J.* **2003**, *39*, 1567–1574. [CrossRef]
84. Plonka, A. Kinetics in condensed media. In *Dispiersive Kinetics*; Springer: Heidelberg, Germany, 2001; pp. 194–195. ISBN 978-94-015-9658-9.
85. Park, B.D.; Kadla, J.F. Thermal degradation kinetics of resole phenol-formaldehyde resin/multi-walled carbon nanotube/cellulose nanocomposite. *Thermochim. Acta* **2012**, *540*, 107–115. [CrossRef]
86. Van Krevelen, D.W.; Te Nijenhuis, K. Thermal Decomposition. In *Properties of Polymers*; Elsevier Science: Rotterdam, The Netherland, 2009; pp. 763–777. ISBN 9780080548197.
87. Saha, T.K.; Purkait, P. Transformer Insulation Materials and Ageing. In *Transformer Ageing: Monitoring and Estimation Techniques*; Wiley-IEEE Press: New York, NY, USA, 2017; pp. 1–34. ISBN 978-1-119-23996-3.

88. Celina, M.; Gillen, K.T.; Assink, R.A. Accelerated aging and lifetime prediction: Review of non-Arrhenius behaviour due to two competing processes. *Polym. Degrad. Stab.* **2005**, *90*, 395–404. [CrossRef]
89. Vyazovkin, S. Activation energies and temperature dependencies of the rates of crystallization and melting of polymers. *Polymers* **2020**, *12*, 1070. [CrossRef] [PubMed]
90. Moon, B.; Kim, K.; Park, K.; Park, S.; Seok, C.S. Fatigue life prediction of tire sidewall using modified Arrhenius equation. *Mech. Mater.* **2020**, *147*. [CrossRef]
91. Tsuji, T.; Mochizuki, K.; Okada, K.; Hayashi, Y.; Obata, Y.; Takayama, K.; Onuki, Y. Time–temperature superposition principle for the kinetic analysis of destabilization of pharmaceutical emulsions. *Int. J. Pharm.* **2019**, *563*, 406–412. [CrossRef] [PubMed]
92. Hulme, A.; Cooper, J. Life prediction of polymers for industry. *Seal. Technol.* **2012**, *9*, 8–12. [CrossRef]
93. Brown, R.P.; Greenwood, J.H. Prediction Techniques. In *Practical Guide to the Assessment of the Useful Life of Plastics*; Smithers Rapra Technology: Shropshire, UK, 2002; pp. 85–94. ISBN 978-1-85957-312-9.
94. Moon, B.; Jun, N.; Park, S.; Seok, C.S.; Hong, U.S. A study on the modified Arrhenius equation using the oxygen permeation block model of crosslink structure. *Polymers* **2019**, *11*, 136. [CrossRef]
95. Whitten, K.; Davis, R.; Peck, L.; Stanley, G. Chemical Kinetics. In *Chemistry*; Mary Finch: Boston, MA, USA, 2018; pp. 606–628. ISBN 978-0-495-39163-0.
96. Xiong, Y.; Chen, G.; Guo, S.; Li, G. Lifetime prediction of NBR composite sheet in aviation kerosene by using nonlinear curve fitting of ATR-FTIR spectra. *J. Ind. Eng. Chem.* **2013**, *19*, 1611–1616. [CrossRef]
97. Madej-Kiełbik, L.; Kośla, K.; Zielińska, D.; Chmal-fudali, E.; Maciejewska, M. Effect of Accelerated Ageing on the Mechanical and Structural Properties of the Material System Used in Protectors. *Polymers* **2019**, *11*, 1263. [CrossRef]
98. Koga, Y.; Arao, Y.; Kubouchi, M. Application of small punch test to lifetime prediction of plasticized polyvinyl chloride wire. *Polym. Degrad. Stab.* **2019**, 109013. [CrossRef]
99. Wang, Y.; Lan, H.; Meng, T. Lifetime prediction of natural gas polyethylene pipes with internal pressures. *Eng. Fail. Anal.* **2019**, *95*, 154–163. [CrossRef]
100. Lee, L.S. Creep and time-dependent response of composites. In *Durability of Composites for Civil Structural Applications*; Karbhari, V.M., Ed.; Woodhead Publishing: Cambridge, UK, 2007; pp. 150–169. ISBN 978-1-84569-035-9.
101. Krauklis, A.E.; Akulichev, A.G.; Gagani, A.I.; Echtermeyer, A.T. Time-temperature-plasticization superposition principle: Predicting creep of a plasticized epoxy. *Polymers* **2019**, *11*, 1848. [CrossRef] [PubMed]
102. Goertzen, W.K.; Kessler, M.R. Creep behavior of carbon fiber/epoxy matrix composites. *Mater. Sci. Eng. A* **2006**, *421*, 217–225. [CrossRef]
103. Ahmed, J.; Ptaszek, P.; Basu, S. Time-Temperature Superposition Principle and its Application to Biopolymer and Food Rheology. In *Advances in Food Rheology and Its Applications*; Woodhead Publishing: Cambridge, UK, 2016; pp. 209–241. ISBN 978-0-08-100431-9.
104. Nakada, M.; Miyano, Y.; Cai, H. Prediction of long-term viscoelastic behavior of amorphous resin based on the time-temperature superposition principle. *Mech. Time Depend. Mater.* **2011**, *15*, 309–316. [CrossRef]
105. Fukushima, K.; Cai, H.; Nakada, M.; Miyano, Y. Determination of Time-Temperature Shift Factor for Long-Term Life Prediction of Polymer Composites. In Proceedings of the ICCM-17 17th International Conference on Composite Materials, Edinburgh, UK, 27–31 July 2009.
106. Yin, W.; Xie, Z.; Yin, Y.; Yi, J.; Liu, X.; Wu, H.; Wang, S.; Xie, Y.; Yang, Y. Aging behavior and lifetime prediction of PMMA under tensile stress and liquid scintillator conditions. *Adv. Ind. Eng. Polym. Res.* **2019**, *2*, 82–87. [CrossRef]
107. Sopade, P.A.; Halley, P.; Bhandari, B.; D'Arcy, B.; Doebler, C.; Caffin, N. Application of the Williams-Landel-Ferry model to the viscosity-temperature relationship of Australian honeys. *J. Food Eng.* **2003**, *56*, 67–75. [CrossRef]
108. Wise, C.W.; Cook, W.D.; Goodwin, A.A. Chemico-diffusion kinetics of model epoxy-amine resins. *Polymer* **1997**, *38*, 3251–3261. [CrossRef]
109. Yildiz, M.E.; Kokini, J.L. Determination of Williams–Landel–Ferry constants for a food polymer system: Effect of water activity and moisture content. *J. Rheol.* **2001**, *45*, 903–912. [CrossRef]
110. Ionita, D.; Cristea, M.; Gaina, C. Prediction of polyurethane behaviour via time-temperature superposition: Meanings and limitations. *Polym. Test.* **2020**, *83*. [CrossRef]

111. Nelson, K.A.; Labuza, T.P. Water activity and food polymer science: Implications of state on Arrhenius and WLF models in predicting shelf life. *J. Food Eng.* **1994**, *22*, 271–289. [CrossRef]
112. Fabre, V.; Quandalle, G.; Billon, N.; Cantournet, S. Time-Temperature-Water content equivalence on dynamic mechanical response of polyamide 6,6. *Polymer* **2018**, *137*, 22–29. [CrossRef]
113. Hu, X.L.; Luo, W.B.; Liu, X.; Li, M.; Huang, Y.J.; Bu, J.L. Temperature and frequency dependent rheological behaviour of carbon black filled natural rubber. *Plast. Rubber Compos.* **2013**, *42*, 416–420. [CrossRef]
114. Yildiz, M.E.; Sozer, N.; Kokini, J.L. Williams-Landel-Ferry (WLF) equation. In *Encyclopedia of Agricultural, Food, and Biological Engineering*; CRC Press: Boca Raton, FL, USA, 2003; pp. 1865–1877. ISBN 9780824709389.
115. Ljubic, D.; Stamenovic, M.; Smithson, C.; Nujkic, M.; Medjo, B.; Putic, S. Time-temperature superposition principle—Application of WLF equation in polymer analysis and composites. *Zast. Mater.* **2014**, *55*, 395–400. [CrossRef]
116. Chang, F.C.; Lam, F.; Kadla, J.F. Application of time–temperature–stress superposition on creep of wood–plastic composites. *Mech. Time-Depend. Mater.* **2013**, *17*, 427–437. [CrossRef]
117. Dan-asabe, B. Thermo-mechanical characterization of banana particulate reinforced PVC composite as piping material. *J. King Saud Univ.-Eng. Sci.* **2016**, 1–9. [CrossRef]
118. Sbirrazzuoli, N. Advanced Isoconversional Kinetic Analysis for the Elucidation of Complex Reaction Mechanisms: A New Method for the Identification ofRate-Limiting Steps. *Molecules* **2019**, *24*, 1683. [CrossRef] [PubMed]
119. Berčič, G. The universality of Friedman's isoconversional analysis results in a model-less prediction of thermodegradation profiles. *Thermochim. Acta* **2017**, *650*, 1–7. [CrossRef]
120. Ozawa, T. A New Method of Analyzing Thermogravimetric Data. *Bull. Chem. Soc. Jpn.* **1965**, *38*, 1881–1886. [CrossRef]
121. Flynn, J.H.; Wall, L.A. A quick, direct method for the determination of activation energy from thermogravimetric data. *J. Polym. Sci. Part B Polym. Lett.* **1966**, *4*, 323–328. [CrossRef]
122. Friedman, H.L. Kinetics of thermal degradation of char-forming plastics from thermogravimetry. Application to a phenolic plastic. *J. Polym. Sci. Part C Polym. Symp.* **2007**, *6*, 183–195. [CrossRef]
123. Akahira, T.; Sunose, T. Joint Convention of Four Electrical Institutes: Method of Determining Activation Deterioration Constant of Electrical Insulating Materials. *Res. Rep. Chiba Inst. Technol.* **1971**, *16*, 22–31.
124. Starink, M.J. Activation energy determination for linear heating experiments: Deviations due to neglecting the low temperature end of the temperature integral. *J. Mater. Sci.* **2007**, *42*, 483–489. [CrossRef]
125. Gao, Z.; Wang, H.; Nakada, M. Iterative method to improve calculation of the pre-exponential factor for dynamic thermogravimetric analysis measurements. *Polymer* **2006**, *47*, 1590–1596. [CrossRef]
126. Vyazovkin, S. Evaluation of activation energy of thermally stimulated solid-state reactions under arbitrary variation of temperature. *J. Comput. Chem.* **1997**, *18*, 393–402. [CrossRef]
127. Vyazovkin, S. Modification of the integral isoconversional method to account for variation in the activation energy. *J. Comput. Chem.* **2001**, *22*, 178–183. [CrossRef]
128. Vyazovkin, S.; Chrissafis, K.; Di Lorenzo, M.L.; Koga, N.; Pijolat, M.; Roduit, B.; Sbirrazzuoli, N.; Suñol, J.J. ICTAC Kinetics Committee recommendations for collecting experimental thermal analysis data for kinetic computations. *Thermochim. Acta* **2014**, *590*, 1–23. [CrossRef]
129. Šimon, P. Isoconversional methods: Fundamentals, meaning and application. *J. Therm. Anal. Calorim.* **2004**, *76*, 123–132. [CrossRef]
130. Criado, J.M.; Sánchez-Jiménez, P.E.; Pérez-Maqueda, L.A. Critical study of the isoconversional methods of kinetic analysis. *J. Therm. Anal. Calorim.* **2008**, *92*, 199–203. [CrossRef]
131. Sánchez-Jiménez, P.E.; Pérez-Maqueda, L.A.; Perejón, A.; Criado, J.M. Generalized master plots as a straightforward approach for determining the kinetic model: The case of cellulose pyrolysis. *Thermochim. Acta* **2013**, *552*, 54–59. [CrossRef]
132. Leroy, V.; Cancellieri, D.; Leoni, E.; Rossi, J.L. Kinetic study of forest fuels by TGA: Model-free kinetic approach for the prediction of phenomena. *Thermochim. Acta* **2010**, *497*, 1–6. [CrossRef]
133. Das, P.; Tiwari, P. Thermal degradation kinetics of plastics and model selection. *Thermochim. Acta* **2017**, *654*, 191–202. [CrossRef]

134. Cui, H.W.; Jiu, J.T.; Sugahara, T.; Nagao, S.; Suganuma, K.; Uchida, H.; Schroder, K.A. Using the Friedman method to study the thermal degradation kinetics of photonically cured electrically conductive adhesives. *J. Therm. Anal. Calorim.* **2014**, *119*, 425–434. [CrossRef]

135. Mittal, A.; Garg, S.; Bajpai, S. Thermal decomposition kinetics and properties of grafted barley husk reinforced PVA/starch composite films for packaging applications. *Carbohydr. Polym.* **2020**, *240*, 116225. [CrossRef] [PubMed]

136. Kropidłowska, A.; Rotaru, A.; Strankowski, M.; Becker, B.; Segal, E. Heteroleptic cadmium(II) complex, potential precursor for semiconducting CDS layers: TTThermal stability and non-isothermal decomposition kinetics. *J. Therm. Anal. Calorim.* **2008**, *91*, 903–909. [CrossRef]

137. Venkatesh, M.; Ravi, P.; Tewari, S.P. Isoconversional kinetic analysis of decomposition of nitroimidazoles: Friedman method vs. Flynn-Wall-Ozawa method. *J. Phys. Chem. A* **2013**, *117*, 10162–10169. [CrossRef] [PubMed]

138. Ali, G.; Nisar, J.; Iqbal, M.; Shah, A.; Abbas, M.; Shah, M.R.; Rashid, U.; Bhatti, I.A.; Khan, R.A.; Shah, F. Thermo-catalytic decomposition of polystyrene waste: Comparative analysis using different kinetic models. *Waste Manag. Res.* **2019**, *38*, 1–11. [CrossRef]

139. Vassiliou, A.A.; Chrissafis, K.; Bikiaris, D.N. In situ prepared PET nanocomposites: Effect of organically modified montmorillonite and fumed silica nanoparticles on PET physical properties and thermal degradation kinetics. *Thermochim. Acta* **2010**, *500*, 21–29. [CrossRef]

140. Benhacine, F.; Yahiaoui, F.; Hadj-hamou, A.S. Thermal Stability and Kinetic Study of Isotactic Polypropylene/Algerian Bentonite Nanocomposites Prepared via Melt Blending. *J. Polym.* **2014**, *2014*, 1–9. [CrossRef]

141. Sbirrazzuoli, N.; Vincent, L.; Mija, A.; Guigo, N. Integral, differential and advanced isoconversional methods. Complex mechanisms and isothermal predicted conversion-time curves. *Chemom. Intell. Lab. Syst.* **2009**, *96*, 219–226. [CrossRef]

142. Lim, A.C.R.; Chin, B.L.F.; Jawad, Z.A.; Hii, K.L. Kinetic Analysis of Rice Husk Pyrolysis Using Kissinger-Akahira-Sunose (KAS) Method. *Procedia Eng.* **2016**, *148*, 1247–1251. [CrossRef]

143. Heydari, M.; Rahman, M.; Gupta, R. Kinetic study and thermal decomposition behavior of lignite coal. *Int. J. Chem. Eng.* **2015**, *2015*, 1–9. [CrossRef]

144. Wang, J.; Laborie, M.G.; Wolcott, M.P. Comparison of model-free kinetic methods for modeling the cure kinetics of commercial phenol—Formaldehyde resins. *Thermochim. Acta* **2005**, *439*, 68–73. [CrossRef]

145. Gabilondo, N.; López, M.; Ramos, J.A.; Echeverría, J.M.; Mondragon, I. Curing kinetics of amine and sodium hydroxide catalyzed phenol-formaldehyde resins. *J. Therm. Anal. Calorim.* **2007**, *90*, 229–236. [CrossRef]

146. Sun, H.; Liu, Y.; Wang, Y.; Tan, H. Curing Behavior of Epoxy Resins in Two-Stage Curing Process by Non-Isothermal Differential Scanning Calorimetry Kinetics Method. *J. Appl. Polym. Sci.* **2014**, *40711*, 1–8. [CrossRef]

147. Vyazovkin, S.; Sbirrazzuoli, N. Isoconversional Analysis of Calorimetric Data on Nonisothermal Crystallization of a Polymer Melt. *J. Phys. Chem. B* **2003**, *107*, 882–888. [CrossRef]

148. Vyazovkin, S. A Unified Approach to Kinetic Processing of Nonisothermal Data. *J. Chem. Kinet.* **1996**, *28*, 95–101. [CrossRef]

149. Vyazovkin, S.; Sbirrazzuoli, N. Mechanism and kinetics of epoxy-amine cure studied by differential scanning calorimetry. *Macromolecules* **1996**, *29*, 1867–1873. [CrossRef]

150. Dunne, R.C.; Sitaraman, S.K.; Luo, S.; Rao, Y.; Wong, C.P.; Estes, W.E.; Gonzalez, C.G.; Coburn, J.C.; Periyasamy, M. Investigation of the curing behavior of a novel epoxy photo-dielectric dry film (ViaLux^{TM} 81) for high density interconnect applications. *J. Appl. Polym. Sci.* **2000**, *78*, 430–437. [CrossRef]

151. Li, S.Y.; Vuorimaa, E.; Lemmetyinen, H. Application of isothermal and model-free isoconversional modes in DSC measurement for the curing process of the PU system. *J. Appl. Polym. Sci.* **2001**, *81*, 1474–1480. [CrossRef]

152. He, G.; Yan, N. Effect of moisture content on curing kinetics of pMDI resin and wood mixtures. *Int. J. Adhes. Adhes.* **2005**, *25*, 450–455. [CrossRef]

153. Polli, H.; Pontes, L.A.M.; Araujo, A.S. Application of model-free kinetics to the study of thermal degradation of polycarbonate. *J. Therm. Anal. Calorim.* **2005**, *79*, 383–387. [CrossRef]
154. Vyazovkin, S.; Dranca, I.; Fan, X.; Advincula, R. Degradation and relaxation kinetics of polystyrene-clay nanocomposite prepared by surface initiated polymerization. *J. Phys. Chem. B* **2004**, *108*, 11672–11679. [CrossRef]

© 2020 by the authors. Licensee MDPI, Basel, Switzerland. This article is an open access article distributed under the terms and conditions of the Creative Commons Attribution (CC BY) license (http://creativecommons.org/licenses/by/4.0/).

Article

Study of Industrial Grade Thermal Insulation at Elevated Temperatures

Amalie Gunnarshaug [1,2], Maria Monika Metallinou [3,*] and Torgrim Log [3]

1. Q Rådgivning AS, 5527 Haugesund, Norway; amg@q-rad.no
2. Department of Physics and Technology, University of Bergen, 5020 Bergen, Norway
3. Fire Disasters Research Group, Department of Safety, Chemistry and Biomedical Laboratory Sciences, Western Norway University of Applied Sciences, 5528 Haugesund, Norway; torgrim.log@hvl.no
* Correspondence: monika.metallinou@hvl.no; Tel.: +47-9882-5104

Received: 21 September 2020; Accepted: 13 October 2020; Published: 16 October 2020

Abstract: Thermal insulation is used for preventing heat losses or heat gains in various applications. In industries that process combustible products, inorganic-materials-based thermal insulation may, if proven sufficiently heat resistant, also provide heat protection in fire incidents. The present study investigated the performance and breakdown temperature of industrial thermal insulation exposed to temperatures up to 1200 °C, i.e., temperatures associated with severe hydrocarbon fires. The thermal insulation properties were investigated using thermogravimetric analysis (TGA), differential scanning calorimetry (DSC) and by heating 50 mm cubes in a muffle furnace to temperatures in the range of 600 to 1200 °C with a 30 min holding time. The room temperature thermal conductivity was also recorded after each heat treatment. Upon heating, the mineral-based oil dust binder was released at temperatures in the range of 300 to 500 °C, while the Bakelite binder was released at temperatures in the range of 850 to 960 °C. The 50 mm test cubes experienced increasing levels of sintering in the temperature range of 700 to 1100 °C. At temperatures above 1100 °C, the thermal insulation started degrading significantly. Due to being heat-treated to 1200 °C, the test specimen morphology was similar to a slightly porous rock and the original density of 140 kg/m^3 increased to 1700 kg/m^3. Similarly, the room temperature thermal conductivity increased from 0.041 to 0.22 W/m·K. The DSC analysis confirmed an endothermic peak at about 1200 °C, indicating melting, which explained the increase in density and thermal conductivity. Recently, 350 kW/m^2 has been set as a test target heat flux, i.e., corresponding to an adiabatic temperature of 1200 °C. If a thin layer of thermally robust insulation is placed at the heat-exposed side, the studied thermal insulation may provide significant passive fire protection, even when exposed to heat fluxes up to 350 kW/m^2. It is suggested that this is further analysed in future studies.

Keywords: industrial thermal insulation; passive fire protection; hydrocarbon fires; thermal conductivity; TGA; DSC; TPS

1. Introduction

Thermal insulation is widely used in several application areas, such as in the building industry, refrigeration plants and in the process industries. In the process industries, examples of typical application areas may be temperature control, personnel protection, humidity condensation prevention or sound attenuation. In the hydrocarbon process industry, thermal insulation may be necessary in order to maintain the required production temperature [1]. A typical example may be distillation columns; in order to obtain good production efficiency and quality of the distilled products, the temperature profiles are carefully designed. These types of process equipment may represent a potential for a major accidental hazard, as it may release large quantities of flammable material upon a potential leak. Ignition of a hydrocarbon leak may lead to severe heat loads to the exposed object, i.e., flame temperatures in the range of 1100 to 1200 °C, corresponding to heat loads in the range of 250 to 350 kW/m² [2,3]. These high heat fluxes will result in a significant temperature increase in exposed production units, which may result in weakening of the steel and a possible escalation of the fire scenario, especially if the steel temperature exceeds 500 °C [4]. The outcome of a process equipment rupture may be a disastrous release, as evidenced by several major accidents during the last decade [5–7]. Hence, these types of units are also often protected with an additional layer of mineral-based passive fire protection.

The fire resistance of the passive fire protection is normally given as a time in minutes until an object reaches a specified critical temperature. In the oil and gas industry, the critical steel temperature is conservatively set to 400 °C, i.e., the protected element should not achieve temperatures above 400 °C within the specified time frame [8]. The fire resistance is typically given in intervals of 15, 30, 60 or 120 min [9].

Previous studies of 50 mm thick thermal insulation (ProRox PSM 971, 50 mm, Rockwool) protecting a 16 mm thick steel wall using small scale testing [10,11] rather than full-scale testing [12] demonstrated that the thermal insulation alone was sufficient to withstand 30 min of jet fire exposure. Even with only 3 mm thick steel walls, the testing showed sufficient fire protection for 20 min. During the testing, the thermal insulation sintered and partially melted in some locations. The sintering and melting of the insulation due to the heat exposure resulted in cracks/openings in the insulation mat, as shown in Figure 1b,c. The previous oven testing up to 1100 °C [11] showed minimal shrinking (less than 25%) of the thermal insulation. Hence, a further examination of the thermal insulation, explaining the observations in the small-scale jet fire testing [10,11] and determining the breakdown temperature of the insulation are the motivations for the present study.

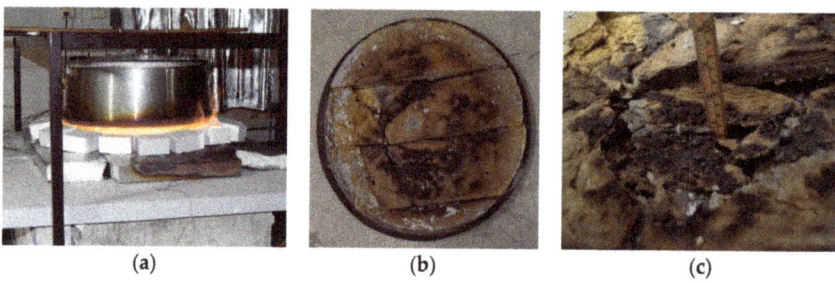

Figure 1. Small-scale jet-fire test setup (**a**), exposed thermal insulation after the jet fire testing (**b**) and melted and sintered remains after high heat flux (350 kW/m²) fire testing (**c**) [10,11].

Sjöström et al. [13] and Olsen et al. [14], performed oven tests of a Rockwool insulation similar to that of Bjørge et al. [11], i.e., recording the temperature in the centre of the insulation. In addition, thermogravimetric analysis (TGA)/differential scanning calorimetry (DSC) tests and transient plane source method (TPS) measurements were performed. However, the scope of their studies [13–17] was limited to temperatures associated with building fires, i.e., at temperatures up to 1000 °C. Their studies focused on the performance of stone wool in such fires [13,14,16] and an analysis of the properties (measured using TGA/DSC/TPS) within the operating range of that thermal insulation [15,17].

Several numerical models have been developed that calculate the fire resistance of insulated walls or columns [18–20]. In order to account for the breakdown of the thermal insulation during fire exposure, the conductivity has, e.g., been adjusted in order to make the model fit with the performed fire tests. This has in some cases overestimated or underestimated the actual conductivity and breakdown of the insulation. The properties of the thermal insulation (Rockwool) at temperatures above the normal operating temperatures are generally missing in the literature.

The present study aimed at investigating the properties of the thermal insulation after being exposed to temperatures up to the breakdown temperature of the insulation. Cubes of the thermal insulation (50 mm) were heat-treated in a muffle furnace at different exposure temperatures up to 1200 °C. To support the findings from the muffle oven tests and further investigate the properties of the thermal insulation, in-depth analyses of the material were performed. To reveal the mass loss at elevated temperatures, thermogravimetric analysis (TGA) was performed at temperatures up to 1250 °C. To examine the melting temperature of the thermal insulation, DSC to 1250 °C was performed. The ambient temperature thermal conductivity of test specimens preheated up to 1200 °C was measured using TPS.

The materials and methods used are explained in Section 2. Section 3 presents the results from the furnace tests, the TGA and DSC analyses and the results from the TPS measurements. Section 4 presents the discussions and Section 5 presents the overall conclusions and suggestions for future studies.

2. Materials and Methods

2.1. The Studied Thermal Insulation

In the present study, industrial-grade pipe section mat (ProRox PSM 971, thickness 50 mm, Trondheim, Norway) delivered by Rockwool, Inc., was studied as a representative industrial thermal insulation, i.e., the same thermal insulation as in previous studies [10,11]. The detailed technical data and thermal conductivity of this thermal insulation up to 350 °C are presented in Appendix A, Tables A1 and A2. The maximum service temperature of the studied insulation, as given by the manufacturer, is 700 °C, which is well below temperatures associated with fires in the oil and gas industry. Temperatures above the service temperature were therefore focused on in the present study.

Chemically, the main components of the thermal insulation are inorganic oxides. The thermal insulation mainly consists of silica, alumina, magnesia, calcium oxide and iron (III) oxide. In addition, there are minor amounts of sodium oxide, potassium oxide, titanium oxide and phosphorous pentoxide. The detailed chemical composition, as received from the supplier, is presented in Appendix A, Table A3.

The thermal insulation is produced by melting the raw materials at 1500 °C before it is cooled and spun into insulation mats [21]. In addition, a dust binder is added (mineral-based oil) to make the material easier to handle when, e.g., cutting and fitting the insulation mat to equipment requiring thermal insulation. Bakelite, i.e., polyoxybenzylmethylenglycolanhydride $(C_6H_6O \cdot CH_2O)_x$, is also added to give some strength to the thermal insulation up to the maximum service temperature.

As the insulation is heated, the mineral oil will gradually pyrolyse/evaporate. The degradation process of the Bakelite is dependent on the actual production conditions and the degradation process may be complicated [22]. The number of molecular cross-links will influence the degradation processes and there may be several reaction paths. Generally, the degradation of Bakelite may be expressed as the following non-balanced reaction:

$$(C_6H_6O \cdot CH_2O)_x \rightarrow CO_2 + CO + H_2O + C_{soot} + \text{other products}. \tag{1}$$

The number of cross-links, in addition to other components mixed into the Bakelite, will have an impact on the degradation temperatures [22].

2.2. Thermal Conductivity

For materials like, e.g., thermal insulation, the thermal conductivity is limited by the pore radiation. Theoretically, it can be shown that the thermal conductivity (k) will be proportional to the absolute temperature to the third power, i.e., T^3 [23]. The thermal conductivity of the virgin thermal insulation as a function of absolute temperature is presented in Figure 2. It can be very well described by Equation (2), which is also presented in Figure 2.

$$k = 0.0304 + 3.11 \times 10^{-10} T^3. \tag{2}$$

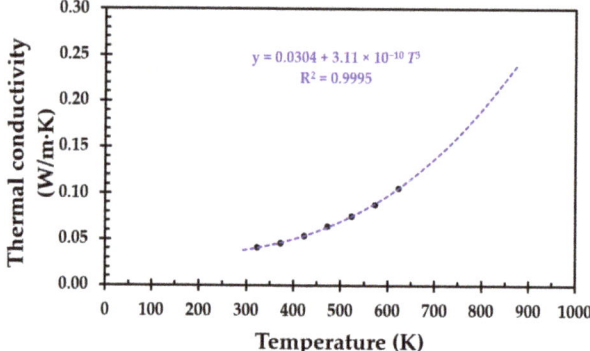

Figure 2. Thermal conductivity of the thermal insulation (ProRox PSM 971, 50 mm) as a function of the absolute temperature. Data from Appendix A, Table A2.

The good fit of Equation (2), i.e., $R^2 = 0.9995$, with a major contribution of the temperature dependency to the third power, i.e., T^3, clearly indicates that the thermal conductivity for this particular thermal insulation is indeed limited by pore radiation. However, when the breakdown of the thermal insulation is significant, the thermal conductivity may no longer be limited by pore radiation and is thus assumed to increase.

When exposing inorganic (ceramic) materials to elevated temperatures, sintering of the material may occur, i.e., an entropy-driven [24] physical process leading to a lower free energy, ΔG. To limit sharp edges and optimise the mix of the material, the atoms in contacting threads will diffuse across the thread boundaries. With time, theoretically, the material may approach a solid state, fusing the threads together to leave a minimum remaining pore fraction. The thermal conductivity of the thermal insulation will depend highly on the temperature exposure and the onset of crystallisation, sintering or melting of the insulation. Hence, the sintering effect will increase the thermal conductivity of the thermal insulation [25].

The sintering process may start at temperatures that are approximately two-thirds of the absolute melting temperature for ceramic materials [26]. Hence, porous ceramic materials may be expected to start the sintering process at temperatures well below the actual melting point.

2.3. Furnace Testing up to 1200 °C

To investigate the thermal insulation dimensional changes and the breakdown temperature, which is defined as a considerable change in physical dimension over a limited temperature range, it was decided that a muffle furnace be used for the heat treatment. In order to minimise any elasticity issues, the thermal insulation test specimens (50 mm cubes) were pre-cut a couple of days before the heat treatment in a muffle furnace (Laboratory Chamber Furnace, Thermconcept GmbH, Bremen, Germany). The maximum temperature of the furnace was 1300 °C, i.e., well above the highest temperature (1200 °C) of interest in the present study.

One thermocouple (type K, mantel, 1.5 mm diameter, Pentronic AB, Västervik, Sweden) was inserted vertically into the centre of the 50 mm cubic test specimens to record the internal test specimen temperature. To record the furnace temperature, a second thermocouple was placed in the upper part of the furnace. The test specimen was placed on a steel plate and lifted approximately 35 mm above the 15 mm thick bottom plate, as shown in Figure 3, to allow for uniform heating of the specimen. In order to minimise any thermal radiation shadowing effects, only one test specimen was placed in the furnace for each heat exposure test.

Figure 3. Test setup in the muffle oven.

The heat treatment of the test specimen was performed for temperatures in the range of 700 to 1200 °C, as presented in Table 1. The heating rate of the oven was set to 15 K/min and the test specimens were kept at the respective holding temperatures for 30 min.

Table 1. Number of tests at each holding temperature.

Holding Temperature (°C)	Number of Tests
700	2
800	2
900	2
1000	2
1100	2
1120	1
1140	2
1160	1
1170	1
1180	2
1190	1
1200	2

After each heat treatment and cooling to below 100 °C, the test specimen was carefully removed from the furnace and the length and width were recorded at three locations at each of the four vertical faces. The average width and height were reported for each test specimen.

2.4. Thermogravimetric Analysis and Differential Scanning Calorimetry

To support the results from the furnace testing and to get more detailed information about the breakdown processes, samples of the thermal insulation were tested in a simultaneous TGA/DSC apparatus (Simultaneous Thermal Analyzer STA 449F3, NETZSCH, Selb, Germany). Prior to the sample preparation, a larger sample was taken from the insulation mat, crushed and mixed well into one large batch, from which each sample was taken. This was done to, as far as possible, even out minor variations in the chemical composition of the different spun layers. The sample mass was approximately 12 mg (±1 mg). The TGA/DSC tests were run at heating rates of 5, 10, 20 and 40 K/min from room temperature to 1250 °C. Three tests were run at each heating rate. The tests were conducted in a nitrogen atmosphere to prevent air oxidation.

2.5. Transient Plane Source Thermal Conductivity Measurements

The thermal conductivity of the virgin thermal insulation was given by the manufacturer, as shown in Table A2. Test specimens for thermal conductivity measurements were also initially 50 mm cubes and were heat-treated in a similar way as previously described. The highest heat treatment temperature for these test specimens was 1200 °C. Post heat treatment, TPS [27,28] was used to record the thermal conductivity of each test specimen at room temperature. However, no thermocouple penetrated these test specimens since that would have left a hole when removed and thus disturbed the TPS thermal conductivity measurements.

3. Results

3.1. Heat Treatment of 50 mm Thermal Insulation Cubes to 1200 °C

During the heat treatment in the muffle furnace, the temperature at the center of the insulation sample was recorded. The recorded temperature as a function of time for each defined holding temperature is presented in Figure 4. The temperature curves show the heating of the thermal insulation, the 30 min holding time and the subsequent cooling of the furnace. Two temperature peaks were observed in all the tests, i.e., two exothermic reactions. The peak temperature in the reaction varied, which may be explained by the variations in the amounts of dust binder and Bakelite in each test cube. The first peak started at about 300 °C, with a peak in temperature between 525 and 587 °C. The second peak started at approximately 870 °C, with a peak temperature between 930 and 990 °C. Thus, the second peak was only observed in test specimens treated at 900 °C and above. As stated in [11], the first peak (exothermic reaction) may be explained by the combustion of the dust binder due to the ambient air atmosphere in the furnace. The second exothermic peak may be explained by the crystallisation of the amorphous silica (SiO_2) in the thermal insulation.

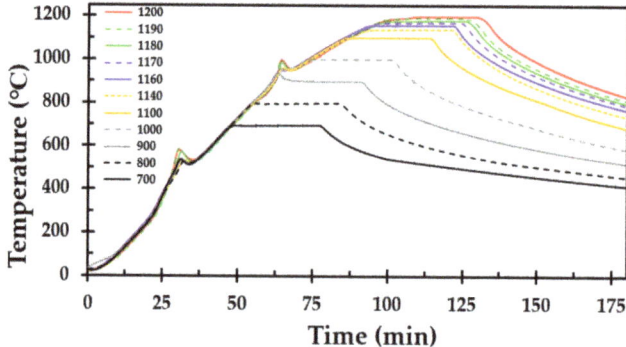

Figure 4. Measured temperature as a function of time, measured at the center of the test specimen, for each holding temperature presented in Table 1.

After the heat treatment, the height of the originally 50 mm high cube of thermal insulation was recorded at three locations at each of the four vertical faces. Similarly, the width of the test specimen was recorded horizontally at three elevations for each of the four vertical faces. There was some variation in height and width in the heat-treated test specimens, hence an average value had to be used. The results from the average measurements from the height (H) and width (W) of the heat-treated thermal insulation are presented in Figure 5.

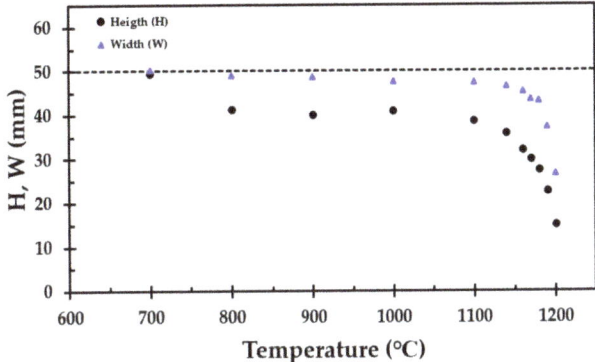

Figure 5. Height (H) of the test specimen (●) and width (W) of the test specimen (▲) after the heat treatment. The height and the width are the average value of three measurements at each side of each of the four vertical faces.

Based on the obtained average height and width of each cube, an estimation of the post-heat-treatment volume was made. The mass of each specimen was also recorded, allowing for the density to be calculated for each cube, as presented in Figure 6.

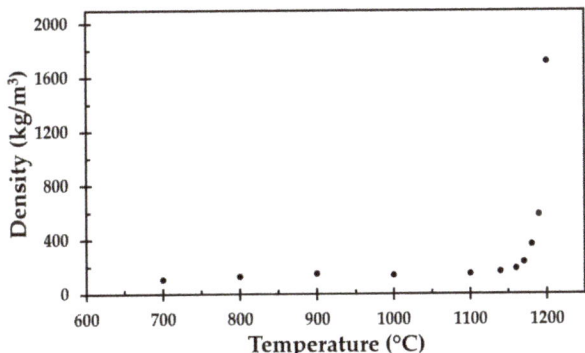

Figure 6. Density as a function of the heat exposure temperature.

A minor decrease in the test specimen height was observed for heat treatment temperatures up to 1100 °C, similar to previously published results [11], which were limited to a maximum temperature of 1100 °C. Above this temperature, the results of the present study clearly show that significant degradation of the thermal insulation started at temperatures just above 1100 °C. There also seemed to be a total breakdown at 1200 °C, as evidenced by a conspicuous increase in the post-heat-treatment density. The virgin test specimen and the test specimen heat-treated up to 1200 °C are presented in Figure 7.

The test specimens after heat treatments up to 1100 °C, 1190 °C and 1200 °C, from left to right, respectively, are shown in Figures 8 and 9. After the heat treatment at 1190 °C, the test specimen had lost 55% of its original height and 25% of its original width, while the heat treatment at 1200 °C resulted in a 76% reduced height and a 46% reduced width. When increasing the heat treatment temperature from 1190 °C to 1200 °C, the thermal insulation material post heat treatment changed in morphology from a chalky consistency to resembling a hard, but still somewhat porous, stone. This was clearly shown in the calculated density, which increased from 589 to 1721 kg/m^3 due to the 10 °C increase in heat treatment temperature from 1190 °C to 1200 °C.

Figure 7. Virgin test specimen (50 mm cube) (left) and heat-treated to 1200 °C (right), including the thermocouple that had to be cut when the specimen was removed from the furnace.

Figure 8. Test specimens after furnace heat treatments up to 1100 °C (left), 1190 °C (middle) and 1200 °C (right), as seen from the side.

Figure 9. Test specimens after furnace heat treatments up to 1100 °C (left), 1190 °C (middle) and 1200 °C (right), as seen from the bottom of the insulation.

3.2. Thermogravimetric Analysis

Thermogravimetric analysis was conducted from ambient temperature up to 1250 °C. The heating rates were 5, 10, 20 and 40 K/min. The samples for the TGA testing were made from the same insulation mat as the muffle furnace testing. The approximate mass loss was between 3 and 4.3%, as shown in Figure 10. The differential thermogravimetric (DTG) analysis, which is the derivative of the TGA curve, is presented in Figure 11.

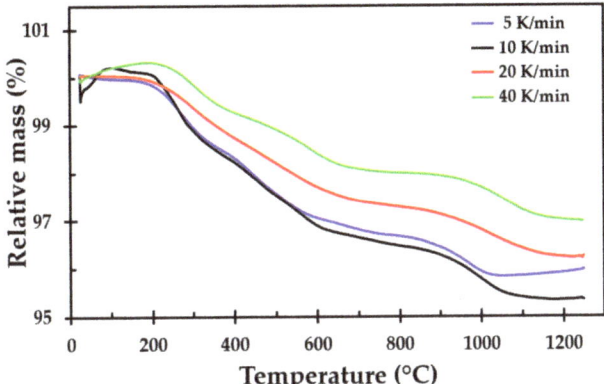

Figure 10. Thermogravimetric analysis of the thermal insulation in a nitrogen atmosphere.

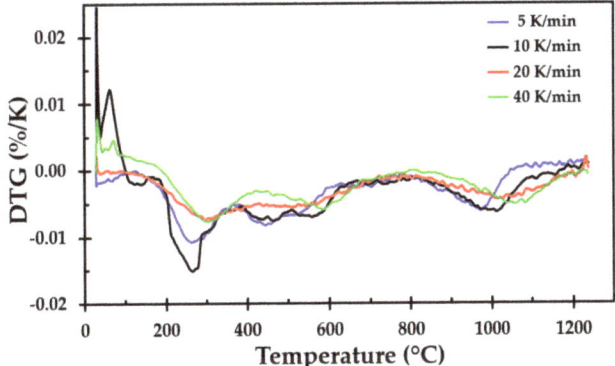

Figure 11. Differential thermogravimetric (DTG) analysis of the results presented in Figure 10.

The mass loss of the insulation started at approximately 180 °C, with a local minimum value between 260 and 290 °C. This may be explained by the release of the dust binder. The mass losses at higher temperatures were most likely due to the Bakelite binder and possibly some released chemically bound water, with the most conspicuous peak observed at about 1000 °C.

3.3. Differential Scanning Calorimetry

Simultaneously with the TGA measurements, DSC analyses were performed from ambient temperature up to 1250 °C at heating rates of 5, 10, 20 and 40 K/min. The results from the DSC analysis are presented in Figure 12. An exothermic reaction started between 800 and 900 °C. An endothermic peak was observed at, or just above, 900 °C. A very conspicuous endothermic reaction was observed starting at approximately 1120 °C, with a maximum local peak between 1170 and 1206 °C.

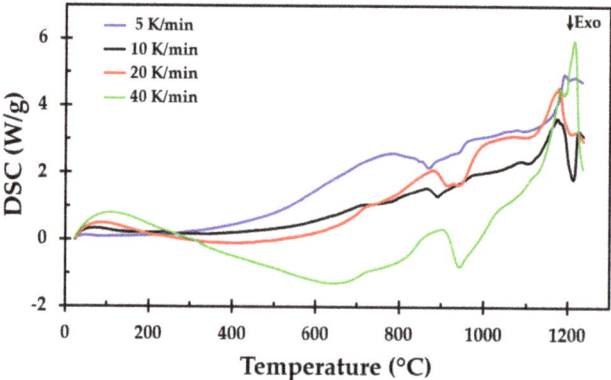

Figure 12. The differential scanning calorimetry (DSC) results as a function of temperature.

The minimum and maximum values from the exothermic and endothermic reactions for the three conducted tests at each heating rate are presented in Table 2, in addition to the heat flow values at the endothermic peaks. There were some variations in the peak value, depending on the heating rate, but there was no clear trend associated with the heating rate and peak temperature.

Table 2. The recorded temperatures at the exothermic ($T_{l,exo}$) and endothermic ($T_{p,endo}$) DSC peaks of Figure 12 and the recorded heat flows at the peaks for each run.

Run	Heating Rate (K/min)	$T_{l,exo}$ (°C)	$T_{p,endo}$ (°C)	DSC (W/g)
1	5	848.0	1210.5	3.18
2	5	873.9	1190.9	4.98
3	5	922.9	1209.9	6.48
Average		881.6	1203.8	4.88
1	10	889.6	1202.6	1.82
2	10	893.9	1174.9	3.65
3	10	917.9	1175.9	4.73
Average		900.4	1184.4	3.40
1	20	920.2	1156.2	1.85
2	20	943.2	1177.2	4.56
3	20	950.0	1175.0	5.59
Average		937.8	1169.4	4.00
1	40	935.1	1216.3	4.47
2	40	942.7	1213.7	5.97
3	40	938.3	1190.3	6.02
Average		938.7	1206.7	5.49

3.4. Thermal Conductivity Measurements

TPS [27,28] was used to record the room temperature thermal conductivity of the heat-treated test specimens. The obtained results as a function of the heat treatment temperature are shown in Figure 13. The thermal conductivity increased with the heat treatment temperature in a similar manner to the recorded density, as presented in Figure 6, i.e., it increased greatly when heat-treated to temperatures above 1150 °C. This was most likely due to the increasing level of sintering and partly due to melting, as evidenced by the endothermic peak in Figure 12 at these high temperatures. The most conspicuous change was observed when the heat treatment temperature was 1200 °C, i.e., where more melting occurred during the heat treatment.

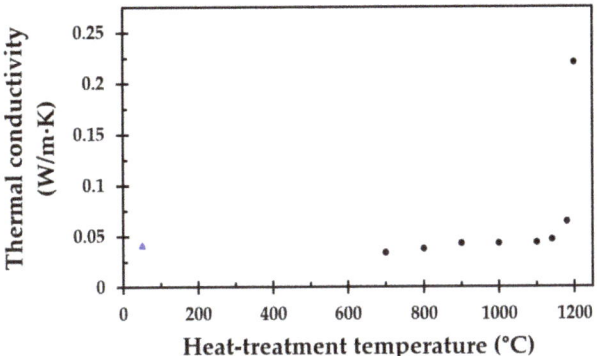

Figure 13. Recorded room temperature thermal conductivity as a function of the test specimen heat treatment temperature. The thermal conductivity at 50 °C (▲) was from Appendix A, Table A2.

4. Discussion

Previous studies have shown that the type of thermal insulation tested in this study survived well when heated in a muffle oven to temperatures up to 1100 °C [11]. The objective of the present study was to investigate the performance of the thermal insulation when exposed to temperatures up to 1200 °C, i.e., temperatures associated with fire heat fluxes of about 350 kW/m². The focus was on finding the breakdown temperature of the thermal insulation. Small scale jet fire testing has proven that the thermal insulation alone may serve as passive fire protection of a 16 mm steel wall [10,11]. The previous jet fire tests showed a complete breakdown of the insulation at the most exposed locations. To determine the breakdown temperature and explain the observations of the insulation after the small-scale jet fire testing, muffle furnace tests up to 1200 °C were performed, as well as TGA/DSC to 1250 °C in a nitrogen atmosphere.

The results from the furnace testing showed the same trend as in [11] up to 1100 °C, which was the upper-temperature limit of that study due to furnace limitations. However, at heat treatment temperatures above 1100 °C, the height of the originally 50 mm cubes started to shrink significantly with increasing heat treatment temperature. From 1160 °C, the width of the test specimens also decreased considerably. The thermal insulation fibers gradually sintered/melted more and more together, and the insulation transformed from being a porous material to a hard, stony consistency when heat-treated to 1200 °C. This was also reflected in the calculated density of the thermal insulation cubes post heat treatment. The density close to tripled due to the heat treatment at 1200 °C compared to 1190 °C. It was clearly shown from the results that heating to 1200 °C is very close to, or even at, the melting point, or eutectic temperature, of the insulation.

Mixtures of inorganic salts, such as the investigated thermal insulation, will not show a defined melting point, but rather an extremely complex phase diagram with several eutectic points. It is therefore expected to gradually melt, without a defined melting temperature. Heat treatment of the test specimen cubes to temperatures above 1200 °C might have resulted in a glass-like substance. This was, however, outside the scope of the present study but may be interesting for future studies.

The thermal conductivity of the virgin thermal insulation was clearly dominated by heat radiation through the pores at moderately elevated temperatures. At higher temperatures, the onset of sintering increased the solid–solid contact phase, improving the true thermal conductivity of the material. This was confirmed by the room temperature thermal conductivity obtained in the present study. However, at a still higher pore fraction, it would be expected that at elevated temperatures, the pore radiation would continue to dominate the effective thermal conductivity. When heat-treated to 1200 °C, the significant increase in density indicated that the pore fraction must be very low. In this stage, the thermal conductivity may not be very dependent on the pore radiation, i.e., not show a very strong dependency on the absolute temperature to the third power. To validate this assumption, the thermal conductivity of the heat-treated test specimens must be recorded at elevated temperatures. This was, however, outside the scope of the present study.

Heat treatment up to 1100 °C revealed some loss in height, approximately 22%. However, little change in the width of the material was observed up to this temperature, i.e., it was not expected that the insulation mat will crack open at temperatures below 1100 °C. When heating to 1180 °C, the loss in width was still below 14%. However, when heating up to 1190 °C, there was a significant loss in width, i.e., 25%.

Due to the heat treatment at 1200 °C, the insulation cube lost 76% of its height and 46% of its original width, explaining the observed cracks and openings in the insulation mat after the small scale jet fire testing presented in [10,11]. In addition to the shrinkage in height (thickness) and the increase in thermal conductivity due to the sintering, in a severe fire scenario, there will be radiant heat transfer through the cracks and openings. Hence, with more and wider cracks, more radiant heat may bypass the thermal insulation, leading to excessive heating of any fire-exposed objects.

In the furnace heat treatment tests, two exothermic peaks were observed. The first of these may be explained by the combustion of the dust binder material at about 300 °C. There were some variations in the peak temperature of the reaction, which may be explained by differences in the chemical compositions between the samples. This exothermic reaction at approximately 300 °C, observed during heat treatment, was not present in the DSC tests. This may be explained by the air access and combustion in the furnace and the inert gas (nitrogen) atmosphere during the DSC tests. In an oxygen atmosphere, both the first and the second peaks were observed [11]. The second peak at about 900 °C may have been due to the Bakelite combustion or a recrystallisation process of the involved inorganic salts.

The TGA showed that only small amounts of the material vaporised during heating, i.e., approximately 4% of the mass was lost. This was also seen when observing the density of the heat-treated test specimens, where there was little change in the mass of the test samples due to the heat treatment, i.e., the density increased as the insulation sintered and finally started to partly melt. The differences in mass loss recorded using the TGA for the different heating rates were also observed in repeated tests at each heating rate. In their study of different stone wool insulations, Livkiss et al. [16] also observed similar discrepancies. They explained these differences using the inhomogeneity of these types of material. We also agree with this assumption since TGA and DSC testing is constrained to test samples that are a few milligrams in size.

The heat treatment in the present study involved a holding time of 30 min at all heat treatment temperatures. It is interesting to notice that if the temperature of the thermal insulation was kept at, or below, 1100 °C, it could stay quite intact for at least the 30 min heat exposure. It started to significantly break down only at temperatures above 1100 °C. Hence, if arranged in a passive fire protection system such that it will not exceed 1100 °C, it may contribute significantly as passive fire protection in addition to its intended function as thermal insulation. A sketch of such an arrangement is presented in Figure 14. The critical point to be kept below is 1100 °C, which is marked on the figure. Unless the object to be protected is internally cooled by, e.g., depressurisation, the temperatures of the system will gradually increase, but these layers of protection may be designed to offer the required protective capacity for the desired time.

It should be noted that different batches of industrial thermal insulation may show slightly different high-temperature performances. Thermal insulation that varies significantly from the chemical composition presented in the present study may show very different high-temperature properties. Care should therefore be taken before using such materials for fire protection. The 50 mm cubes that were heat-treated in the muffle furnace and the TGA/DSC analysis both provided results supporting the conclusions in the present study. When considering industrial thermal insulation for that also supplies some passive protection in fire situations, it may in the future be sufficient to use only one of these methods for a preliminary evaluation of the potential passive fire protection capability.

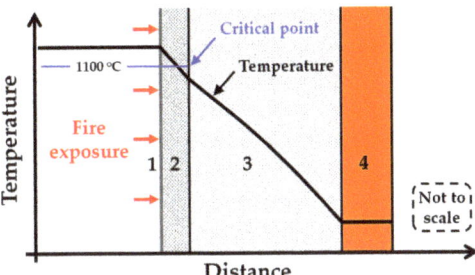

Figure 14. Sketch of fire exposure, weather protection cladding (1), the layer of heat-resistant insulation (2), thermal insulation (3) and the object to be protected from fire exposure (4). The critical point to be kept below is a temperature of 1100 °C, which is marked on the figure.

In the future, it would be beneficial to do further fire testing of, e.g., small-scale jet fire tests [10,11] with an additional protective layer, as demonstrated in Figure 14. When properly chosen, this layer may then keep the exposed thermal insulation below the breakdown temperatures, thereby ensuring that it may contribute towards providing significantly prolonged fire protection.

Heat treatment tests in a muffle furnace that test other types of insulation, e.g., mineral-based passive fire protection or different types of Rockwool insulation, could give more information about future possibilities. It would also be beneficial to measure the thermal conductivity of the thermal insulation at elevated temperatures. This could give the information required for developing a numerical model of the thermal insulation performance when exposed to fires, with or without a protective layer, as indicated in Figure 14.

5. Conclusions

Except for a 20% vertical (z-direction) shrinkage at 800 °C, the present study showed that the properties of the analysed thermal insulation did not change much when heat-treated to 1100 °C, which is associated with pool fire heat flux levels (250 kW/m^2). However, when heat treated to 1200 °C, which is associated with jet fire heat flux levels (350 kW/m^2), the thermal insulation changed greatly. The density and room temperature thermal conductivity increased from 140 to 1700 kg/m^3 and from 0.041 W/m·K to 0.22 W/m·K, respectively. The horizontal (x- and y-direction) shrinkage that took place at heat treatment temperatures above 1180 °C created gaps in the insulation, i.e., allowed for unrestricted radiant heat flow at exposed locations. However, if a thin layer of thermal insulation that is robust to temperatures above 1200 °C is placed at the heat exposed side, the studied thermal insulation may provide significant passive fire protection. It is recommended that this is tested in future studies.

Author Contributions: Conceived the project idea, T.L.; guided the study, A.G. and M.M.M.; selected the test materials, performed the muffle furnace heat-treatment tests and prepared samples for TGA and DSC analysis and thermal conductivity, A.G.; did all the data preparation, A.G.; did the data analysis, A.G., T.L. and M.M.M.; prepared the artwork and wrote the first draft paper, A.G.; wrote the final paper, A.G., T.L. and M.M.M. All authors have read and agreed to the published version of the manuscript.

Funding: The project was supported by Gassco Inc., Norway. A.G. was supported by a Research Council Norway industrial PhD scholarship, grant no. 305336.

Acknowledgments: The authors would like to acknowledge Gisle Kleppe for technical support, Leif Inge Larsen for supplying test materials and Emil Hallberg, Rise Fire Research, Sweden, for performing the TGA, DSC and TPS measurements. Technical data for the thermal insulation, which was supplied by Søren Nyborg Rasmussen, Rockwool, is also much appreciated.

Conflicts of Interest: None of the participants in the study have any connections to specific equipment or materials being mentioned in the paper.

Appendix A

The technical data for the thermal insulation is given in Tables A1 and A2. The chemical composition of the thermal insulation is given in Table A3.

Table A1. Technical data for the Rockwool pipe section mat thermal insulation [29].

Name	Description	
Material	Stone Wool	
Operating Range	−40 to 700 °C	
Name	Performance	Norms
Maximum Service Temperature	700 °C	EN 14706
Reaction to Fire	Euroclass A1	EN 13501-1
Nominal Density	140 kg/m^3	EN 1602
Water Absorption	≤1 kg/m^2	EN 1609
	≤20 kg/m^3	BP 172
Water Vapor Diffusion Resistance	Sd > 200 m	EN 12086
Air Flow Resistivity	>60 kPa·s/m^2	
Designation Code	MW EN 14303-T4-ST(+)700-WS1-MV2	EN 14303

Table A2. Thermal conductivity of the thermal insulation studied (Rockwool ProRox PSM 971, 50 mm) [29].

Temperature (°C)	Thermal Conductivity (W/m·K)
50	0.041
100	0.046
150	0.054
200	0.064
250	0.075
300	0.088
350	0.106

Table A3. Data for the thermal insulation studied (Rockwool ProRox PSM 971. 50 mm) [30].

Name	Product	Percentage
Dust Binder [1]	Oil product	<0.5%
Binder [1]	$(C_6H_6O \cdot CH_2O)_N$	2.5% (±0.4%)
Bulk Oxide	SiO_2	40.6–44.6%
Bulk Oxide	Al_2O_3	17.4–20.4%
Bulk Oxide	$MgO + CaO$	23.9–27.9%
Bulk Oxide	Fe_2O_3	5.5–8.5%
Bulk Oxide	$Na_2O + K_2O$	1.3–4.3%
Bulk Oxide	TiO_2	0.6–2.6%
Bulk Oxide	P_2O_5	Max. 1.2%

[1] The binder calorific value is 27 MJ/kg according to ISO 1716.

References

1. Bahadori, A. *Thermal Insulation Handbook for the Oil, Gas, and Petrochemical Industries*; Elsevier BV: Amsterdam, The Netherlands, 2014.
2. Scandpower. *Guidelines for the Protection of Pressurised Systems Exposed to Fire*; Report No. 27.207.291/R1; Version 2; Scandpower: Kjeller, Norway, 2004.

3. Norsk Standard. *S-001 Technical Safety. NORSOK Standard*, 5th ed.; Norwegian Technology Standards Institution: Oslo, Norway, 2018; Available online: http://www.standard.no (accessed on 13 June 2020).
4. Sintef Byggforsk. *Brannbeskyttelse av Stålkonstruksjoner*; 520.315; Sintef Byggforsk: Trondheim, Norway, 2004.
5. Kletz, T. *What Went Wrong? Case Histories of Process Plant Disasters and How They Could Have Been Avoided*, 5th ed.; Institution of Chemical Engineers: London, UK, 2009; ISBN 13-978-1-85617-531-9.
6. U.S. Chemical Safety and Hazard Investigation Board. *Investigation Report Executive Summary*; Drilling Rig Explosion and Fire at the Macondo Well, Report No. 2010-10-I-OS; U.S. Chemical Safety and Hazard Investigation Board: Washington, DC, USA, 2010.
7. Murray, J.A.; Sander, L.C.; Wise, S.A.; Reddy, C.M. *Gulf of Mexico Research Initiative 2014/2015 Hydrocarbon Intercalibration Experiment: Description and Results for SRM 2779, Gulf of Mexico Crude Oil and Candidate SRM 2777 Weathered Gulf of Mexico Crude Oil*; NISTIR 8123; National Institute of Standards and Technology: Gaithersburg, MD, USA, 2016.
8. Standard Norge. *Petroleum and Natural Gas Industries—Control and Mitigation of Fires and Explosions on Offshore Production Installations—Requirements and Guidelines*; NS-EN ISO 13702; Standard Norge: Lysaker, Norway, 2015.
9. International Organization for Standardization. *Determination of the Resistance to Jet Fires of Passive Fire Protection Materials—Part 1: General Requirements*; ISO 22899-1; International Organization for Standardization: Geneva, Switzerland, 2007.
10. Bjørge, J.S.; Metallinou, M.-M.; Kraaijeveld, A.; Log, T. Small Scale Hydrocarbon Fire Test Concept. *Technologies* **2017**, *5*, 72. [CrossRef]
11. Bjørge, J.S.; Gunnarshaug, A.; Log, T.; Metallinou, M.-M. Study of Industrial Grade Thermal Insulation as Passive Fire Protection up to 1200 °C. *Safety* **2018**, *4*, 41. [CrossRef]
12. Jet Fire Test Working Group. *The Jet-Fire Resistance of Passive Fire Protection Materials*; HSE Report OTI 95 634; Health & Safety Executive—Offshore Technology Report: Sheffield, UK, 1995; ISBN 0-11-322731-0.
13. Sjöström, J.; Jansson, R. Measuring thermal material properties for structural fire engineering. In Proceedings of the 15th International Conference on Experimental Mechanics ICEM15, Porto, Portugal, 22–27 July 2012.
14. Olsen, H.; Sjöström, J.; Jansson, R.; Anderson, J. Thermal Properties of Heated Insulation Materials. In Proceedings of the 13th International Fire and Engineering Conference INTERFLAM, London, UK, 24–26 June 2013.
15. Al-Ajlan, S.A. Measurements of thermal properties of insulation materials by using transient plane source technique. *Appl. Therm. Eng.* **2006**, *26*, 2184–2191. [CrossRef]
16. Livkiss, K.; Andres, B.; Bhargava, A.; Van Hees, P. Characterization of stone wool properties for fire safety engineering calculations. *J. Fire Sci.* **2018**, *36*, 202–223. [CrossRef]
17. Moesgaard, M.; Pedersen, H.; Yue, Y.; Nielsen, E. Crystallization in stone wool fibres. *J. Non-Cryst. Solids* **2007**, *353*, 1101–1108. [CrossRef]
18. Keerthan, P.; Mahendran, M. Thermal Performance of Composite Panels Under Fire Conditions Using Numerical Studies: Plasterboards, Rockwool, Glass Fibre and Cellulose Insulations. *Fire Technol.* **2012**, *49*, 329–356. [CrossRef]
19. Thomas, G.C. Fire Resistance of Light Timber Framed Walls and Floors. Ph.D. Thesis, University of Canterbury, Christchurch, New Zeeland, 1996.
20. Schleifer, V. Zum Verhalten von Raumabschliessenden Mehrschichtigen Holzbauteilen im Brandfall. Ph.D. Thesis, ETH Zurich University, Zurich, Germany, 2009.
21. Rockwool. Available online: https://www.rockwool.com/west-virginia/factory-operations-and-production (accessed on 1 October 2020).
22. Solyman, W.S.; Nagiub, H.M.; Alian, N.A.; Shaker, N.O.; Kandil, U.F. Synthesis and characterization of phenol/formaldehyde nanocomposites: Studying the effect of incorporating reactive rubber nanoparticles or Cloisite-30B nanoclay on the mechanical properties, morphology and thermal stability. *J. Radiat. Res. Appl. Sci.* **2017**, *10*, 72–79. [CrossRef]
23. Kingery, W.D. Thermal Conductivity: XII. Temperature Dependence of Conductivity for Single-Phase Ceramics. *J. Am. Ceram. Soc.* **1955**, *38*, 251–255. [CrossRef]
24. Pozzoli, V.; Ruiz, M.; Kingston, D.; Razzitte, A. Entropy Production during the Process of Sintering. *Procedia Mater. Sci.* **2015**, *8*, 1073–1078. [CrossRef]

25. Log, T.; Jackson, T.B. Simple and Inexpensive Flash Technique for Determining Thermal Diffusivity of Ceramics. *J. Am. Ceram. Soc.* **1991**, *74*, 941–944. [CrossRef]
26. Log, T.; Cutler, R.A.; Jue, J.F.; Virkar, A.V. Polycrystalline t'-ZrO$_2$(Ln$_2$O$_3$) formed by displacive transformations. *J. Mater. Sci.* **1993**, *28*, 4503–4509. [CrossRef]
27. Log, T.; Gustafsson, S.E. Transient plane source (TPS) technique for measuring thermal transport properties of building materials. *Fire Mater.* **1995**, *19*, 43–49. [CrossRef]
28. International Organization for Standardization. *Plastics—Determination of Thermal Conductivity and Thermal Diffusivity—Part 2: Transient Plane Heat Source (Hot Disc) Method*; ISO 22007:2; International Organization for Standardization: Geneva, Switzerland, 2015.
29. Rockwool. Available online: https://www.rockwool.co.uk/product-overview/hvac/pipe-section-mat-psm-en-gb/?selectedCat=downloads (accessed on 9 September 2020).
30. Rasmussen, S.N.; (ROCKWOOL Inc., Hedehusene, Denmark). Personal communication, 20 February 2018.

Publisher's Note: MDPI stays neutral with regard to jurisdictional claims in published maps and institutional affiliations.

© 2020 by the authors. Licensee MDPI, Basel, Switzerland. This article is an open access article distributed under the terms and conditions of the Creative Commons Attribution (CC BY) license (http://creativecommons.org/licenses/by/4.0/).

Article

Investigation and Possibilities of Reuse of Carbon Dioxide Absorbent Used in Anesthesiology

Bartłomiej Rogalewicz *, Agnieszka Czylkowska, Piotr Anielak and Paweł Samulkiewicz

Institute of General and Ecological Chemistry, Faculty of Chemistry, Lodz University of Technology, Zeromskiego 116, 90-924 Lodz, Poland; agnieszka.czylkowska@p.lodz.pl (A.C.); piotr.anielak@p.lodz.pl (P.A.); psamulkiewicz@swspiz.pl (P.S.)
* Correspondence: 211150@edu.p.lodz.pl

Received: 14 October 2020; Accepted: 5 November 2020; Published: 9 November 2020

Abstract: Absorbents used in closed and semi-closed circuit environments play a key role in preventing carbon dioxide poisoning. Here we present an analysis of one of the most common carbon dioxide absorbents—soda lime. In the first step, we analyzed the composition of fresh and used samples. For this purpose, volumetric and photometric analyses were introduced. Thermal properties and decomposition patterns were also studied using thermogravimetric and X-ray powder diffraction (PXRD) analyses. We also investigated the kinetics of carbon dioxide absorption under conditions imitating a closed-circuit environment.

Keywords: soda lime; carbon dioxide; anesthesiology; absorbent; thermogravimetric analysis; PXRD analysis

1. Introduction

Soda lime is one of the most popular carbon dioxide absorbents used in order to maintain a safe level of this gas. Its composition has slightly changed over time; however, calcium hydroxide is still the main component. Often an indicator signaling its consumption is added, as well as small amounts of sodium (or potassium) hydroxide since NaOH is more reactive than $Ca(OH)_2$. Moreover, the hygroscopic properties of NaOH reduce interphase mass transfer barriers and speed up the CO_2 sorption process. Soda lime is most commonly used in environments characterized by reduced or no connection with fresh air, like anesthetic and diving apparatus or spacecraft. Such environments are commonly called "closed" and "semi-closed" circuit environments. It is an important issue since humans are there to the most degree, exposed to increased levels of CO_2. Some of the most important examples of such systems, where soda lime and similar absorbents are used are anesthetic and diving apparatus, submarines, spacecraft and mine refuge chambers [1,2]. Breathing in such environments results in increased carbon dioxide concentrations. This may cause occurrence of several symptoms or even lead to death. In order to prevent it, absorbents based on alkali hydroxides are used to capture carbon dioxide, with soda lime being the most common one. Despite relatively low costs of production and simple operating principle, soda lime also has numerous flaws and limitations. It may undergo reaction with some of the gases used in general anesthesia, especially with sevoflurane, desflurane, isoflurane and enflurane to form a number of degradation products [3–5]. One of the gases, fluoromethyl 2,2-difluoro-1-(trifluoromethyl)vinyl ether, often abbreviated in the literature as *Compound A*, has been proven nephrotoxic to rats [6–8]. It may also contain viruses derived from the exhaled air, including rhinovirus, respiratory syncytial virus, parainfluenza virus, adenovirus, coronavirus, human metapneumovirus and influenza virus [9–12]. Under certain conditions, soda lime may also support the formation of carbon monoxide, one of the most toxic gases. Some of the factors that increase the possibility of carbon monoxide formation are the use of volatile anesthetic agents in question, their concentration and flow rate, dryness, type of used absorbent and the temperature in which absorption takes place [13–16]. Another complication is the need to control the absorbent exhaustion level since an excess of CO_2 can cause hypercapnia. In order to signalize soda lime consumption, indicators like ethyl violet or ethyl orange are added. It does not, however, provide the necessary level of safety since sometimes these indicators may return to their previous color despite absorbent exhaustion [17]. Thus, capnometers are used in order to control the level of carbon dioxide in a patient's organism. These devices are, however, relatively expensive and are not used in less developed areas. Soda lime with additions of indicators has also been withdrawn from U.S. Navy Fleet since it was suspected of releasing harmful compounds [18]. During the absorption process, extensive amounts of heat are produced, especially when baralyme (modification of soda lime in which calcium hydroxide is replaced with barium hydroxide) is used. Indicator color change monitoring and extensive heat emission additionally complicate soda lime use in diving apparatus. Furthermore, soda lime dust inhalation was observed, which may contribute to the occurrence of airway diseases in divers [19]. There are also no reports of reliable recycling methods of exhausted soda lime, which is most commonly considered medical waste. Taking into consideration all the mentioned problems and limitations, soda lime requires rigorous and careful handling. It is also the basis for seeking new alternative absorbents that would be more reliable and versatile. Some of them are carbonaceous materials, solid and liquid organic amines, mixtures of metals peroxides, hyperoxides or superoxides with water, membranes and zeolites [20]. There is no doubt that carbon dioxide plays one of the most significant roles taking into consideration both biological and environmental issues.

In this paper, we present a critical evaluation of soda lime performance as carbon dioxide absorbent. We have also investigated composition (volumetric analysis, photometric analysis) and thermal properties (thermogravimetric analysis, X-ray powder diffraction analysis) of two soda lime commercial samples, as well as proposed its possible recycling method.

2. Experimental

2.1. Materials and Analysis

All chemicals used during analysis were purchased from Avantor Performance Materials, Gliwice, Poland S.A.: pure calcium hydroxide, pure calcium carbonate, hydrochloric acid (1 mol·L^{-1}), sodium hydroxide (4 mol·L^{-1}), EDTA (0.05 mol·L^{-1}), phenolphthalein, methyl orange, Patton and Reeder's indicator. Analyzed soda lime samples came from the company producing carbon dioxide absorbents. The first sample was a fresh, unused sample, while the second one was used and considered exhausted prior to the research. The samples were marked as follows:

SL (F)—fresh soda lime sample;
SL (U)—used soda lime sample.

Chemical composition and thermal decomposition curves of samples were investigated using volumetric, photometric, thermogravimetric analysis. In order to better understand thermal decomposition pathways and to exclude the theoretical presence of other products of decomposition, a PXRD analysis was performed for sinters of samples SL (F) and SL (U) prepared at 950 °C. The sinters were obtained by heating the samples to the temperature defined from the thermal curves.

In the second part, we conducted an experiment under conditions imitating carbon dioxide absorption in closed circuit anesthetic apparatus, which allowed us to draw conclusions about the kinetics of carbon dioxide absorption and soda lime performance as a carbon dioxide absorbent. Chemical composition and thermal destruction ways after absorption were investigated in the same way as in the first part of our study. These samples were marked in the following way:

SL (5 min)—fresh soda lime sample after 5 min of carbon dioxide absorption;
SL (15 min)—fresh soda lime sample after 15 min of carbon dioxide absorption.
SL (30 min)—fresh soda lime sample after 30 min of carbon dioxide absorption.

The experimental setup is schematically shown in Figure 1. A compressor was used to flow atmospheric air through a water bubbler and then a packed bed of sorbent. The carbon dioxide concentration in the inlet air (C_o) was around 4% (average concentration of carbon dioxide in exhaled air). The experiment was conducted at room temperature (23–25 °C), and relative humidity was maintained at around 55% during the experiment.

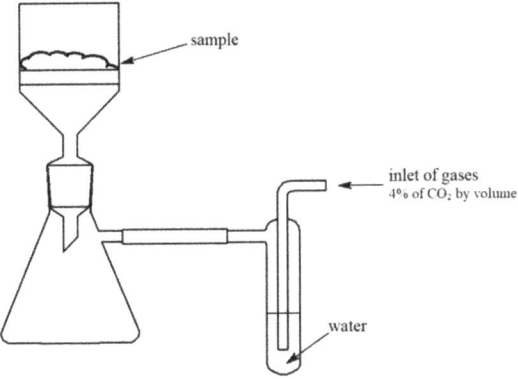

Figure 1. Experimental setup for carbon dioxide absorption under conditions imitating closed-circuit environment.

2.2. Methods and Instruments

A bigger batch of each sample was ground in a mortar. After homogenization, around 1.0 grams of each sample was stirred in 1 L of distilled water on a magnetic stirrer for 24 h. Suspensions prepared this way were then investigated using volumetric analysis. In order to ensure repeatability of results, no less than three portions per each sample were collected and titrated. Powders resulting from grinding in a mortar were investigated using thermogravimetric analysis. We have also performed thermal decompositions of two main soda lime components—calcium hydroxide $Ca(OH)_2$ and calcium carbonate $CaCO_3$. Volumetric analysis was performed using automatic burettes at room temperature (23–25 °C). P alkalinity and M alkalinity determinations were performed using 2 mol·L^{-1} hydrochloric acid solution in the presence of phenolphthalein and methyl orange, respectively. Calcium ion concentration determinations were performed using 0.05 mol·L^{-1} EDTA solution in the presence of Patton and Reeder's indicator. Photometric analysis was performed using BWB-XP flame photometer (BWB Technologies, Newbury, England). The content of sodium in samples was measured at an analytical spectral line 589 nm with the limit of detection 0.02 ppm. Thermal behavior and decomposition patterns of samples were investigated using IRIS 209 (Netzsch, Selb, Germany) in the temperature range 25–980 °C at a heating rate of 4°·min^{-1} in flowing dynamic nitrogen atmosphere (v = 30 mL·min^{-1}) using platinum crucibles; as reference material, platinum crucibles were used. PXRD analysis was performed using a X'Pert Pro MPD diffractometer (PANalytical, Malvern, England) in the Bragg–Brentano reflection geometry using CuK$_\alpha$ radiation in the 2θ range 5–90° with a step of 0.0167° and exposure per step of 50 s.

3. Results and Discussion

The performed analyses allowed us to determine the composition of the investigated samples SL (F), SL (U), SL (5 min), SL (15 min), SL (30 min). Volumetric and thermogravimetric analyses allowed us to calculate the percentage of the contents of calcium hydroxide and calcium carbonate for each sample. The contents of water were derived from thermal decomposition, as dehydration is the first process of thermolysis. Photometric analysis allowed us to establish the content of sodium hydroxide in investigated samples. Table 1 presents the collected data.

Table 1. Composition of investigated samples.

Sample	Percentage Content [% (m/m)]			
	Ca(OH)$_2$	CaCO$_3$	H$_2$O	NaOH
SL (F)	96.78	0	0.89	2.50
SL (U)	35.85	60.50	1.93	2.10
SL (5 min)	63.23	34.09	2.08	2.28
SL (15 min)	42.34	55.09	2.74	2.15
SL (30 min)	31.41	64.45	2.95	2.04

Determination of each component's content was done separately using different analytical techniques, which may have caused propagation of error. This is the reason why, in some samples, the summed contents of components may exceed 100%. The biggest error occurred in SL (15 min) sample (contents sum up to 102.32%); however, it was still within the error tolerance.

3.1. Thermal Decomposition of Samples: SL (F), SL (U), Samples After Absorption: SL (5 min), SL (15 min) and SL (30 min)

Figure 2 presents the thermal decomposition of calcium hydroxide and calcium carbonate. Thermolysis began at 280 °C for calcium hydroxide (DTA peak at 430 °C) and at 560 °C for calcium carbonate (DTA peak at 740 °C). Mass losses and decomposition curves were consistent with reactions that took place during the process. For calcium hydroxide, it was the release of one molecule of water, and for calcium carbonate, it was the release of a molecule of carbon dioxide. The final product of decomposition in both cases was pure calcium oxide.

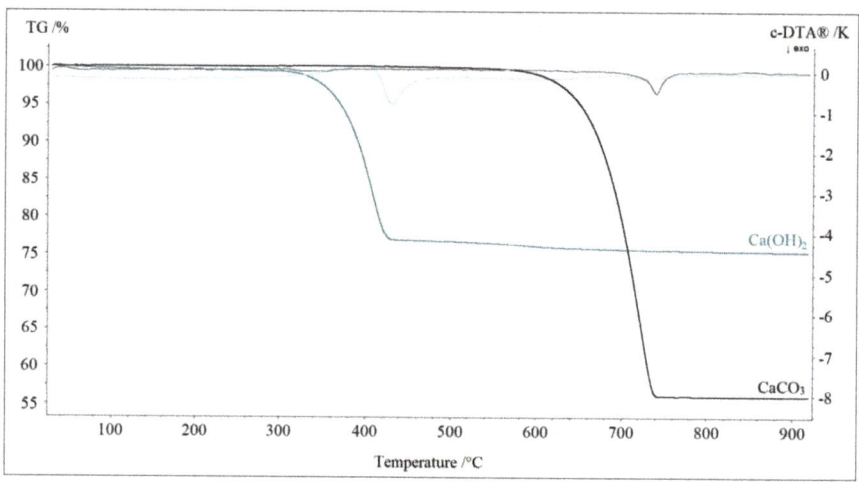

Figure 2. TG-DTA curves of decomposition of calcium hydroxide and calcium carbonate in nitrogen.

Figure 3 presents TG-DTA curves of the investigated fresh soda lime sample SL (F) and the used sample SL (U) that came from a hospital and were considered exhausted. It is clear that the first step of decomposition was the dehydration of samples. For SL (F), the sample mass loss related to this process was 0.89% at a temperature range 25–275 °C, while for the SL (U) sample, it was 1.93% at a temperature range 25–300 °C. In the second step, one of the samples' components decomposed—calcium hydroxide. For the SL (F) sample, it took place above 275 °C (DTA peak at 405 °C), while for the SL (U) sample—at a temperature range of 300–400 °C (DTA peak at 390 °C). It was clearly visible that above 400 °C, for the SL (U) sample, decomposition of calcium carbonate took place (DTA peak at 675 °C).

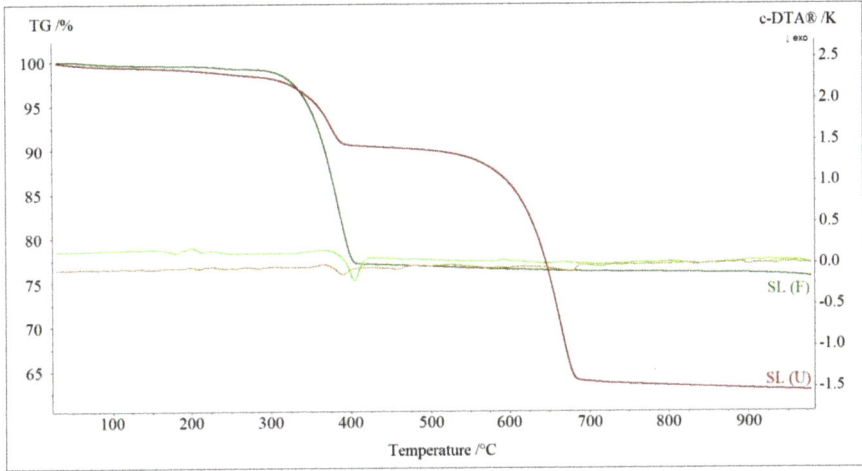

Figure 3. TG-DTA curves of decomposition of fresh soda lime sample (SL (F)) and used soda lime sample (SL (U)) in nitrogen.

Figure 4 presents the TG curves of investigated soda lime samples after 5, 15 and 30 min of carbon dioxide absorption, as well as the TG curve of the SL (F) sample (after 0 min of absorption). The decomposition path was analogous in all cases. The first step (up to 300 °C) was associated with dehydration. It is clear that along with the increasing time of CO_2 absorption, the content of water increased. We could also observe how the second step of decomposition, associated with decomposition of calcium hydroxide, shortened, while the last step, associated with decomposition of calcium carbonate, increased along with time. These curves also show that the process of absorption slowed down with time.

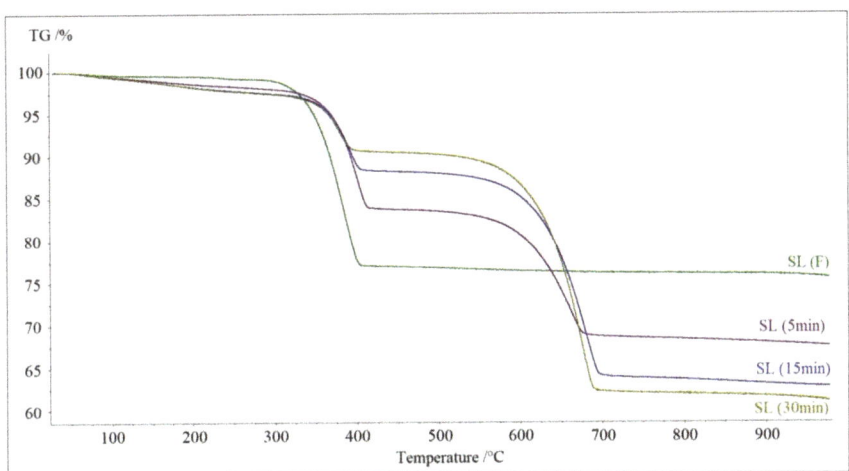

Figure 4. TG curves of decomposition of SL (F) sample and samples after carbon dioxide absorption: SL (5 min), SL (15 min) and SL (30 min) in nitrogen.

Soda lime is a mixture of different chemicals, and thus its thermal decomposition is a multistage process. In all cases, the first step is dehydration. Later, decomposition of calcium hydroxide and calcium carbonate takes place. In order to thoroughly investigate the thermal properties of such absorbents, we decided to study the composition of two sinters prepared at the end of the decomposition process (950 °C). Both the fresh sample's (SL (F)) and the used sample's (SL (U)) sinters were prepared. Their X-ray powder diffraction patterns are shown in Figure 5. These patterns correlate very well with calcium oxide, proving it is a final product of the thermal decomposition.

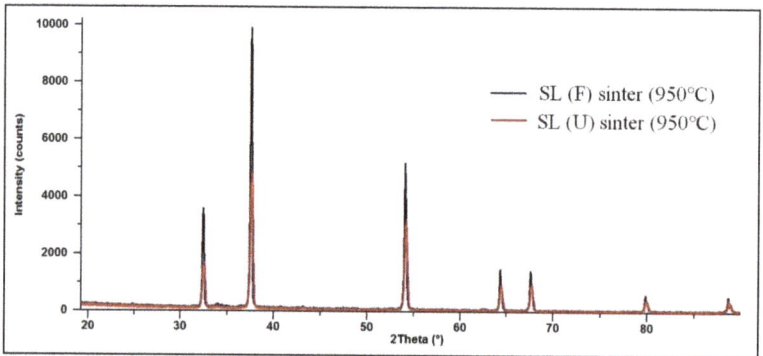

Figure 5. X-ray powder diffraction patterns of analyzed SL (F) and SL (U) sinters prepared at 950 °C.

3.2. Chemical Kinetics of Carbon Dioxide Absorption by SL (U) Sample

Three samples of SL (F) were exposed to CO_2 absorption for 5 min, 15 min and 30 min, and then analyzed in the same way as SL (F) and SL (U) samples. The absorption of carbon dioxide and water is a multistage process. The first stage involves the formation of carbonic acid from CO_2 and water. Then, NaOH (or KOH) added in small amounts acts as an activator to speed up the process through the formation of sodium (or potassium) carbonate. It can also be concluded that the absorption and the hydration of CO_2 and the formation of CO_3^{2-} are rapid steps, and the dissolution of $Ca(OH)_2$ is the slowest step of the carbonation process [21,22]. Calcium hydroxide reacts with the carbonates within minutes to form an insoluble precipitate of calcium carbonate as well as results in a regeneration of NaOH [23]. Some carbon dioxide may also react directly with $Ca(OH)_2$ to form calcium carbonates, but this reaction is much slower. In addition, calcium bicarbonate may be formed on the surface of the sorbent particles. The higher solubility of bicarbonate enhances CO_2 diffusion through the bulk of the particle [24]. Soda lime is exhausted when all hydroxides become carbonates.

Figure 6 shows the experimental results of CO_2 sorption on soda lime as a relationship between conversion rate α and time t. It clearly indicates that the conversion of the sorbent was incomplete and would be difficult to reach under the typical working conditions. According to the results shown in the figure, about 20–30 min from the beginning of the experiment, the carbonation rate slowed down noticeably. We can observe that the curve is composed of two sections. The initial upslope of the curve depicts the fastest rate of carbonation; its initial rate was 3.3 min^{-1}. After 30 min, the reaction slowed down and reached a rate of 0.17 min^{-1}. It was a result of significant limitations of CO_2 transport from the surface to the bulk of the sorbent particles, and differentiation between kinetics-controlled and diffusion-controlled ranges occurs.

The reaction rate of a solid-state process, $\frac{d\alpha}{dt}$, can be related to the process temperature, T, and to the fraction reacted, α, by means of the following general equation [25]:

$$\frac{d\alpha}{dt} = k \cdot f(\alpha) \qquad (1)$$

where k is a constant rate.

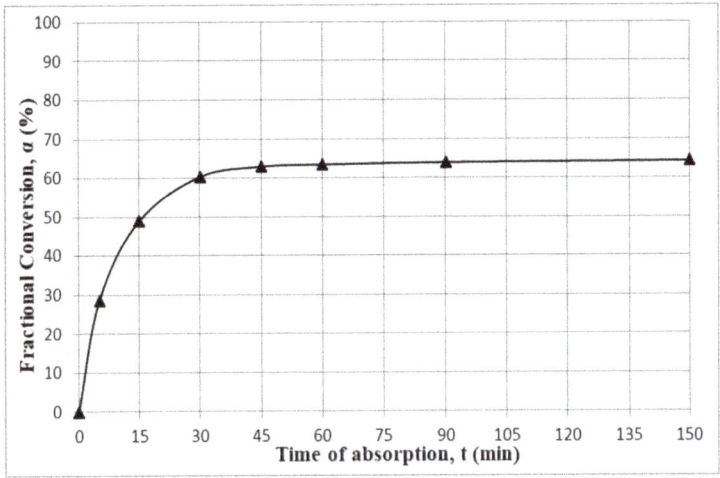

Figure 6. Relationship between the fractional conversion of soda lime and time of carbon dioxide absorption.

The kinetic curve of CO_2 absorption of soda lime can be described by the pseudo-first or pseudo-second order kinetic equation [26,27]. In the first section of the kinetic curve, the carbonation is controlled by the surface reaction, whereas in the second section, a heterogeneous system is controlled mainly by diffusion [28]. Assuming a driving force of CO_2 removal to be proportional to the difference between its concentrations in sorbent at any time prior to equilibrium and its concentration at equilibrium, we can use the equation:

$$\frac{d\alpha}{dt} = k_n(\alpha_e - \alpha)^n \tag{2}$$

The fittings of the experimental data to the linear form of the two kinetic models, i.e., pseudo-first order and pseudo-second order, are shown in Figure 7.

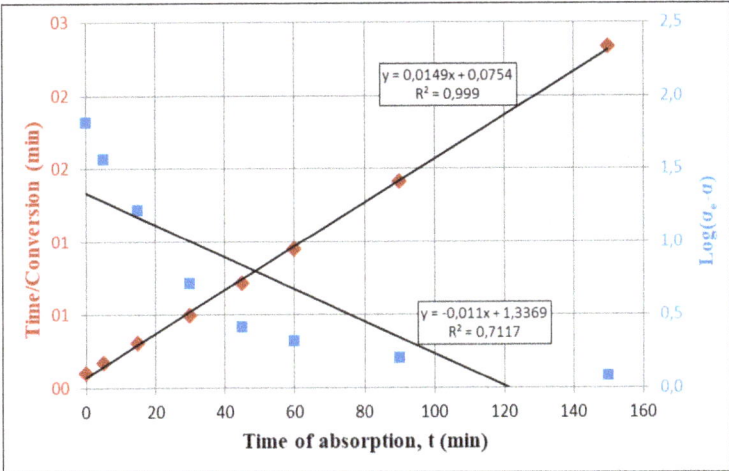

Figure 7. Linearized equation of the pseudo-first (right axis) and pseudo-second (left axis) order kinetics models.

The values of the correlation coefficient for linear forms of both kinetic equations are significantly different. The pseudo-second order model describes the kinetic data better than pseudo-first order model when the process is diffusion-limited [29]. The obtained values of the correlation coefficients were, therefore, 0.999 and 0.712, respectively. Thus, pseudo-first order model does not cover both stages of the CO_2 sorption, i.e., the chemical reaction and the diffusion process. However, the carbonation rate constant determined using first order reaction was greater than for the second order reaction and amounted to 0.011 min^{-1} and 0.0022 min^{-1}, respectively. Experimental data have shown that the carbonation process ends before all lime is converted into a calcium carbonate [23]. On the other hand, the first, fast absorption stage is completed within one hour, and the experimental and calculated values of fractional conversion (Figure 6) were in good agreement with values calculated for both kinetic models: 65.5% and 67.1%, respectively [27].

One of the possible ways to express soda lime exhaustion rate is a relationship between calcium carbonate content and calcium hydroxide content $\frac{\alpha\ CaCO_3}{\alpha\ Ca(OH)_2}$ in the bed and time t. This relationship is presented in Figure 8.

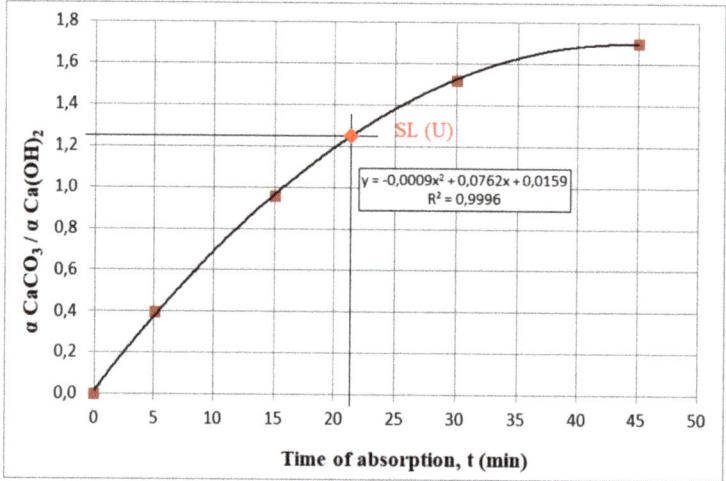

Figure 8. Relationship between calcium carbonate amount and calcium hydroxide amount ratio and time.

We used this relationship to determine what time of absorption corresponds with the chemical composition of the SL (U) sample. For example, a bed exhaustion rate of 1.247 could be obtained after 21.7 min.

4. Conclusions

Using various analytical techniques, we determined the chemical composition of several soda lime samples: fresh sample, exhausted sample after use in hospital and three samples after carbon dioxide absorption under conditions imitating semi-closed circuit apparatus. Thermogravimetric and XRD analyses comprehensively described the thermal properties and decomposition ways of the investigated samples. This product decomposed in a stable manner, releasing water and carbon dioxide. It is possible to recycle and reuse soda lime in different forms; however, the calcination process would require relatively high temperatures. On the other hand, high temperatures would ensure the biological neutrality of recycled soda lime. Calcium oxide itself could be reused in many different areas, e.g., in absorption and desiccation, in the construction industry or in the manufacturing of chemicals.

Soda lime is a fairly efficient carbon dioxide absorbent that has been used for a long time. It has, however, some limitations and drawbacks that require further investigations, as it is a product used in environments where dependability is a factor of great importance. One of the issues that should be addressed is possible interactions between absorbent and anesthetic gases, which can lead to the release of harmful compounds. Another limitation is the speed of carbon dioxide absorption, which is the highest at the beginning of the process and slows down relatively fast. On the surface of the soda lime granules, water forms a film less than three molecular layers thick, and the reaction rate is reduced [21]. As the carbonation proceeds, the product particles precipitate on the surface of $Ca(OH)_2$ and cover it a thicker, porous deposit layer, which inhibits the exchange of reacting species between the surface of calcium hydroxide and bulk solution. Therefore, the diffusion rate of reacting species is an important factor affecting the final stage of carbonation. The carbonation of $Ca(OH)_2$ was observed to stop before one hour. However, carbonation may go on by diffusion through the covering layer, but its rate is too slow to be detected in the range of carbonation time used [28]. Thorough research on soda lime properties is an important step in the evaluation of its performance as an agent responsible for preventing carbon dioxide poisoning. Lack of data concerning the kinetics of this process causes this problem to be still very interesting and important in both anesthetic and medical science. Such data would provide a reliable tool to compare different types of absorbents and thus would allow proper absorbent choice taking into consideration all other aspects of the environment or apparatus. Our research, only to some extent, covers the main problems of soda lime use. Additional further studies must be performed in order to ensure the required level of safety and efficiency and to determine the best recycling method.

Author Contributions: Conceptualization, A.C., P.A., P.S. and B.R.; methodology, A.C., P.A., P.S., B.R.; validation, A.C., P.A. and P.S.; formal analysis, P.A, B.R.; investigation, A.C., P.A., P.S., B.R.; resources, A.C., P.S.; data curation, P.A., B.R.; writing—original draft preparation, P.A., B.R.; writing—review and editing, A.C., P.A., P.S., B.R.; visualization, P.A., B.R.; supervision, A.C.; project administration, A.C., B.R; All authors have read and agreed to the published version of the manuscript.

Funding: This research received no external funding.

Conflicts of Interest: The authors declare no conflict of interest.

References

1. Issa, M.C.; Yusoff, N.H.N.; Yati, M.S.D.; Muhammad, M.M.; Salleh, N.A.; Nor, M.F.M.; Minal, A.; Nain, H.; Nor, I.M. Characterisation od Carbon Dioxide Absorbent Material For Enclosed Space Applications. *Def. S T Tech. Bull. Malays.* **2012**, *5*, 1–10.
2. Gai, W.-M.; Deng, Y.-F.; Du, Y. Adsorption Properties of Modified Soda Lime for Carbon Dioxide Removal within the Closed Environment of a Coal Mine Refuge Chamber. *Ind. Eng. Chem. Res.* **2016**, *55*, 10794–10802. [CrossRef]
3. Fang, Z.X.; Kandel, L.; Laster, M.J.; Ionescu, P.; Eger, E.I. Factors Affecting Production of Compound A from the Interaction of Sevoflurane with Baralyme and Soda Lime. *Anesth. Analg.* **1996**, *82*, 775–781. [CrossRef]
4. Morio, M.; Fujii, K.; Satoh, N.; Imai, M.; Kawakami, U.; Mizuno, T.; Kawai, Y.; Ogasawara, Y.; Tamura, T.; Negishi, A.; et al. Reaction of Sevoflurane and Its Degradation Products with Soda Lime. Toxicity of the Byproducts. *Anesthesiology* **1992**, *77*, 1155–1164. [CrossRef]
5. Frink, E.J.; Malan, T.P.; Morgan, S.E.; Brown, E.A.; Malcomson, M.; Brown, B.R. Quantification of the Degradation Products of Sevoflurane in Two CO_2 Absorbants during Low-flow Anesth. in Surgical Patients. *Anesthesiology* **1992**, *77*, 1064–1069. [CrossRef] [PubMed]
6. Gonsowski, C.T.; Laster, M.J.; Eger, E.I.; Ferrell, L.D.; Kerschmann, R.L. Toxicity of Compound A in Rats. Effect of increasing duration of administration. *Anesthesiology* **1994**, *80*, 566–573. [CrossRef] [PubMed]
7. Kandel, L.; Laster, M.J.; Ii, E.I.E.; Kerschmann, R.L.; Martin, J. Nephrotoxicity in Rats Undergoing a One-Hour Exposure to Compound A. *Anesth. Analg.* **1995**, *81*, 559–563. [CrossRef] [PubMed]
8. Keller, K.A.; Callan, C.; Prokocimer, P.; Delgado-Herrera, L.; Friedman, M.B.; Hoffman, G.M.; Wooding, W.L.; Cusick, P.K.; Krasula, R.W. Inhalation Toxicity Study of a Haloalkene Degradant of Sevoflurane, Compound A (PIFE), in Sprague-Dawley Rats. *Anesthesiology* **1995**, *83*, 1220–1232. [CrossRef] [PubMed]

9. Charlson, E.S.; Diamond, J.M.; Bittinger, K.; Fitzgerald, A.S.; Yadav, A.; Haas, A.R.; Bushman, F.D.; Collman, R.G. Lung-enriched Organisms and Aberrant Bacterial and Fungal Respiratory Microbiota after Lung Transplant. *Am. J. Respir. Crit. Care Med.* **2012**, *186*, 536–545. [CrossRef] [PubMed]
10. Piters, W.A.A.D.S.; Sanders, E.A.M.; Bogaert, D. The role of the local microbial ecosystem in respiratory health and disease. *Philos. Trans. R. Soc. B Biol. Sci.* **2015**, *370*, 20140294. [CrossRef]
11. Nguyen, L.D.N.; Viscogliosi, E.; Delhaes, L. The lung mycobiome: An emerging field of the human respiratory microbiome. *Front. Microbiol.* **2015**, *6*, 89. [CrossRef] [PubMed]
12. Yi, H.; Yong, D.; Lee, K.; Cho, Y.-J.; Chun, J. Profiling bacterial community in upper respiratory tracts. *BMC Infect. Dis.* **2014**, *14*, 583. [CrossRef] [PubMed]
13. Fang, Z.X.; Eger, E.I.; Laster, M.J.; Chortkoff, B.S.; Kandel, L.; Ionescu, P. Carbon Monoxide Production from Degradation of Desflurane, Enflurane, Isoflurane, Halothane, and Sevoflurane by Soda Lime and Baralyme. *Anesth. Analg.* **1995**, *80*, 1187–1193. [CrossRef]
14. Coppens, M.J.; Versichelen, L.F.M.; Rolly, G.; Mortier, E.P.; Struys, M.M.R.F. The mechanisms of carbon monoxide production by inhalational agents. *Anaesth.* **2006**, *61*, 462–468. [CrossRef] [PubMed]
15. Holak, E.J.; Mei, D.A.; Dunning, M.B.; Gundamraj, R.; Noseir, R.; Zhang, L.; Woehlck, A.H.J. Carbon Monoxide Production from Sevoflurane Breakdown: Modeling of Exposures Under Clinical Conditions. *Anesth. Analg.* **2003**, *96*, 757–764. [CrossRef] [PubMed]
16. Knolle, E.; Heinze, G.; Gilly, H. Carbon Monoxide Formation in Dry Soda Lime is Prolonged at Low Gas Flow. *Anesth. Analg.* **2001**, *93*, 488–493. [CrossRef]
17. Pond, D.; Jaffe, R.A.; Brock-Utne, J.G. Failure to Detect CO_2-absorbent Exhaustion: Seeing and Believing. *Anesthesiology* **2000**, *92*, 1196. [CrossRef]
18. Lillo, R.S.; Ruby, A.; Gummin, D.D.; Porter, W.R.; Caldwel, J.M. Chemical Safety of U.S. Navy Fleet Soda Lime. *Undersea Hyperb Med.* **1996**, *23*, 43.
19. Neubauer, B.; Mutzbauer, T.S.; Tetzlaff, K. Exposure to Soda-Lime Dust in Closed and Semi-Closed Diving Apparatus. *Aviat Space Environ Med.* **2000**, *71*, 1248.
20. Holloway, A.M. Possible Alternatives to Soda Lime. *Anaesth. Intensiv. Care* **1994**, *22*, 359–362. [CrossRef] [PubMed]
21. Shih, S.-M.; Ho, C.-S.; Song, Y.-S.; Lin, J.-P. Kinetics of the Reaction of $Ca(OH)_2$ with CO_2 at Low Temperature. *Ind. Eng. Chem. Res.* **1999**, *38*, 1316–1322. [CrossRef]
22. Van Balen, K. Carbonation reaction of lime, kinetics at ambient temperature. *Cem. Concr. Res.* **2005**, *35*, 647–657. [CrossRef]
23. Freeman, B.S.; Berger, J.S. *Anesthesiology Core Review: Part One Basic Exam*; Springer: Berlin/Heidelberg, Germany, 2014; Chapter 17.
24. Samari, M.; Ridha, F.; Manovic, V.; Macchi, A.; Anthony, E.J. Direct capture of carbon dioxide from air via lime-based sorbents. *Mitig. Adapt. Strat. Glob. Chang.* **2019**, *25*, 25–41. [CrossRef]
25. Pérez-Maqueda, L.A.; Criado, A.J.M.; Sánchez-Jiménez, P.E. Combined Kinetic Analysis of Solid-State Reactions: A Powerful Tool for the Simultaneous Determination of Kinetic Parameters and the Kinetic Model without Previous Assumptions on the Reaction Mechanism. *J. Phys. Chem. A* **2006**, *110*, 12456–12462. [CrossRef]
26. Bos, M.; Kreuger, T.; Kersten, S.; Brilman, D. Study on transport phenomena and intrinsic kinetics for CO_2 adsorption in solid amine sorbent. *Chem. Eng. J.* **2019**, *377*, 120374. [CrossRef]
27. Singh, V.K.; Kumar, E.A. Comparative Studies on CO_2 Adsorption Kinetics by Solid Adsorbents. *Energy Procedia* **2016**, *90*, 316–325. [CrossRef]
28. Nikulshina, V.; Gálvez, M.; Steinfeld, A. Kinetic analysis of the carbonation reactions for the capture of CO_2 from air via the $Ca(OH)_2$–$CaCO_3$–CaO solar thermochemical cycle. *Chem. Eng. J.* **2007**, *129*, 75–83. [CrossRef]
29. Simonin, J.-P. On the comparison of pseudo-first order and pseudo-second order rate laws in the modeling of adsorption kinetics. *Chem. Eng. J.* **2016**, *300*, 254–263. [CrossRef]

Publisher's Note: MDPI stays neutral with regard to jurisdictional claims in published maps and institutional affiliations.

© 2020 by the authors. Licensee MDPI, Basel, Switzerland. This article is an open access article distributed under the terms and conditions of the Creative Commons Attribution (CC BY) license (http://creativecommons.org/licenses/by/4.0/).

Article

Comparison of the Properties of Natural Sorbents for the Calcium Looping Process

Krzysztof Labus

Department of Applied Geology, Silesian University of Technology, 2 Akademicka St., 44-100 Gliwice, Poland; krzysztof.labus@polsl.pl

Abstract: Capturing CO_2 from industrial processes may be one of the main ways to control global temperature increases. One of the proposed methods is the calcium looping technology (CaL). The aim of this research was to assess the sequestration capacity of selected carbonate rocks, serpentinite, and basalt using a TGA-DSC analysis, thus simulating the CaL process. The highest degrees of conversion were obtained for limestones, lower degrees were obtained for magnesite and serpentinite, and the lowest were obtained for basalt. The decrease in the conversion rate, along with the subsequent CaL cycles, was most intense for the sorbents with the highest values. Thermally pretreated limestone samples demonstrated different degrees of conversion, which were the highest for the calcium-carbonate-rich limestones. The cumulative carbonation of the pretreated samples was more than twice as low as that of the raw ones. The thermal pretreatment was effective for the examined rocks.

Keywords: TGA-DSC; calcium looping; CO_2 capture; mineral carbonation; natural sorbents; carbonate rock; serpentine

Citation: Labus, K. Comparison of the Properties of Natural Sorbents for the Calcium Looping Process. *Materials* **2021**, *14*, 548. https://doi.org/10.3390/ma14030548

Academic Editors: Sergey V. Ushakov

Received: 1 January 2021
Accepted: 20 January 2021
Published: 24 January 2021

Publisher's Note: MDPI stays neutral with regard to jurisdictional claims in published maps and institutional affiliations.

Copyright: © 2021 by the author. Licensee MDPI, Basel, Switzerland. This article is an open access article distributed under the terms and conditions of the Creative Commons Attribution (CC BY) license (https://creativecommons.org/licenses/by/4.0/).

1. Introduction

The prevailing view today is that anthropogenic and geogenic greenhouse gases (GHGs) emitted into the atmosphere are the main cause of global warming. Energy production and use may be responsible for almost two-thirds of global greenhouse gas emissions (e.g., [1]). CO_2 is considered to be the greenhouse gas with the greatest contribution to global warming, and global CO_2 emissions are used as a clear indicator of global fossil energy consumption (e.g., [2]). Given the above assumptions, it can be concluded that capturing CO_2 from industrial processes may be one of the main ways to control global temperature increases.

Mineral carbonation (MC) is considered to be one of the safest technologies for reducing CO_2 emissions into the atmosphere and is used to capture and store CO_2 in situ—in geological formations—or ex situ—as a potential solution for CO_2 sequestration from smaller emitters where geological sequestration is not a viable option [3].

The advantage of mineral carbonation is the permanent storage of CO_2 in the form of thermodynamically stable and environmentally friendly carbonates (e.g., [4–6]). This process is exothermic, and the raw materials for its operation are widely available (which is advantageous from an economic point of view) [7]. In the MC process, appropriately selected mineral substrates react with CO_2 and form thermodynamically stable carbonates. This prevents emissions and ensures permanent CO_2 sequestration [8,9]. One of proposed ex situ methods is the calcium looping technology—CaL [10].

Calcium looping (CaL) systems have been proposed as a less expensive method of CO_2 capture for conventional power plants. In this process, a key role is played by calcium sorbent, which is used in alternating calcination and carbonation processes. The efficiency of the process varies depending on the properties of the sorbents used, which are expressed, inter alia, in the effects observed for the decreasing efficiency of gas capture with increasing number of CaL cycles. It is believed that this phenomenon is related to the reduction of the active surface of the sorbent due to sintering and, possibly, the decrease in the chemical activity resulting from the reaction with sulfur oxides competing with carbonation. The reaction in which sulfur compounds are involved is largely similar to carbonation; however, it is irreversible under CaL conditions. It takes place in pores of small dimensions, and its products are deposited on the sorbent surface, which, in turn, makes carbonation difficult. The environments of carbonation are meso- and micropores, and especially in the latter ones, rapid filling with reaction products can take place [11].

The aim of this work was to assess the sequestration capacity of selected rocks using a simultaneous TGA-DSC analysis, thus simulating the calcium looping process. Such a method is suitable for small samples (e.g., drill cuttings, rock fragments), which are easy to obtain even at a very early stage of the raw material deposit recognition. Moreover, such tests do not require extensive reactors, and, in a relatively quick and simple way, they allow for the characterization of the material or screening of samples in terms of their suitability for CaL.

The calcium looping process—shown in Figure 1—uses a reversible chemical reaction,

$$CaO + CO_2 = CaCO_3, \tag{1}$$

between lime (CaO) and CO_2, to capture CO_2 from waste gas streams [10]. CO_2 in the gas stream reacts with CaO in an exothermic carbonation reaction, forming calcium carbonate ($CaCO_3$) at temperatures in the range of 600–700 °C. The $CaCO_3$ from the carbonizer is then sent to a separate device called a calciner, where the calcination reaction takes place at a high temperature (around 900 °C). In these conditions, high-purity CO_2 is released, which is suitable for transport to the sequestration site. The CaO produced is then sent back to the carbonator, closing the loop. Many researchers have proposed the oxy-combustion of coal in a calciner as a heat source for the calcination reaction [10,12]. The heat can be recovered from the exothermic carbonation reaction as well as from high-temperature gas and solid waste streams to generate electricity. As a result, CaL CO_2 capture technology can be less energy intensive and more economical than the amine-based chemical absorption process.

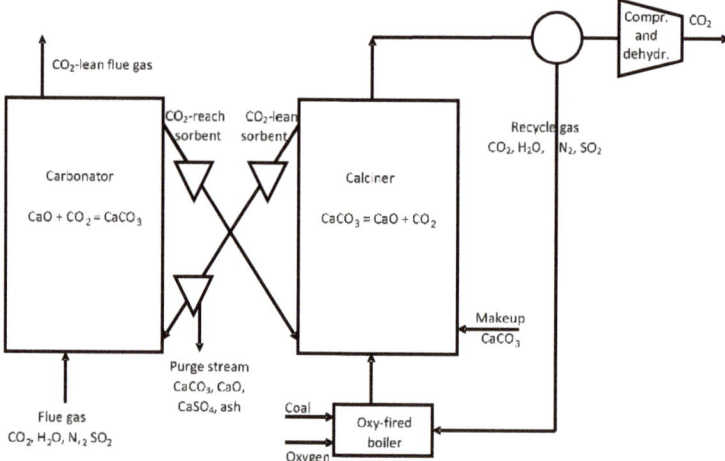

Figure 1. Diagram of the calcium looping (CaL) process for CO_2 capture, according to [13].

The efficiency of CO_2 capture (carbonation) by the sorbent and its regeneration (calcination) depends on the reaction kinetics, the sorbent grain size, its specific surface area, and the pore space characteristics. The cyclicity of the CaL process is accompanied by a decrease in the active surface of the sorbent particles due to the tight packing of CaO (regular system) in comparison to CaCO$_3$ (trigonal system). Capture of CO$_2$ by the CaO phase occurs in two stages. The first stage, in which the surface of the sorbent is covered with calcium carbonate, is characterized by a fast reaction rate and is strongly dependent on the partial pressure of CO$_2$ [14,15]. The second stage is slow—the contact of CO$_2$ with the sorbent depends on diffusion through the CaCO$_3$ coating. Therefore, CaL installations should be based on the use of the first stage [16]. The capture process can be described by, among others, the shrinking core model (SCM), in which both stages are included, as shown in Figure 2. According to this model, the reaction proceeds at a narrow front that moves into the solid particle. The reactant is completely converted as the front passes by [17].

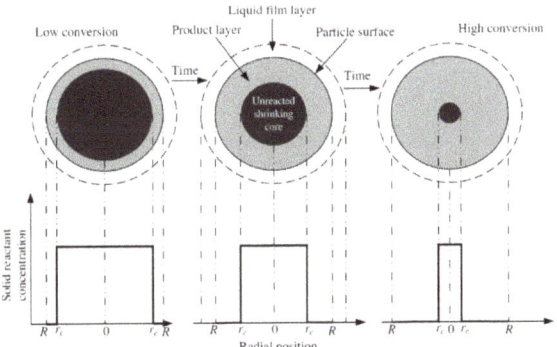

Figure 2. Schematic diagram of the shrinking core model (SCM), according to [17].

The conversion of sorbent can be expressed on the basis of mass change as follows:

$$X = \left(\frac{m - m_0}{m_0}\right) \cdot \frac{M_{CaO}}{M_{CO_2}} \tag{2}$$

where m is the sorbent mass at the time t, m_0 is the initial mass of the sorbent, and M_{CaO} and M_{CO_2} are the molar masses of CaO and CO_2.

According to [17], given the spherical shape of sorbent particles, the relationship of sorbent conversion with particle radius is given by the formula:

$$1 - X = \left(\frac{volume\ of\ unreacted\ core}{total\ volume\ of\ particle}\right) = \frac{\frac{4}{3}\pi r_c^3}{\frac{4}{3}\pi R^3} = \left(\frac{r_c}{R}\right)^3 \tag{3}$$

Let the time for complete conversion of a particle be τ (s), and t is the time of the carbonation reaction (s). Then, in terms of the fractional time for the carbonation reaction, the conversion of the sorbent is given by:

$$X = 1 - \left(\frac{r_c}{R}\right)^3 = \frac{t}{\tau}. \tag{4}$$

Thus, we obtain the general relationship of time with the radius and with the conversion, and the progression of the chemical reaction in terms of the fractional conversion becomes:

$$1 - (1 - X)^{\frac{1}{3}} = \frac{t}{\tau}, \tag{5}$$

while the progression of diffusion is:

$$1 - 3(1 - X)^{\frac{2}{3}} + 2(1 - X) = \frac{t}{\tau}. \tag{6}$$

2. Materials and Methods

In this work, rock samples representing the listed lithological types were studied as potential sorbents in the CaL process (Table 1).

Table 1. Samples studied as potential sorbents in the CaL process.

Sample	Rock Type	Site/Age	Sample Mass [mg]	Mass Loss [%]	$CaCO_3$ [%]
1	limestone	Stramberk(Czechia)/Jurassic	20.02	41.57	90.1
2	limestone	Podlesie (Poland)/Devonian	19.58	39.00	84.5
3	limestone	Butkov (Slovakia)/Cretaceous	20.32	31.81	68.9
4	bituminous limestone	Dębnik (Poland)/Devonian	14.72	34.51	78.5
5	limestone	Saint Anne Mountain (Poland)/Triassic	15.46	42.88	97.5
6	limestone	Gorazdze (Poland)/Triassic	13.9	42.09	95.7
7	dolomite	Olkusz (Poland)/Triassic	14.99	46.06	94.4 [1]
8	marl	Cisownica (Poland)/Cretaceous	15.09	29.75	67.7
9	basalt (nefelinite)	Saint Anne Mountain (Poland)/Tertiary	14.55	1.92	-
10	magnesite	Braszowice (Poland)/Tertiary	14.91	50.91	97.5 [2]
11	serpentinite	Jordanów (Poland)/older than UpperDevonian	14.46	26.97	-

[1] % $CaMg(CO_3)_2$, [2] % $MgCO_3$.

The rock samples were mechanically ground using a PM 100 CM ball mill (Retsch, Haan, Germany) to a fraction of less than 0.08 mm. CO_2 sorption studies were carried out on the STA 449 F3 Jupiter (Netzsch, Selb, Germany) apparatus. Samples weighing about 15 mg were placed in an Al_2O_3 crucible. First, the mass change of the analyzed samples was measured (Table 1); this allowed the determination of the share of calcite (or dolomite) in carbonate rocks. Measurements were made in N_2 atmosphere at a temperature of up to 1030 °C and a heating rate of 10 K/min. Next, the simulation of the CaL process was carried out using a temperature program (Figure 3), which consisted in heating (calcination) of the primary sample from a temperature of 40 °C to about 900 °C at a rate of 20 °C/min in a N_2 atmosphere with a flow of 25 mL/min. Then, a 10 min isothermal section was introduced, and the temperature was lowered to about 650 °C. Then, the carbonation process was carried out, keeping the sample at this temperature for 10 min with the attached CO_2 flow at the rate of 25 mL/min. The test was carried out in 10 cycles of alternating calcination and carbonation. During the measurements, changes in the mass of the sample over time (TGA) and heat flow (DSC) were recorded.

In this work, we the tested raw sorbents (unmodified) as well as selected limestone sorbents that were thermally modified by pre-heating the sample for one hour at 1000 °C in 100% N_2 atmosphere.

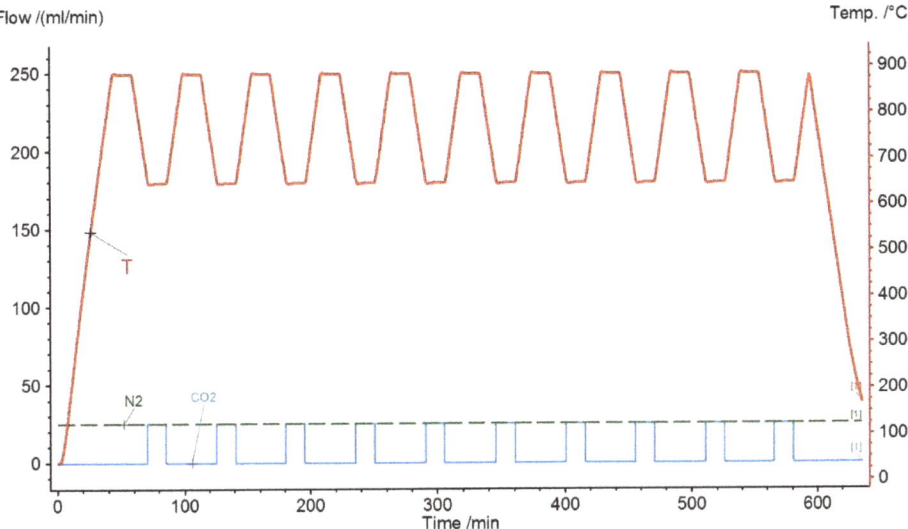

Figure 3. Example of a temperature program sequence (purple), and gas flow (green dashed line—N_2, blue solid line—CO_2).

3. Results

3.1. Raw Sorbents

3.1.1. Dolomite

The simulation of the CaL process for dolomite showed an initial weight loss of the sorbent (calcination) amounting to 45.06% of its weight (Figure 4), which means that the sorbent was composed of almost pure $CaMg(CO_3)_2$. The dolomite derivatogram reveals that two reactions registered on the DSC curve as two adjacent endothermic effects at 740 and 870 °C. The first is responsible for the CO_2 release from $MgCO_3$, and the second from $CaCO_3$; the sample mass losses are 23.59 and 21.47%, respectively. During carbonation, the CO_2 capture was obtained, causing the sample mass changes in the range from 18.77% (first cycle) to 10.30% (10th cycle) (Figure 5). The gas capture efficiency decreased with increasing number of CaL cycles, which may be related to the decreasing active surface of the sorbent due to sintering.

It is worth noting that the carbonation process was not completed within the assumed time of 10 min. This is evident in the sample mass change (TGA) graph, which shows the rapid rise (reaction step—line 1) smoothly moving through the transition (line 2) towards the diffusion step (line 3). According to the SCM model, such effects are connected with the increasing thickness of the $CaCO_3$ layer surrounding the unreacted CaO core. However, this last step was not fully completed, as shown by the line 3, which is tangent to a portion of the mass loss curve still deviates from the horizontal position. This means that in the case of dolomite sorbent, carbonation could be carried out for a longer time than assumed in the analyzed simulation. This is justified by the observation of the occurrence of a segment typical for the diffusion that was visible at the time of about 10 min after closing the CO_2 flux to the reaction chamber of the furnace, when the atmosphere was not fully replaced with the protective gas (N_2). In this case, the extension of the carbonation time may be associated with a potential reduction in the economics of the capture process, as the recorded increase in uptake by diffusion was only 0.29% in the first cycle, and in the next, it was about 1% of the sorbent sample mass (Figure 5). This issue would require further tests with extended carbonation time in order to calculate the amount of CO_2 bound by the sorbent through the diffusion process.

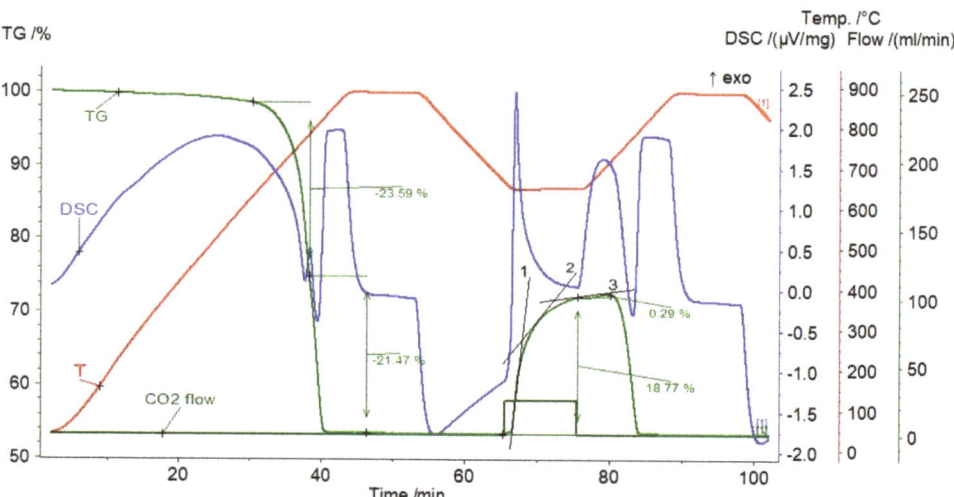

Figure 4. TGA and DSC curves for dolomite.

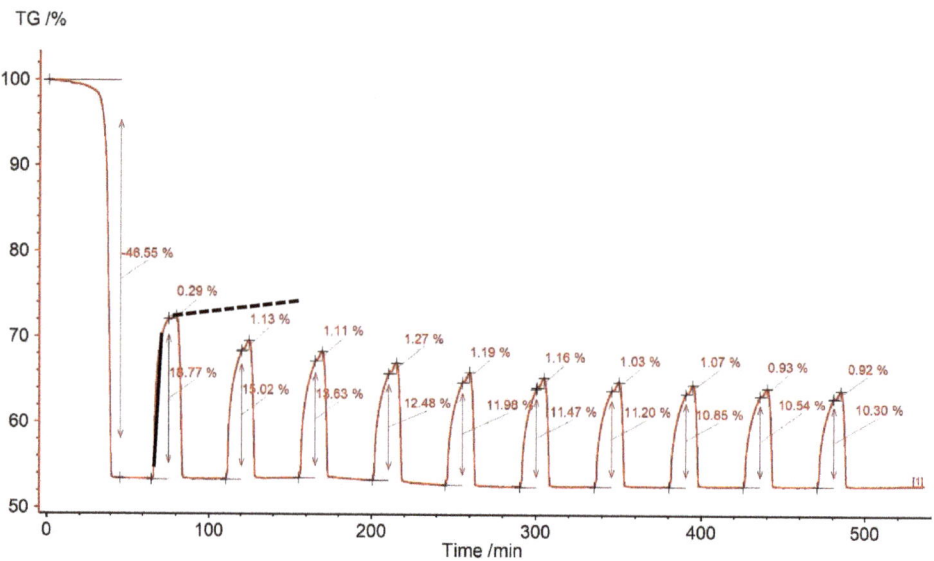

Figure 5. Relative mass changes for dolomite in the CaL process (the solid black line marks the section of the surface reaction, and the dashed line marks the section possibly corresponding to diffusion).

3.1.2. Saint Anne Mountain Limestone

The initial weight loss of the sorbent was observed (dehydration and calcination) to amount to 42.89% of its weight (Figure 6), which means that the sorbent was composed of almost pure calcium carbonate. Carbonation caused the sample mass changes in the range from 30.49% (first cycle) to 12.42% (10th cycle) (Figure 7).

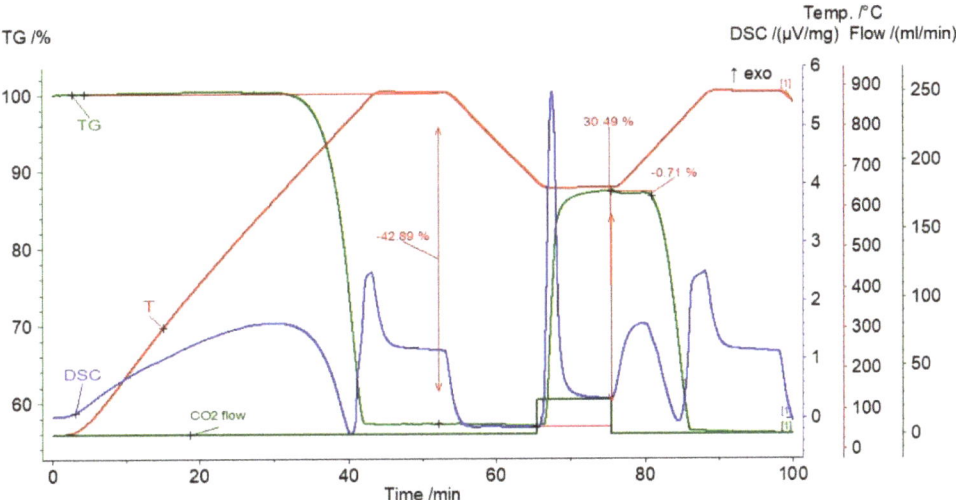

Figure 6. TGA and DSC curves for limestone from Saint Anne Mountain.

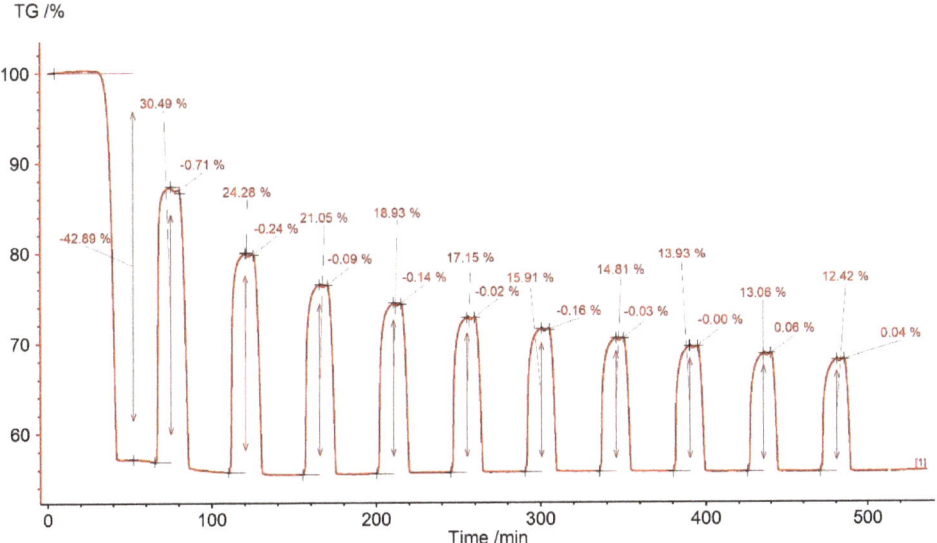

Figure 7. Relative mass changes for the limestone from Saint Anne Mountain.

The gas capture efficiency clearly decreased after the first cycle—to 24.28%—which may be related to the reduction in the active sorbent surface due to sintering. In subsequent cycles, this decline showed a downward trend. It should be noted that the reaction segment (visible as a steep mass increase) was shortened more and more in subsequent steps. At the same time, the elongation of the transition section was noticeable—this proves the increasing role of diffusion and confirms the sintering phenomenon. At the end of the assumed carbonation stage, a slightly inclined section of the sample mass increase was formed, which proves the diffusion process at that time (Figure 6). In addition, within about 5 min after closing the CO_2 supply, small, unsystematic fluctuations in the mass of the sample are revealed (from −0.71% to +0.06%), followed by rapid calcination.

3.1.3. Marl

A decrease in the weight of the sorbent was observed, amounting to 29.75%, of which about 4% corresponded to dehydration and dehydroxylation of clay minerals, and the remaining 25.73% corresponded to carbonate decomposition (Figure 8). Carbonation caused the sample mass changes within the range from 3.52% (first cycle) to 1.67% (10th cycle) (Figure 9). The efficiency of CO_2 gas uptake decreased at a decreasing rate, which may be related to the reduction in the active surface of the sorbent due to sintering (Figure 12). The reaction segment was shortened in successive stages, and instead, the elongation of the transition section became noticeable. This proves the increasing role of diffusion and confirms the sintering phenomenon. Within about 5 min after closing the CO_2 valve, slight fluctuations in the mass of the sample were revealed, with a clearly decreasing character with subsequent cycles. They were followed by rapid calcination.

3.1.4. Nephelinite

The initial weight loss of the sorbent (poorly marked dehydration and calcination) was observed, amounting to 1.90% of its weight (Figure 10) and showing a small, potential share of carbonates (most likely filling cracks or voids formed during degassing of basaltic lava). During carbonation, the gas capture efficiency showed no systematic variability, and in most cycles, it ranged from 0.63 to 0.68%. In only the first cycle, the reaction section (steep mass increase fragment) was short, followed by the diffusion section. In subsequent stages, the reaction segments were higher. They were followed by slight fluctuations in mass, lasting until the end of the assumed carbonation stage.

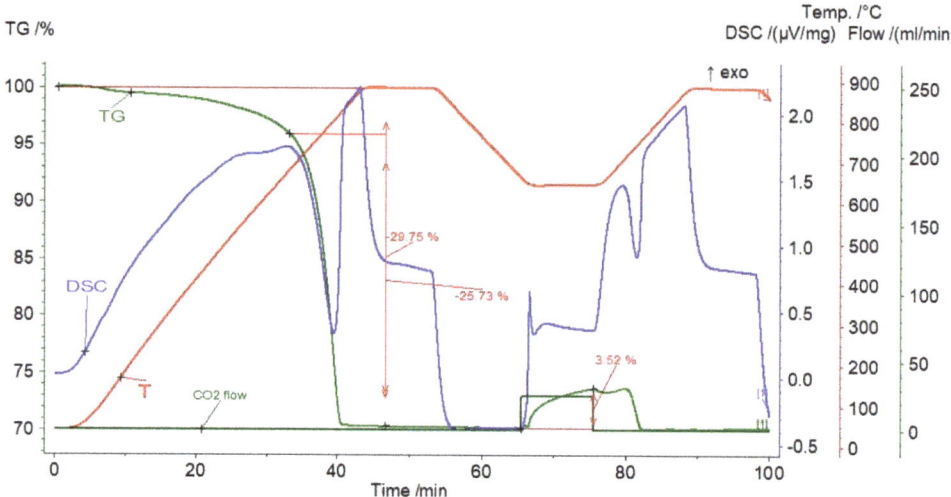

Figure 8. TGA and DSC curves for marl.

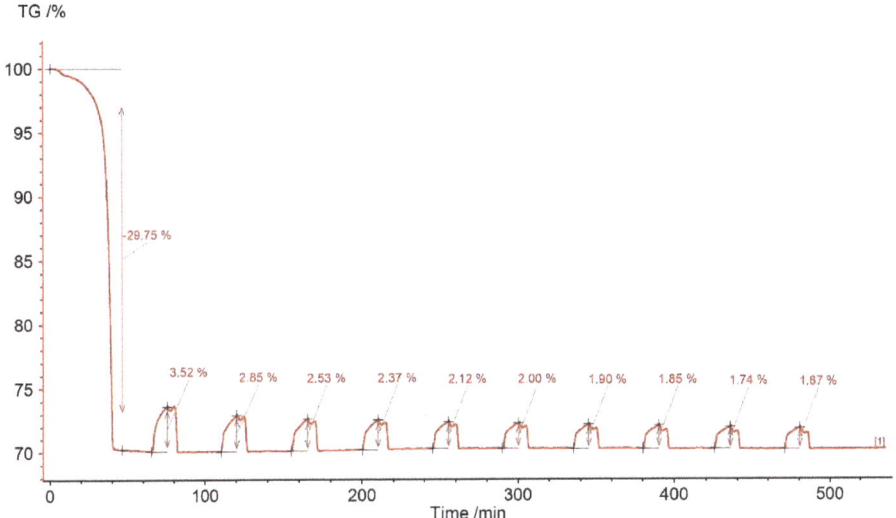

Figure 9. Relative mass changes for the marl.

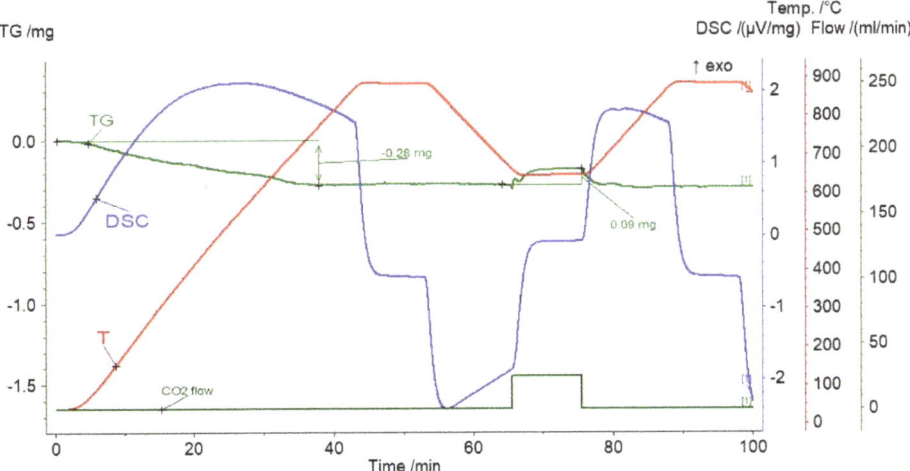

Figure 10. TGA and DSC curves for nephelinite.

3.1.5. Magnesite

The significant initial weight loss of the sorbent amounted to 50.92% of its weight (Figure 11), which proves the negligible potential contributions of other carbonates. The gas capture efficiency showed a decreasing trend, burdened with a non-systematic component, and ranged from 0.85% (cycle 1) to 0.72% (cycle 10). As in the case of the previously described nephelinite, in only the first cycle, the reaction section was short, followed by the diffusion process. In the next stages, the reaction segments were higher. Fluctuations in mass lasted until the end of the carbonation stage, and were followed by relatively slow calcination.

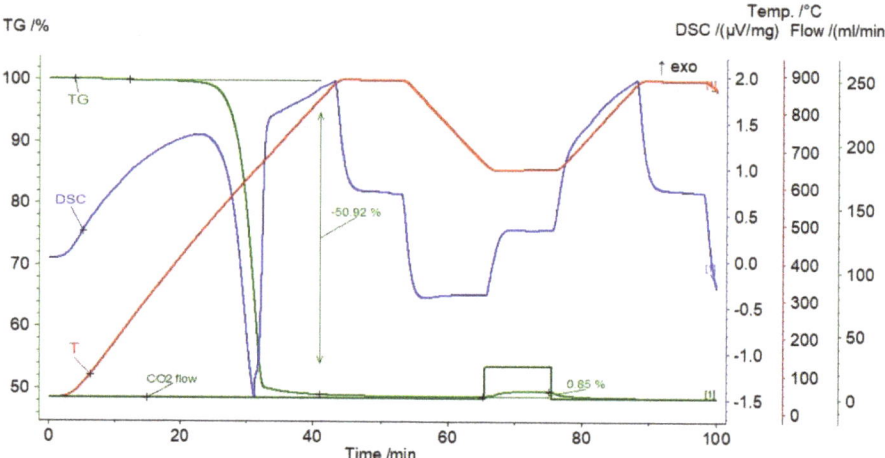

Figure 11. TGA and DSC curves for magnesite.

3.1.6. Serpentinite

The result of the DTA–TG analysis of serpentine sample is shown in Figure 12. Two endothermic peaks were found at temperatures of 623.1 and 701.9 °C due to the release of structural water. At a temperature of 834.1 °C, a large exothermic peak was visible, which represents the destruction of the serpentine crystalline structure and the formation of forsterite, enstatite, and clinoenstatite. For temperatures higher than 750 °C, the TGA analysis showed no significant weight variation. During carbonation, the sample mass changes reached from 1.42% (first cycle) to 0.95% (10th cycle). The efficiency of gas capture decreased at a decreasing rate. The reaction segment was shortened in successive stages, while the diffusion segment became more apparent, which confirms the sintering of the sample. About 5 min after the CO_2 shut off, there was a slight decrease in the mass of the sample—typical for the first-order reaction—followed by a sharp but slight weight loss due to calcination.

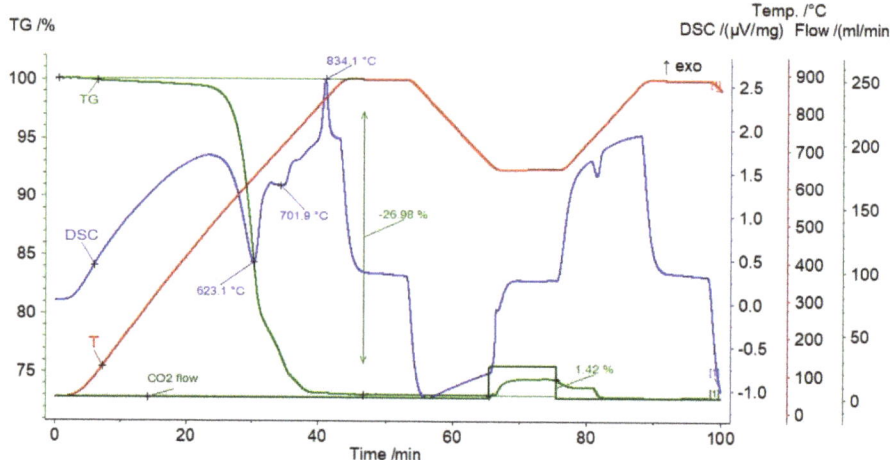

Figure 12. TGA and DSC curves for serpentinite.

3.2. Thermal Pretreatment of Sorbents

Calcium sorbents are characterized by a decreasing activity of capturing CO_2 in each subsequent carbonation cycle. After approximately one hundred cycles, the asymptotically decreasing sorbent yield ranges from 7 to 15% for a 10 min carbonation time. It is believed that this is the result of changes in the sorbent's morphology, during which its specific surface area decreases and the micropores disappear. In order to improve the activity of sorbents, the following enhancement techniques are used:

- Doping—aimed at postponing or avoiding sintering of sorbent in order to moderate sintering and abrasion of the sorbent (e.g., [18,19]). The effectiveness of doping depends on the concentration of the substrate used. Too low of a concentration will have no effect, while too high of a concentration may block the pores [12,20];
- Chemical treatment—to obtain a better sintering performance and more favorable pore area (e.g., [21,22]). Although the chemical treatment presents reactivity benefits, it has two drawbacks: the cost and availability of the acid and the marginal increase in CO_2 uptake [23],
- Thermal pretreatment—to improve the conversion of CaO in long series of cycles and to stabilize the sorption capacity (e.g., [24,25]).

Research performed by Manovic and Anthony [25] and Manovic et al. [26] demonstrated that thermal pre-treatment could be an important method of improving conversion of CaO over long series of cycles. Such a phenomenon might be explained by a theory proposed by Lysikov et al. [27] and developed by Manovic and Anthony [25], according to which the repetitive carbonation/calcination cycles enhance the formation of a skeleton of interconnected CaO. This skeleton acts as the outer layer of the reactive CaO layer and stabilizes the sorption capacity.

The tests by Manovic and Anthony [25] showed that the particles were strongly sintered and that carbonization occurred on the surface of the solid particle. The pretreatment resulted in the formation of an internal skeleton of the sorbent particles and protection of their integrity. When the sorbents are preheated, after the decomposition of $CaCO_3$, ion diffusion continues and stabilizes the skeleton, which, due to its porous structure, is able to maintain significant carbonation (Figure 13).

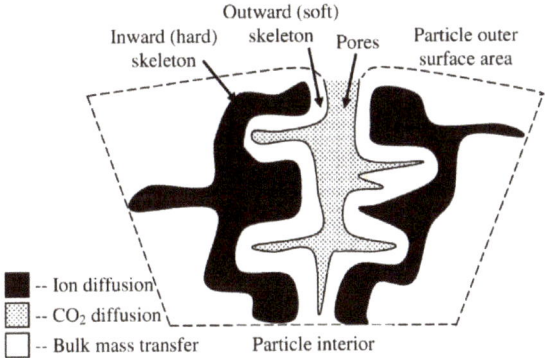

Figure 13. Schematic representation of the pore–skeleton model [25].

Manovic and Anthony [20] studied the improvement of sorbent properties (Kelly Rock limestone) with 54.39% CaO content and loss on ignition (44.20%) through steam reactivation, thermal pre-treatment, and addition of calcium-aluminate-based pellets. The most promising results were obtained for powdered Kelly Rock limestone samples (<50 μm). For the preheating temperatures of 1000 °C, the sorbent was self-reactivated for the next 30 carbonation cycles. The highest conversions were obtained at 1000 °C, for which 49% conversion was achieved in the last cycle, with an average value of ∼45% for 30 cycles.

Lower conversion values were obtained for samples pretreated at 1100 and 1200 °C. However, taking into account the self-reactivation effect, these results may also be promising, especially since heating the sorbent at these temperatures (in combination with granulation) gives better mechanical properties that could prevent its undesirable attrition. Assuming that the pretreatment time is an important parameter, a 6 h thermal activation study was also carried out on Kelly Rock powdered samples at temperatures of 900–1100 °C. It was confirmed that the pretreatment time influences the properties of the sorbent, and that shorter thermal treatment times can positively influence the effectiveness of the sorbent [25].

In this work, based on the results of the preliminary tests, three limestone samples from Stramberk (Czechia), Podlesie (Poland), and Butkov (Slovakia) (Table 1) were selected for further testing. They were characterized by high, medium, and the lowest weight loss, respectively. The tested sorbents were thermally modified by pre-heating the sample for one hour at 1000 °C in 100% N_2 atmosphere. A further research cycle was carried out according to the procedure described earlier.

The effect of thermal pretreatment is presented for the example of limestone from Stramberk (Czechia) (Figure 14A,B). During the carbonation of the raw sample, CO_2 capture was achieved, causing the sample mass to change within a range from 31.36% (first cycle) to 13.68% (10th cycle). The gas capture efficiency decreased with increasing number of CaL cycles. After the first cycle, the gas uptake efficiency decreased to 25.36%, which may be due to the reduction of the active sorbent surface due to sintering. It is noticeable that the reaction time (visible as a steep section of the mass increase) got shorter in subsequent cycles. Simultaneously with the reduction of the reaction section, the extension of the transition section was noticeable, which proves the increasing role of diffusion and confirms the phenomenon of sintering of the sample.

It is noteworthy that, within the assumed time of 10 min, the carbonation process was not completed. This is evident in the sample mass change graph, which shows a sharp rise (reaction stage) moving smoothly (transition stage) towards the diffusion stage. The latter, however, is not observed (no element approaching the horizontal line, mass growth curve). This means that in the case of the Stramberk limestone, carbonation could be carried out for a longer time than assumed in the analyzed simulation. The extension of the carbonation time in this case may be associated with a potential reduction in the economics of the capture process. This issue would require further tests with extended carbonation time in order to calculate the amount of CO_2 bound by the sorbent through the diffusion process.

The comparison of the relative mass changes for Stramberk limestone without modification and thermally pretreated samples proves that the end of the reaction stage occurs at a similar temperature—around 644.5 °C (Figure 15). The calcination time varies, however, and is shorter for the unmodified sorbent, which is related to the lower content of CO_2 blocked in this sample.

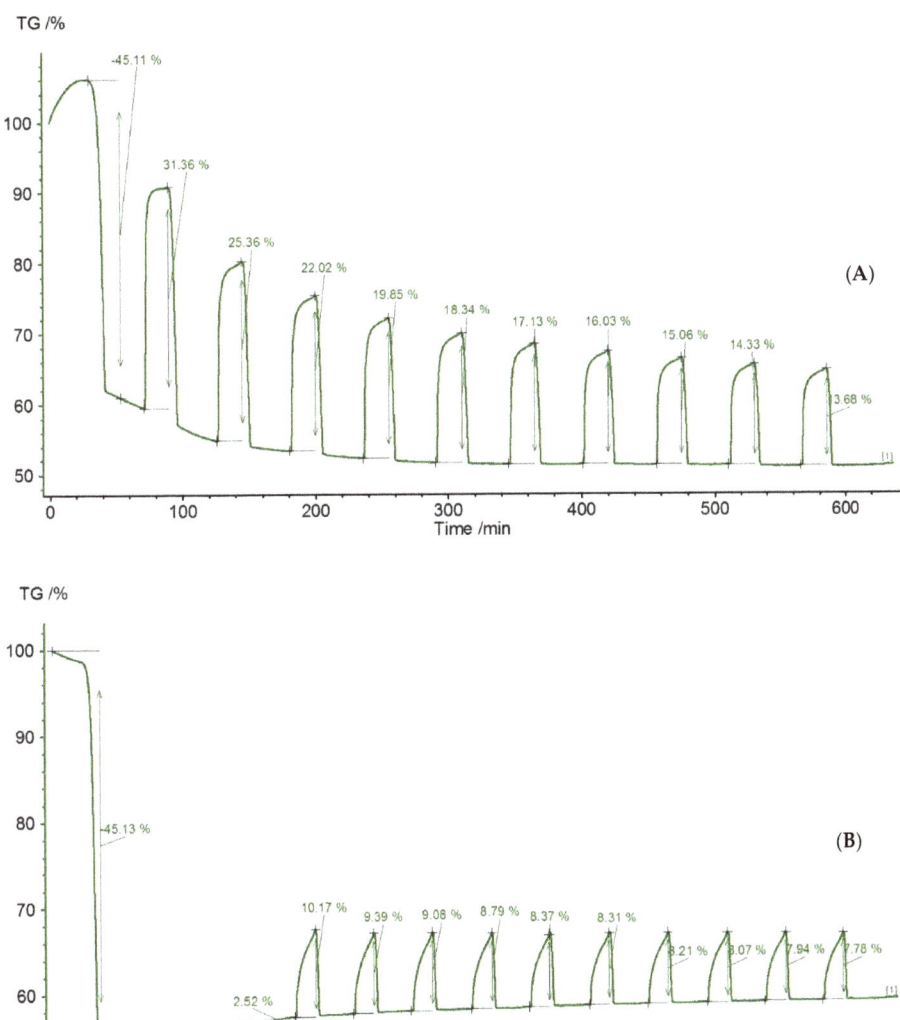

Figure 14. Relative mass changes for Stramberk limestone: (**A**) unmodified; (**B**) thermally pretreated for 1 h at 1000 °C.

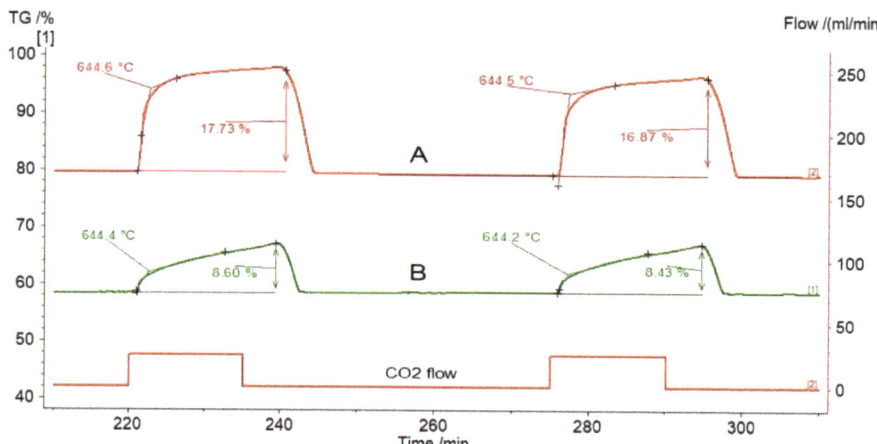

Figure 15. Relative mass changes for Stramberk limestone at the fifth and sixth calcium looping cycles: (**A**) without modification; (**B**) pretreated for 1 h at 1000 °C.

4. Discussion

The CaL simulations showed different degrees of conversion for the tested rock sorbents (Figures 16 and 17). As expected, they were the highest for carbonate rocks (except for bituminous limestone); intermediate values were found for marl and bituminous limestone, and the values were lower by an order of magnitude for the remaining sorbents: magnesite and serpentinite. The lowest degree of conversion was determined for basalt.

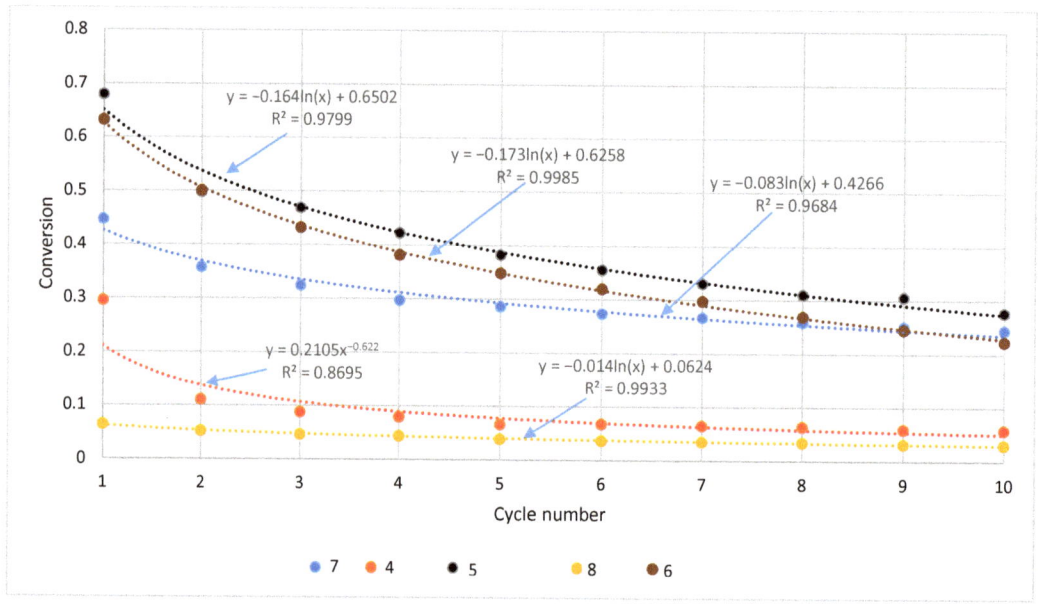

Figure 16. The conversion of carbonate rocks in the CaL process.

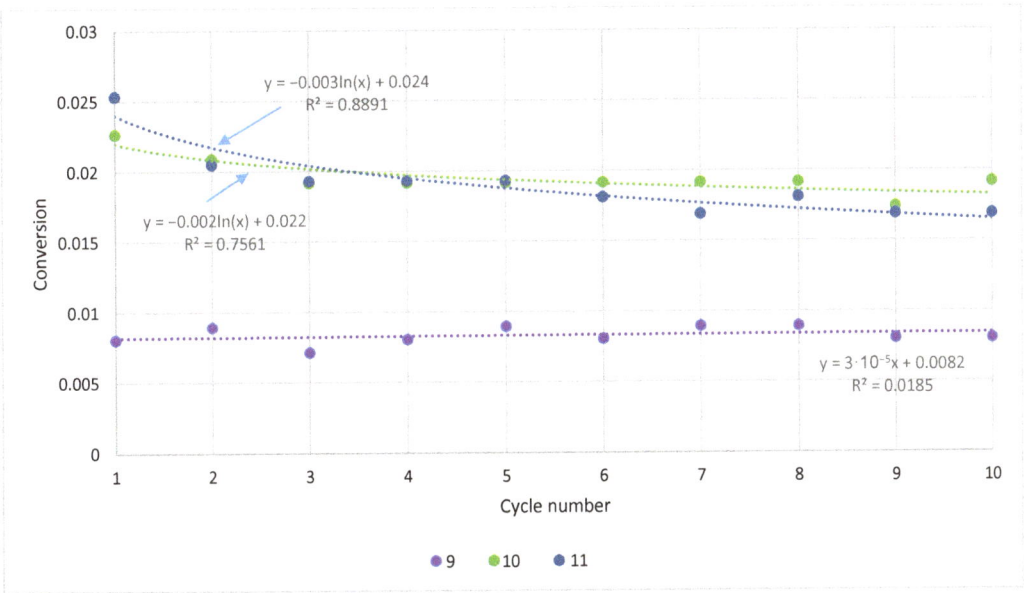

Figure 17. The conversion of magnesite and non-carbonate rocks in the CaL process.

The decrease in the conversion rate with subsequent CaL cycles was most pronounced for the sorbents with the highest values of this parameter, again with the exception of bituminous limestone (for which a significant decrease was noted between the first and second cycles; Figure 16). In the case of basalt, there were no significant changes in the conversion rate during the simulated cycles.

From the standpoint of the efficiency of the CaL process, an important indicator is the parameter called "cumulative carbonation relative to the initial mass of the sorbent", which, for the 10 analyzed cycles ($n = 10$), could be defined as the degree of carbonation for n cycles.

It represents the multiplicity of the captured CO_2 relative to the initial sorbent mass. In the course of the analyzed cycles, the cumulative carbonation shows regularities that are similar to the degree of conversion (Figures 18 and 19). This parameter, exceeding 1.0, was also the highest for carbonate rocks (limestones: Stramberk—1.93, Saint Anne Mountain—1.81, Gorazdze—1.65, Podlesie—1.62, and dolomite of Olkusz—1.26); values that were lower by an order of magnitude were achieved for marl, bituminous limestone, and serpentine, and values lower by two orders of magnitude were achieved for magnesite and basalt.

The CaL simulations performed for the thermally pretreated samples also demonstrated different degrees of conversion for the tested rock sorbents (Table 2). As predicted, they were the highest for the Stramberk limestone sample, which was characterized by the highest proportion of calcium carbonate. The decrease in conversion with subsequent CaL cycles was most apparent for the sorbents with the highest values of this parameter.

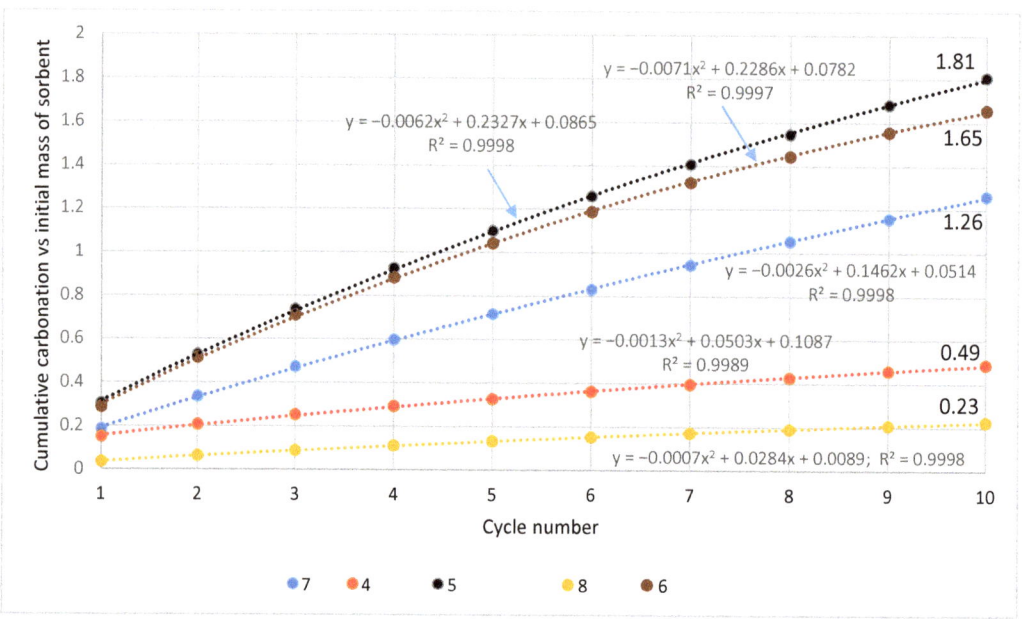

Figure 18. Cumulative carbonation relative to the initial mass of the sorbent—carbonate rocks.

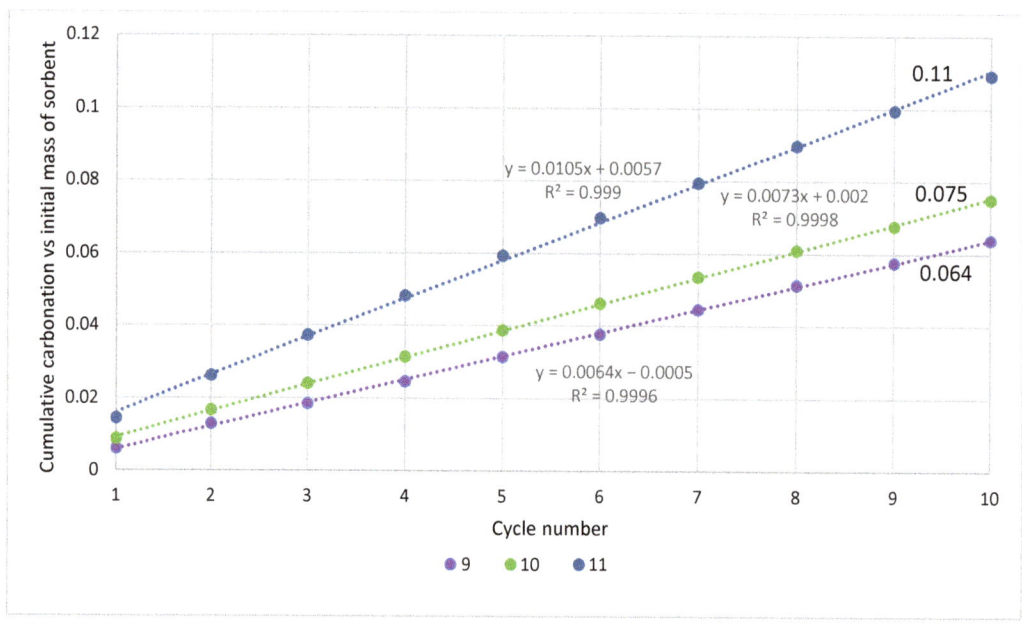

Figure 19. Cumulative carbonation relative to the initial mass of the sorbent—non-carbonate rocks.

Table 2. Conversion in raw and pretreated limestones.

Cycle	Sample					
	Štramberk		Podlesie		Butkov	
	90.1% $CaCO_3$		84.5% $CaCO_3$		68.9% $CaCO_3$	
	Untreated	Pretreated	Untreated	Pretreated	Untreated	Pretreated
1	0.73	0.24	0.61	0.14	0.28	0.09
2	0.59	0.22	0.49	0.12	0.17	0.08
3	0.51	0.21	0.42	0.12	0.13	0.08
4	0.46	0.20	0.38	0.11	0.11	0.07
5	0.43	0.19	0.34	0.11	0.10	0.07
6	0.40	0.19	0.31	0.11	0.09	0.07
7	0.37	0.19	0.28	0.11	0.09	0.06
8	0.35	0.19	0.25	0.10	0.09	0.06
8	0.33	0.18	0.23	0.10	0.09	0.06
10	0.32	0.18	0.21	0.10	0.08	0.06

The cumulative carbonation in the course of the analyzed cycles shows a pattern similar to that of the degree of conversion (Table 3); however, the value of this parameter was more than two times lower than for the raw samples of Stramberk (0.86) and Butkov (0.35). The greatest relative decrease was recorded for the Podlesie limestone, the cumulative carbonation of which decreased by more than three times (from 1.62 for raw sample to 0.5 after thermal treatment) despite the relatively high content of $CaCO_3$.

Table 3. Cumulative carbonation relative to the initial weight of the sorbent for raw and pretreated limestones.

Cycle	Sample					
	Štramberk		Podlesie		Butkov	
	Untreated	Pretreated	Untreated	Pretreated	Untreated	Pretreated
1	0.31	0.10	0.28	0.06	0.14	0.05
2	0.57	0.20	0.50	0.12	0.23	0.09
3	0.79	0.29	0.70	0.17	0.29	0.13
4	0.99	0.37	0.87	0.22	0.35	0.16
5	1.17	0.46	1.03	0.27	0.40	0.20
6	1.34	0.54	1.17	0.32	0.45	0.23
7	1.50	0.62	1.30	0.37	0.50	0.26
8	1.65	0.70	1.41	0.41	0.54	0.29
8	1.80	0.78	1.52	0.46	0.58	0.32
10	1.93	0.86	1.62	0.50	0.62	0.35

The thermal pretreatment was not effective for the examined limestones, as also reported by Manovic et al. [20]; it is believed to be efficient for only some types of natural materials. It is likely that different types of limestone require different pretreatment conditions due to differences in impurities and internal structures [28]. However, this treatment has clear advantages: It is simple and relatively inexpensive compared to other techniques. On the other hand, it should be underlined that this would require additional energy to heat up the sorbent prior to its final use. This may result in a reduction in the power output of a CaL-equipped power generation system. Nonetheless, several studies proved that even if the pretreated limestone shows lower values of initial sorption capacity, this capacity increases over many cycles due to the softening of the hard skeleton. The disadvantage of this refining technique is that although the reactivity increases, the attrition of the particles significantly increases [29].

5. Summary and Conclusions

The suitability of selected rocks—limestone, dolomite, magnesite, marl, serpentinite, and basalt—was tested for the purpose of CO_2 sequestration in the CaL process. TGA-DSC tests were carried out based on a temperature program designed for this purpose. The tests were performed in 10 cycles of alternating calcination and carbonation. During the measurements, changes in the mass of the sample over time (TGA) and heat flow (DSC) were recorded.

CaL simulations showed various degrees of conversion for the tested rock sorbents—the highest values were achieved for carbonate rocks (except for bituminous limestone), intermediate values were achieved for marl and bituminous limestone, lower ones were achieved for the remaining sorbents (magnesite and serpentinite), and the lowest were achieved for basalt.

The decrease in the conversion rate with subsequent CaL cycles was most intense for the sorbents with the highest values of this parameter. In the case of basalt, no significant changes in the conversion rate were observed. The decrease in gas capture efficiency with an increasing number of CaL cycles may be related to the decreasing active sorbent surface due to sintering.

The values of the parameter called "cumulative carbonation relative to the initial mass of the sorbent" corresponded to the multiplicity of the captured CO_2 relative to the initial sorbent mass. This parameter, exceeding a value of 1.0, was the highest for carbonate rocks; it achieved lower values for marl, bituminous limestone, and serpentine, as well as—by two orders of magnitude—for magnesite and basalt.

In most of the analyzed samples, the carbonation process was not completed within the assumed time of 10 min. In practice, however, extending the carbonation time could reduce the economics of the capture process.

The simulations of the thermally pretreated samples also demonstrated different degrees of conversion for the tested rock sorbents, which were the highest for the calcium-carbonate-rich Stramberk limestone. The cumulative carbonation of the pretreated samples was more than two times lower than that of the raw ones. The largest relative decrease was recorded in the case of Podlesie limestone, the cumulative carbonation of which decreased by more than three times, despite the relatively high $CaCO_3$ content.

Author Contributions: Conceptualization, methodology, validation, formal analysis, investigation, data curation, writing—original draft preparation, writing—review and editing, visualization, K.L. All authors have read and agreed to the published version of the manuscript.

Funding: This research received no external funding.

Institutional Review Board Statement: Not applicable.

Informed Consent Statement: Not applicable.

Data Availability Statement: The data presented in this study are available on request from the corresponding author.

Conflicts of Interest: The author declare no conflict of interest.

References

1. International Energy Agency. *Energy and Climate Change—World Energy Outlook—Special Briefing for COP21*; International Energy Agency: Paris, France, 2015.
2. Olivier, J.G.J.; Janssens-Maenhout, G.; Muntean, M.; Peters, J.A.H.W. *Trends in Global CO_2 Emissions-2015 Report*; No: JRC98184; PBL Netherlands Environmental Assessment Agency: The Hague, The Netherlands, 2015.

3. Sanna, A.; Uibu, M.; Caramanna, G.; Kuusik, R.; Maroto-Valer, M.M. A review of mineral carbonation technologies to sequester CO_2. *Chem. Soc. Rev.* **2014**, *43*, 8049. [CrossRef] [PubMed]
4. Gislason, S.R.; Wolff-Boenisch, D.; Stefansson, A.; Oelkers, E.H.; Gunnlaugsson, E.; Sigurdardottir, H.; Sigfusson, B.; Broecker, W.S.; Matter, J.M.; Stute, M.; et al. Mineral sequestration of carbon dioxide in basalt: A pre-injection overview of the carbfix project. *Int. J. Greenh. Gas Contr.* **2010**, *4*, 537–545. [CrossRef]
5. McKelvy, M.J.; Chizmeshya, A.V.G.; Diefenbacher, J.; Béarat, H.; Wolf, G. Exploration of the role of heat activation in enhancing serpentine carbon sequestration reactions. *Environ. Sci. Technol.* **2004**, *38*, 6897–6903. [CrossRef] [PubMed]
6. Olajire, A.A. A review of mineral carbonation technology in sequestration of CO_2. *J. Pet. Sci. Eng.* **2013**, *109*, 364–392. [CrossRef]
7. Werner, M.; Verduyn, M.; van Mossel, G.; Mazzotti, M. Direct flue gas CO_2 mineralization using activated serpentine: Exploring the reaction kinetics by experiments and population balance modelling. *Energy Procedia* **2014**, *4*, 2043–2049. [CrossRef]
8. Huijgen, W.J.J.; Comans, R.N.J. *Carbon Dioxide Sequestration by Mineral Carbonation: Literature Review*; ECN—Clean Fossil Fuels Environmental Risk Assessment: Petten, The Netherlands, 2003; p. 112.
9. Kirsch, K. CO_2-Induced Metal Release from Sandstones: Implications for Geologic Carbon Sequestration. Master's Thesis, Hydrologic Science and Engineering Faculty, School of Mines, Golden, CO, USA, 2013.
10. Blamey, J.; Anthony, E.J.; Wang, J.; Fennell, P.S. The calcium looping cycle for large-scale CO_2 capture. *Prog. Energy Combust. Sci.* **2010**, *36*, 260–279. [CrossRef]
11. Laursen, K.; Duo, W.; Grace, J.R.; Lim, J. Sulfation and reactivation characteristics of nine limestones. *Fuel* **2000**, *79*, 153–163. [CrossRef]
12. Dean, C.C.; Blamey, J.; Florin, N.H.; Al-Jeboori, M.J.; Fennell, P.S. The calcium looping cycle for CO_2 capture from power generation, cement manufacture and hydrogen production. *Chem. Eng. Res. Des.* **2011**, *89*, 836–855. [CrossRef]
13. Mantripragadaa, H.C.; Rubin, E.S. Calcium looping cycle for CO_2 capture: Performance, cost and feasibility analysis. *Energy Procedia* **2014**, *63*, 2199–2206. [CrossRef]
14. Lee, D. An apparent kinetic model for the carbonation of calcium oxide by carbon dioxide. *Chem. Eng. J.* **2004**, *100*, 71–77. [CrossRef]
15. Oakeson, W.G.; Cutler, I.B. Effect of CO_2 Pressure on the Reaction with CaO. *J. Am. Ceram. Soc.* **2006**, *62*, 556–558. [CrossRef]
16. Butler, J.; Lim, J.; Grace, J. Kinetics of CO_2 absorption by CaO through pressure swing cycling. *Fuel* **2014**, *127*, 78–87. [CrossRef]
17. Levenspiel, O. *Chemical Reaction Engineering*, 2nd ed.; John Wiley and Sons: New York, NY, USA, 1972.
18. Salvador, C.; Lu, D.; Anthony, E.J.; Abanades, J.C. Enhancement of CaO for CO_2 capture in an FBC environment. *Chem. Eng. J.* **2003**, *96*, 187–195. [CrossRef]
19. González, B.; Blamey, J.; McBride-Wright, M.; Carter, N.; Dugwell, D.; Fennell, P. Calcium looping for CO2 capture: Sorbent enhancement through doping. *Energy Procedia* **2011**, *4*, 402–409. [CrossRef]
20. Manovic, V.; Anthony, E.J. Lime-Based sorbents for high-temperature CO_2 capture—A review of sorbent modification methods. *Int. J. Environ. Res. Public Health* **2010**, *7*, 3129–3140. [CrossRef]
21. Li, Y.J.; Zhao, C.S.; Duan, L.B.; Liang, C.; Li, Q.Z.; Zhou, W. Cyclic calcination/carbonation looping of dolomite modified with acetic acid for CO_2 capture. *Fuel Process. Technol.* **2008**, *89*, 1461–1469. [CrossRef]
22. Ridha, F.N.; Manovic, V.; Macchi, A.; Anthony, E.J. The effect of SO_2 on CO_2 capture by CaO-based pellets prepared with a kaolin derived $Al(OH)_3$ binder. *Appl. Energy* **2012**, *92*, 415–520. [CrossRef]
23. Erans, M.; Manovic, V.; Anthony, E.J. Calcium looping sorbents for CO_2 capture. *Appl. Energy* **2016**, *180*, 722–742. [CrossRef]
24. Albrecht, K.O.; Wagenbach, K.S.; Satrio, J.A.; Shanks, B.H.; Wheelock, T.D. Development of a CaO-based sorbent with improved cyclic stability. *Ind. Eng. Chem. Res.* **2008**, *47*, 7841–7848. [CrossRef]
25. Manovic, V.; Anthony, E.J. Thermal Activation of CaO-Based Sorbent and Self-Reactivation during CO_2 Capture Looping Cycles. *Environ. Sci. Technol.* **2008**, *42*, 4170–4174. [CrossRef]
26. Manovic, V.; Anthony, E.J.; Grasa, G.; Abanades, J.C. CO_2 Looping Cycle Performance of a High-Purity Limestone after Thermal Activation/Doping. *Energy Fuels* **2008**, *22*, 3258–3264. [CrossRef]
27. Lysikov, A.I.; Salanov, A.N.; Okunev, A.G. Change of CO_2 carrying capacity of CaO in isothermal recarbonation-decomposition cycles. *Ind. Eng. Chem. Res.* **2007**, *46*, 4633–4638. [CrossRef]
28. Arias, B.; Grasa, G.S.; Alonso, M.; Abanades, J.C. Post-combustion calcium looping process with a highly stable sorbent activity by recarbonation. *Energy Environ. Sci* **2012**, *5*, 7353–7359. [CrossRef]
29. Chen, Z.; Song, H.S.; Portillo, M.; Lim, C.J.; Grace, J.R.; Anthony, E.J. Long-term calcination/carbonation cycling and thermal pretreatment for CO_2 capture by limestone and dolomite. *Energy Fuels* **2009**, *23*, 1437–1444. [CrossRef]

Article

Measurements of Density of Liquid Oxides with an Aero-Acoustic Levitator

Sergey V. Ushakov [1,*], Jonas Niessen [2,*], Dante G. Quirinale [3,*], Robert Prieler [2], Alexandra Navrotsky [1] and Rainer Telle [2]

1. School of Molecular Sciences and Center for Materials of the Universe, Arizona State University, Tempe, AZ 85287, USA; anavrots@asu.edu
2. Institut fuer Gesteinshuettenkunde/Mineral Engineering, RWTH Aachen University, 52062 Aachen, Germany; prieler@ghi.rwth-aachen.de (R.P.); telle@ghi.rwth-aachen.de (R.T.)
3. Neutron Scattering Division, Oak Ridge National Laboratory, Oak Ridge, TN 37830, USA
* Correspondence: sushakov@asu.edu (S.V.U.); j.niessen@ghi.rwth-aachen.de (J.N.); quirinaledg@ornl.gov (D.G.Q.)

Abstract: Densities of liquid oxide melts with melting temperatures above 2000 °C are required to establish mixing models in the liquid state for thermodynamic modeling and advanced additive manufacturing and laser welding of ceramics. Accurate measurements of molten rare earth oxide density were recently reported from experiments with an electrostatic levitator on board the International Space Station. In this work, we present an approach to terrestrial measurements of density and thermal expansion of liquid oxides from high-speed videography using an aero-acoustic levitator with laser heating and machine vision algorithms. The following density values for liquid oxides at melting temperature were obtained: Y_2O_3 4.6 ± 0.15; Yb_2O_3 8.4 ± 0.2; $Zr_{0.9}Y_{0.1}O_{1.95}$ 4.7 ± 0.2; $Zr_{0.95}Y_{0.05}O_{1.975}$ 4.9 ± 0.2; HfO_2 8.2 ± 0.3 g/cm^3. The accuracy of density and thermal expansion measurements can be improved by employing backlight illumination, spectropyrometry and a multi-emitter acoustic levitator.

Keywords: levitation; rare earth oxides; zirconia; hafnia; melting; thermodynamics

Citation: Ushakov, S.V.; Niessen, J.; Quirinale, D.G.; Prieler, R.; Navrotsky, A.; Telle, R. Measurements of Density of Liquid Oxides with an Aero-Acoustic Levitator. *Materials* **2021**, *14*, 822. https://doi.org/10.3390/ma14040822

Academic Editor: Franz Saija

Received: 30 December 2020
Accepted: 27 January 2021
Published: 9 February 2021

Publisher's Note: MDPI stays neutral with regard to jurisdictional claims in published maps and institutional affiliations.

Copyright: © 2021 by the authors. Licensee MDPI, Basel, Switzerland. This article is an open access article distributed under the terms and conditions of the Creative Commons Attribution (CC BY) license (https://creativecommons.org/licenses/by/4.0/).

1. Introduction

Most metal alloys are produced by melt processing, and thermodynamic and thermophysical properties of metallic melts have been systematically investigated for over a century. The developed ab initio and Calphad-based computational tools show spectacular results for the prediction of crystallization pathways and equilibrium phases for metal alloys [1]. Refractory oxide ceramics are usually produced by sintering, and application-driven incentives to study high temperature oxide melts have been largely limited to metallurgical slugs and glasses. The situation has changed with the application of additive manufacturing techniques to ceramic materials [2–7]. These techniques often involve laser melting, and their advance is hampered by a lack of data on oxide melts.

A plethora of techniques is available for high temperature study of metal alloys. The "exploding wire" technique [8,9] has been in development for more than 300 years and has also been adapted for electrically conductive carbides and nitrides [10–13]. In this method, pulse discharge through a metallic wire or conducting ceramic coating provides instantaneous heating and excludes any contamination from the container.

Electromagnetic [14] and electrostatic [15] levitation have been successfully used for contactless high temperature studies on metal alloys for decades [16–22]. Electromagnetic levitation was also applied to semiconductors after preheating; liquid silicon was studied extensively with this technique [20,23]. Image-based density measurements using levitation have been reported since the 1960s [24], and have been further refined using modern machine vision algorithms and applying Legendre polynomial fitting to the results of edge detection routines for volume calculation [23,25–28]. Modulated laser calorimetry on electromagnetically levitated melts was developed by Fukuyama et al. in 2007 [29]. Data on excess volume from image processing combined with data on excess heat capacities of mixing [30] provide a thermodynamic foundation for constructing realistic solution models for metallic alloys.

However, many refractory oxides are dielectrics and cannot be studied using exploding wire or electromagnetic levitation. Electrostatic levitation can be applied to dielectric materials, but it relies on surface charges and is challenging in terrestrial conditions. Most of the work on the application of electrostatic levitation to oxides was accomplished by the group at Tsukuba Space Center in Japan [31–34]. It culminated in the design of an electrostatic levitation furnace (ELF), which is currently in operation at the International Space Station [35]. The first results on the density of liquid Er_2O_3 and Gd_2O_3 were published in 2020 [36,37].

Aerodynamic levitation in a conical nozzle (CNL) [38–41] has been used extensively for oxide melts; however, it has limitations of limited sample visibility and large thermal gradient. The development of an aero-acoustic levitator (AAL) was funded by NSF and NASA and built by Intersonics Inc. in 1990 [42]. In this method, the sample is stabilized above the gas jet by acoustic forces that allow unimpeded access for multi-beam laser heating, pyrometer aiming and video recording.

Only two AAL instruments were commercially produced [43]: the first one with analog controls [44], which was operated in Japan [45–47]. The second one, used in this work, was built for RWTH Aachen University [48]. It enabled the first direct observation of liquid immiscibility between zirconia and silica-rich melts in the ZrO_2-SiO_2 system [49]. To the authors' knowledge, this is the only instrument of its kind in operation to date. However, it will not be for long. Marzo et al. [50] made openly accessible a new acoustic levitator design using mass-produced acoustic transducers. This drastically reduces the cost of the development of new generation AAL. We anticipate that this innovation will result in a wider application of this technique to study oxide melts. In this work, we present measurements of the density of Y_2O_3, Yb_2O_3, YSZ, and HfO_2 melts with an aero-acoustic levitator using machine vision algorithms developed for metal alloys.

2. Materials and Methods

2.1. Sample Synthesis

HfO_2, Y_2O_3, and Yb_2O_3 oxide spheroids 2–3 mm in diameter were prepared by the melting of oxide powders obtained from Alfa Aesar (Ward Hill, MA, USA) with metals purity 99.98% or higher. The powders were sintered at 1500 °C in air for 5 h, then placed into a copper hearth and melted with a 400 W CO_2 laser beam into irregularly shaped pieces surrounded by an unmelted powder bed. The resulting solid pieces were remelted in a conical nozzle aerodynamic levitator in Ar flow. Experiments were also performed on laser melted $Y_{0.05}Zr_{0.95}O_{1.975}$ and $Y_{0.1}Zr_{0.9}O_{1.95}$ samples prepared and characterized earlier for neutron diffraction experiments [51].

2.2. Measurement Procedure

The design of the aero-acoustic levitator used in this work (Figure 1) was described in detail by Nordine et al. [48]. The sample for levitation was positioned above the alumina tube heated to 550 °C, which serves as a gas jet. The levitation of the sample above the jet is stabilized by six acoustic transducers controlled with levitator software using a positioning system with three low power 808 nm solid state lasers. Experiments with Y_2O_3 were performed using N_2 or Ar gas jets; due to high density of HfO_2 and Yb_2O_3 samples their levitation was only possible using an Ar gas jet.

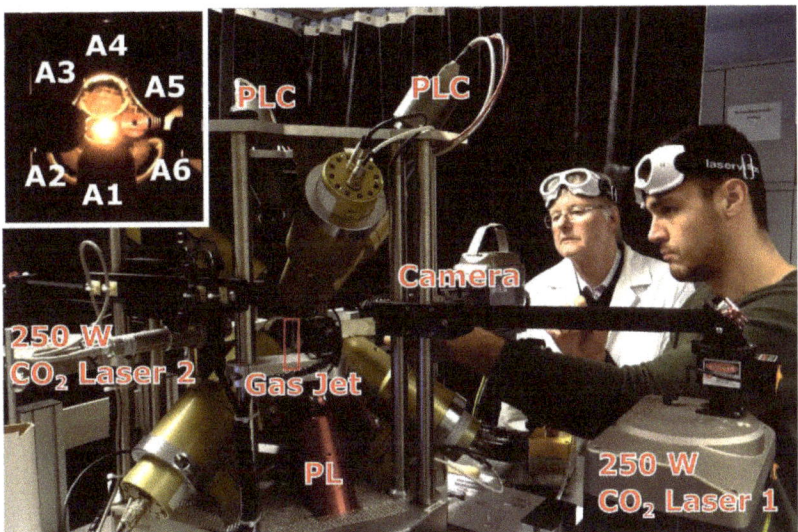

Figure 1. Operation of the aero-acoustic levitator at RWTH Aachen. J.N. is positioning the bead using an air pick above the gas jet for levitation. Inset: the sample bead heated with the dual laser beam, A1–A6 are frequency matched 22.2 kHz acoustic transducers controlled with input from positioning lasers and cameras (labeled PL and PLC, respectively).

After levitation was established, the sample was heated to its melting temperature with two antiparallel 240 W CO_2 laser beams (Synrad, WA, USA). The video was recorded using a Phantom V9.1 camera from Vision Resarch, Inc. (Wayne, NJ, USA) with an acquisition interval of 0.5–1 ms and exposure time of 20–200 µs. The temperature was recorded with a narrow band 650 nm Exactus pyrometer from BASF Corporation (Florham Park, NJ, USA). The pyrometer was operated with a 1-ms acquisition interval; the measurement spot size was set to 0.8 mm, and emissivity was set to 1. It was possible to record videos and cooling traces for the crystallization of Y_2O_3, $Zr_{0.95}Y_{0.05}O_{1.975}$, and $Zr_{0.9}Y_{0.1}O_{1.95}$ samples. This allowed correlation of the spike in density trace with recalescence peaks on crystallization, obtaining density values at melting temperature, and evaluating volume thermal expansion of the liquids. Levitated HfO_2 and Yb_2O_3 melts became unstable after turning off the lasers and fell out of the field of view before recalescence peaks were captured with a pyrometer. This is attributed to their higher density. Levitation stability in AAL is discussed in detail by Nordine et al. [48].

2.3. Video Processing

Volume was calculated using a modification of the algorithm developed by Bradshaw et al. [25] as implemented by Bendert et al. [52]. As the videos were not filtered, the large temperature range and surface features of the molten samples made shape determination using edge detection routines difficult, so the video contrast was post-processed using open source software [53] to provide better definition. The numerical routines were used to calculate volume assumed symmetry about the vertical axis. This is not necessarily the case with aero-acoustically levitated droplets, which may be slightly asymmetric and experience a slight precession, introducing some additional uncertainty into the measurement. The implementation of multi-emitter acoustic levitator design [50] may reduce this uncertainty.

Camera calibration was performed by imaging commercially obtained machined Al_2O_3 spheres 3.27 mm in diameter levitated without laser heating. The video was processed using the same procedure as for the laser-heated molten oxides. The variation in calculated volume from the machine vision algorithm did not exceed 1%. The main contribution to calibration uncertainty comes from measurements of diameter and sphericity of the Al_2O_3 sphere (taken as ±0.025 mm). The total uncertainty in volume from camera calibration was estimated as ±3%, from ±2.3% uncertainty in volume of the calibration standard and ±0.3% variation in volume from video edge detection procedure.

Correlation of the pyrometer trace with the video recording and density curve obtained from video analysis was performed manually. The moment of turning off the lasers is clearly observed from the disappearance of the bright spot on the molten sample. The onset of crystallization was evident from the video from a sudden increase in sample brightness—sample "flash" or recalescence, caused by reheating the sample on crystallization by released heat of fusion.

3. Results and Discussion

In experiments performed on Y_2O_3, $Zr_{0.95}Y_{0.05}O_{1.975}$, and $Zr_{0.9}Y_{0.1}O_{1.95}$, it was possible to record recalescence on cooling traces and videos. This allowed accurate temperature correlation of density values and evaluation of thermal expansion of the liquid. Stable levitation of HfO_2 and Yb_2O_3 through recalescence was challenging, and we did not succeed in recording videos of recalescence. However, considering 3% uncertainty from calibration, we attribute the measured values for HfO_2 and Yb_2O_3 to the melting temperatures. Matched temperature–density profiles are shown in Figures 2 and 3. The obtained density and thermal expansion data are summarized in Table 1, together with relevant reference values.

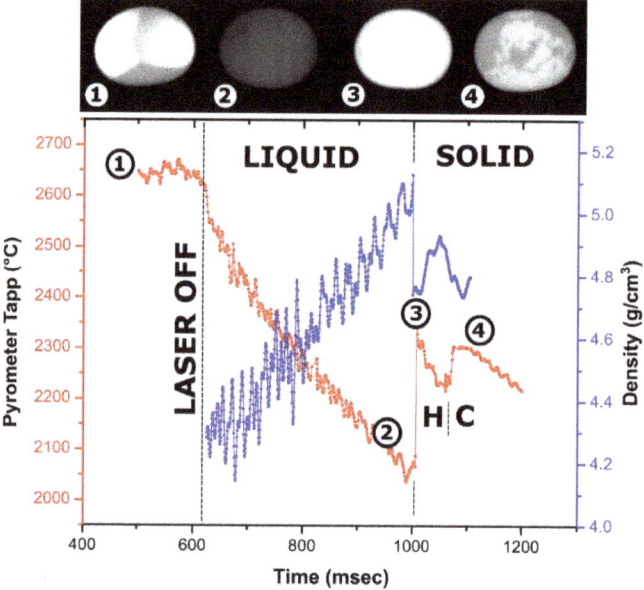

Figure 2. Cooling trace of a 59-mg Y_2O_3 sample (~2.5 mm in diameter) melted in an Ar jet in an aeroacoustic levitator with overlayed density measurements and video frames (lateral view): (1)—molten droplet heated laterally with two laser beams which are visible as bright spots; (2)—undercooled liquid before crystallization; (3)—recalescence or "flash" on crystallization; (4)—phase transformation from high temperature hexagonal to cubic bixbyite phase. The brightness of images (2–4) was adjusted by the same degree for visibility. A video fragment is provided in Supplementary Materials (Video S1).

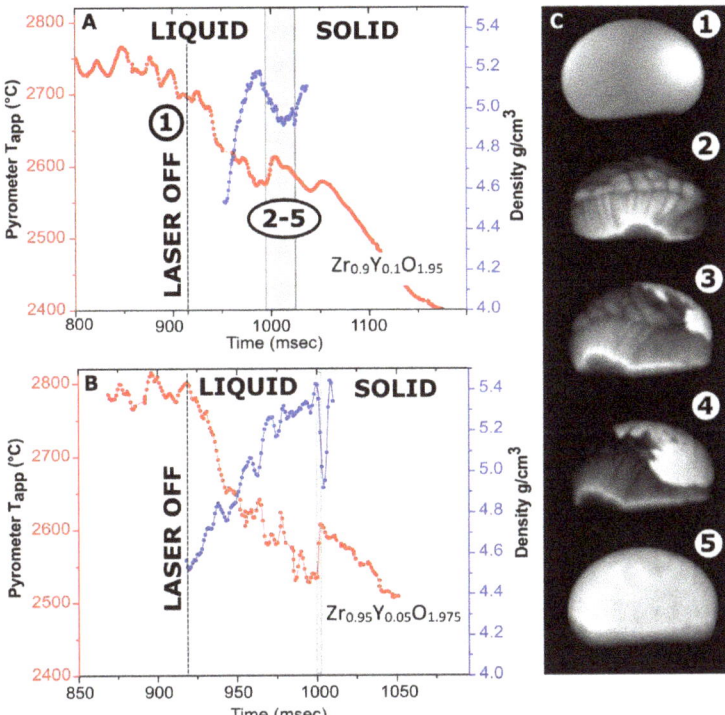

Figure 3. Cooling traces with overlayed density measurements. (**A**) 119.5-mg $Zr_{0.9}Y_{0.1}O_{1.95}$ sample levitated in Ar; (**B**) a 32-mg $Zr_{0.95}Y_{0.05}O_{1.975}$ sample levitated in nitrogen; (**C**) video frames for the $Zr_{0.9}Y_{0.1}O_{1.95}$ sample (~3.5 mm in diameter): (1)—before turning off the lasers (visible as bright spot); (2–5)—113, 132, 135 and 151 ms after turning off the laser. The brightness was adjusted individually for every frame. A video fragment is provided in Supplementary Materials (Video S2).

3.1. Measurements on Y_2O_3

The density profile overlayed with the cooling trace of the Y_2O_3 droplet is shown in Figure 2. The ~300 °C temperature rise on recalescence gives the magnitude of the observed undercooling of liquid Y_2O_3. Recalescence can be pinpointed as a flash on the video, and the density profile shows a sharp decrease due to reheating of the sample. After crystallization onset, the density calculations from the bead dimension are not meaningful since the surface of the droplets crystallizes first, and cavities are formed on further crystallization of core parts of the sample.

The first phase to crystallize from Y_2O_3 melt is known to be a hexagonal phase, common to lanthanide oxides [54–57]. It is stable in a narrow ~100 °C temperature range and undergoes a transition to cubic bixbyite, which is stable at room temperature. This transition is clearly seen on the cooling trace as a second peak with a smaller ~100 °C rise corresponding to undercooling on hexagonal (H-type) to cubic phase transformation and a plateau at a temperature ~100 °C lower than the recalescence peak. Apart from the "flash" of the bead due to temperature increase, crystallization of the H phase is not clearly distinguished in videos, indicating its emissivity is similar to that of the liquid.

Table 1. Density and volumetric thermal expansion coefficient (TEC) for liquid oxides at melting temperatures (T_m) measured in this work compared with previously published data.

Composition	T_m, °C [Ref.]	Density g/cm^3	TEC 10^{-4} K^{-1}	Method †	Ref.
Y$_2$O$_3$	2431 [58]	4.6 ± 0.15	3 ± 1	AAL	This work
		4.42	1.9	CNL↓	Granier 1988 [59]
		4.15 ‡	4.5	AI MD	Kapush 2017 [60]
HfO$_2$	2800 [61]	8.2 ± 0.3	-	AAL	This work
		8.16		PDF	Gallington 2017 [62]
		8.7 *		AI MD	Hong 2018 [63]
Zr$_{0.95}$Y$_{0.05}$O$_{1.975}$	2730 [64]	4.9 ± 0.2	2 ± 1	AAL	This work
Zr$_{0.9}$Y$_{0.1}$O$_{1.95}$	2750 [64]	4.7 ± 0.2	3 ± 1	AAL	This work
ZrO$_2$	2710 [61]	4.9	0.9	AI MD	Hong 2018 [63]
		5.05	1.8	CNL→	Kohara 2014 [65]
		4.69 ± 0.23	0.9	CNL→	Kondo 2019 [66]
Yb$_2$O$_3$	2434 [67]	8.4 ± 0.2	-	AAL	This work
		7.94	0.9	CNL↓	Granier 1988 [59]
		8.75	4.5	AI MD	Fyhrie 2019 [68]
Er$_2$O$_3$	2417 [67]	8.17 ± 0.16	1.0	ESL(ISS)	Koyama 2020 [37]
		7.60	0.4	CNL↓	Granier 1988 [59]
Gd$_2$O$_3$	2420 [67]	7.24 ± 0.14	0.7 ± 0.2	ESL(ISS)	Ishikawa 2020 [36]
		6.93	1.5	CNL↓	Granier 1988 [59]

† Methods abbreviation: AAL—aero-acoustic levitation; CNL ↓—conical nozzle aerodynamic levitation top view; CNL →—idem., side view; PDF—refined from experimental pair distribution function; ESL(ISS)—electrostatic levitation at international space station; AI MD—ab initio molecular dynamic computations. ‡ The density value from calculations at 2377 °C. * The density value from calculations at 2827 °C.

On turning off the laser, the density of liquid Y$_2$O$_3$ increases from 4.3 g/cm^3 at 2650 °C to 5.1 g/cm^3 at 2100 °C. The density of the liquid at the melting temperature is estimated at 4.6 ± 0.15 g/cm^3. For the experiment shown in Figure 2, observed fluctuation in refined density is likely due to the non-symmetrical oscillation of the molten sample in the acoustic field, which ceases after crystallization. In addition to the experiment shown in Figure 2, density was refined from experiments on three more Y$_2$O$_3$ samples, 56–80 mg in weight, using Ar and N$_2$ for levitation. The refined values varied from 4.3 to 4.7 g/cm^3. In several experiments, the scatter in volume from video processing was as low as ±0.01 g/cm^3. We were not able to correlate other measurements with the temperature trace on cooling; however, the observed variation is consistent with the density value from Figure 2. Volume thermal expansion of liquid Y$_2$O$_3$ at melting temperature was estimated as $(3 \pm 1) \times 10^{-4}$ K^{-1}.

3.2. Measurements on Zr$_{0.9}$Y$_{0.1}$O$_{1.95}$ and Zr$_{0.95}$Y$_{0.05}$O$_{1.975}$

The density profiles overlayed with cooling traces from Y-doped zirconia samples are shown in Figure 3. Undercooling on the crystallization of Zr$_{0.9}$Y$_{0.1}$O$_{1.95}$ and Zr$_{0.95}$Y$_{0.05}$O$_{1.975}$ does not exceed 50 °C and 80 °C, respectively. The recalescence step is much more pronounced for a smaller sample. In previous experiments on these compositions in an aerodynamic levitator, no recalescence peaks were detected [51], likely due to the larger sample gradient.

The video stills from the crystallization of Zr$_{0.9}$Y$_{0.1}$O$_{1.95}$ (Figure 3C) have interesting features. The Marangoni flows on the cooling of Zr$_{0.9}$Y$_{0.1}$O$_{1.95}$ are more pronounced than for Y$_2$O$_3$ but are not observed at the bottom of the sample. The Marangoni flows are caused by temperature or composition-related gradients in surface tension. Calorimetry experiments indicated the possibility of oxygen dissolution in liquid ZrO$_2$ and HfO$_2$ [63]. In our experiments, the melting of Zr$_{0.9}$Y$_{0.1}$O$_{1.95}$ was accomplished in an air–argon mixture, with argon flow provided by an auxiliary gas jet. The lower oxygen fugacity at the bottom of the sample impinged upon by the argon jet could be a plausible reason for this behavior.

In the case of congruent crystallization, one would expect that a solid phase would first appear at the bottom surface of the sample due to additional cooling by the argon jet. This is the case for the C–H transformation in Y_2O_3 (Figure 2). However, this is not what happens in $Zr_{0.9}Y_{0.1}O_{1.95}$, in which the bottom of the bead seems to crystallize last. This supports the hypothesis of variable oxygen content in the melt. The densities of $Zr_{0.95}Y_{0.05}O_{1.975}$ and $Zr_{0.9}Y_{0.1}O_{1.95}$ at melting temperatures were estimated at 4.9 and 4.7 g/cm^3, respectively. The difference between compositions is within the assigned experimental uncertainty of $\pm 0.2\ g/cm^3$.

3.3. Comparison with Previously Reported Density Values

In Table 1 the density and thermal expansion coefficient values for liquid oxides measured in this work are listed together with four types of previously published density data: (i) measurements in aerodynamic levitator by Granier and Heurnault's [59], Kohara et al. [65], and Kondo et al. [66]; (ii) measurements with the electrostatic levitator on board of International Space Station [36,37]; (iii) refinement from pair distribution function analysis (PDF) of synchrotron X-ray scattering [62]; (iv) ab initio molecular dynamic computations [60,63].

Granier and Heurnault reported the density of liquid alumina, yttria, and several lanthanide sesquioxides [59,69]. They performed the measurements on photographs of laser-heated droplets levitated in a conical nozzle aerodynamic levitator (CNL). Their value for Al_2O_3 density at melting temperature is about 10% lower than most of the previous measurements [69]. Granier's values for Y_2O_3 and for Yb_2O_3 are 4–5% lower than those measured in this work.

Density measurements for Gd_2O_3 and Er_2O_3 were recently performed by a Japanese group with an electrostatic levitator furnace (ELF) at the International Space Station (ISS) [35–37]. The electrostatic levitation in microgravity conditions ensures the absence of disturbances by acoustic waves or by gas flow, resulting in a perfectly spherical shape of the levitating droplet. Notably, Granier's values for Gd_2O_3 and Er_2O_3 from CNL are also 4–7% lower than measured in ELF at ISS (Table 1).

There is a simple explanation for this discrepancy if one considers how measurements were performed in Granier's and Heurnault's study [59]. In the early version of the aerodynamic levitator they used, the sample was completely surrounded by the nozzle. The photographs were taken from the top, and density values were calculated assuming spherical shape of the sample. However, the shape of the levitated droplet is not an ideal sphere but an oblate spheroid. This can be clearly seen from the lateral view of molten samples in an aero-acoustic levitator (Figure 2). Using top view and assumption of spherical shape would overestimate the volume and underestimate the density. In terrestrial measurements, the degree of oblateness of the spheroid depends on the surface tension of the melt. Melting temperatures of Y_2O_3, Yb_2O_3, Gd_2O_3, and Er_2O_3 are ~400 °C higher than Al_2O_3, and they are expected to have higher surface tension. For these oxides, the bead of the same dimension will be closer to spherical shape than in the case of Al_2O_3. This is consistent with better agreement of measurements for these oxides with Granier's values.

Kohara et al. [65] also used CNL for measurements of density for liquid ZrO_2, however they employed a very shallow nozzle which allowed the side view of the sample. Their value for density and thermal expansion of ZrO_2 (5.05 g/cm^3 and $1.8 \times 10^{-4}\ K^{-1}$) is the same within uncertainty as our results for $Zr_{0.95}Y_{0.05}O_{1.975}$ ($4.9 \pm 0.2\ g/cm^3$ and $(2 \pm 1) \times 10^{-4}\ K^{-1}$). The value measured for liquid HfO_2 density in this work ($8.2 \pm 0.3\ g/cm^3$) is in excellent agreement with the density reported by Gallington et al. [62] (8.16 g/cm^3) from refinement of total synchrotron X-ray scattering on liquid HfO_2.

Liquid ZrO_2 density from ab initio molecular dynamic (AI MD) calculations [63] coincides with our value; however, the Y_2O_3 value from computations is 10% lower than measured, and for HfO_2 and Yb_2O_3, it is 4–6% higher. The comparison of absolute density values from computations with experiment is compromised by uncertainty in absolute temperature in AI MD simulations and underlying assumptions such as choices of exchange-correlation functionals and size of the simulation system.

3.4. Temperature Measurements

In this work, we correlated high-speed video recording with recalescence peak to provide density and thermal expansion values in the proximity of melting temperature. This approach assumes that peak temperature on recalescence is close to the melting point and does not require knowledge of absolute values of the sample temperature. Thus, our measurements did not require knowledge of emissivity, but relied on known melting temperatures of measured oxides. The only assumption made about emissivity is that it does not change substantially for measured liquid oxides around the melting temperature. However, temperature calibration is a paramount issue for measurement of thermal expansion above the melting temperature and for determination of unknown temperatures of melting and phase transformations. Below we discuss methods of estimation of emissivity values and effective emissivities calculated for measured samples.

Temperature measurements with a single-color pyrometer require knowledge of spectral emissivity and its temperature dependence. These data for refractory oxides above 2000 °C are fragmentary [70]. For opaque materials, the sum of emissivity and reflectivity at given wavelength must be equal to one. From careful measurements of reflectivities at 650 nm in a solar furnace, Yamada and Noguchi [71] obtained emissivities for Y_2O_3 and ZrO_2 at melting temperatures as 0.92 ± 0.005, and 0.89 ± 0.005. In an earlier study [72], the same group reported 650 nm emissivity values for HfO_2 and Al_2O_3 at freezing points as 0.91 and 0.96, but without estimation of the uncertainties.

Nordine et al. [48] suggested that reflectivity of opaque melts can be estimated from the refraction index of the solid and emissivity can be calculated using the following relationship:

$$\epsilon(\lambda) = 1 - r(\lambda) = 1 - \frac{(n-1)^2}{(n+1)^2} \qquad (1)$$

where r is the reflectivity and n is the refractive index at the given wavelength λ This approach neglects temperature dependence of emissivity and changes in emissivities between solid and liquid phases. However, in the case of Al_2O_3 for which high temperature data are available [73] these changes are small.

The videos indicate that the melt is opaque at visible wavelengths, and therefore the transmittance can be neglected. Emissivity for different materials is calculated in Table 2 based on Equation (1). The differences from the available values reported by Noguchi [71] from direct measurements for liquid oxides do not exceed 3%. Thus, this approach can be used for future measurements.

Table 2. Calculated emissivity based on the reported values for the refractive index (n).

Composition	T, °C	n	λ, nm	ϵ_{calc}	Reference for (n) Value
Al_2O_3	25	1.78	632.8	0.92	Krishnan 1991 [73]
Y_2O_3	25	1.92	650	0.90	Nigara 1968 [74]
HfO_2	25	2.08	600	0.88	Hu 2003 [75]
Yb_2O_3	25	1.94	643.8	0.90	Medenbach 2001 [76]
ZrO_2 12 mol% Y_2O_3	25	2.15	650	0.87	Wood 1982 [77]

When the melting temperatures are known or independently measured, the common approach is to estimate effective emissivity ϵ_λ at melting temperature using Wien's approximation:

$$\frac{1}{T} = \frac{1}{T_A} + \frac{\lambda}{C_2} ln(\epsilon_\lambda) \qquad (2)$$

where T is the melting temperature, T_A is the apparent melting temperature (measured by pyrometer), and C_2 is Planck's second radiation constant. The temperature dependence of emissivity is usually neglected, and the obtained value for effective emissivity can be used to correct the apparent temperature of the melt in the range of the measurements.

In dynamic measurements of phase transformations, metastability is common on cooling but not on heating. Thermal arrests on melting cannot be unambiguously distinguished on heating with a continuous wave (CW) laser. On cooling of oxide melts, crystallization onset is usually below equilibrium melting temperature. This undercooling is more pronounced in levitated samples in the absence of any solid–liquid interfaces. The crystallization of undercooled melt results in a peak in temperature-recalescence which can be visually observed as a "flash". If on recalescence the sample is reheated to the equilibrium melting temperature, the true thermal arrest can be observed. Ideally, its temperature should be used for emissivity calculation at the melting point.

In our experiments, we do not observe thermal arrest on recalescence. It is not surprising for 2–3 mm beads with melting temperatures above 2400 °C. Due to the relatively large surface to volume ratio and high temperatures, heat transfer by radiation does not allow sample reheating to melting temperature by released heat of fusion.

Table 3 lists emissivity values calculated from Equation (2), taking the maximum temperature of recalescence peak as the apparent melting temperature. The values calculated based on this method will always underestimate emissivity when the melting point is not reached; therefore, they should be seen as a lower boundary. Measurements on a curved surface will also underestimate emissivity; thus, these differences are expected. While single-color pyrometry remains the fastest and most sensitive technique, it must be noted that approaches which do not require knowledge of emissivity for temperature measurements are well established, such as direct measurements of reflectivity [78] and spectropyrometry [79–81].

Table 3. Emissivity values calculated from assumption that the recalescence peak reaches the melting temperature (T_m).

Composition	T_m °C [Ref.]	ε_{calc} (at T_m 650 nm)	Ref.
Y_2O_3	2431 [58]	0.8	This work
$Zr_{0.95}Y_{0.05}O_{1.975}$	2730 [64]	0.68	This work
$Zr_{0.9}Y_{0.1}O_{1.95}$	2750 [64]	0.65	This work

4. Conclusions and Future Directions

This work demonstrates that reasonable values for density of liquid oxides can be obtained from high-speed videography measurements with an aero-acoustic levitator. Volume change on melting cannot be directly obtained from these experiments. While videography can be used on solids, samples in this work were prepared by laser melting and contained cavities formed on solidification. However, volume change on melting can be derived by combining density data for liquids with thermal expansion data on solids from X-ray diffraction [51,57,82].

The accuracy of the measurements can be significantly improved by back illumination of the levitated samples, combined with appropriate filters on the camera [27,34], which is known to aid edge detection in image processing. The multi-emitter single-axis acoustic levitator introduced by Marzo et al. [50] allows levitation of non-spherical samples with density up to 6.5 g/cm^3. The adaptation of a new multi-emitter design for laser heating and density measurements can simplify levitation, decrease the deformation of the liquid sample by acoustic waves, and eliminate or drastically reduce the need for auxiliary aerodynamic support of the sample.

The measurements of change in density with melt composition can be used to obtain the excess volume of mixing in the liquid state for multicomponent systems and derive realistic thermodynamic mixing models, as demonstrated by Fukuyama et al. [30] for metal alloy systems. Such measurements for key refractory oxide systems will be the subject of future studies.

Supplementary Materials: The following are available online at https://www.mdpi.com/1996-1944/14/4/822/s1, Video S1: Crystallization of 58.79 mg Y_2O_3 sample (500 μs acquisition interval, 200 μs exposure, exported to AVI format at 50 frames per second), Video S2: Crystallization of 119.50 mg Y_2O_3 sample (1000 μs acquisition interval, 40 μs exposure, exported to AVI at 50 frames per second).

Author Contributions: Conceptualization, S.V.U., A.N. and R.T.; experiments, J.N., R.P., S.V.U., R.T.; video processing, D.G.Q.; emissivity analysis, J.N.; writing—original draft preparation, S.V.U., D.G.Q. and J.N.; writing—review and editing, A.N., R.P. and R.T. All authors have read and agreed to the published version of the manuscript.

Funding: This research was funded by the U.S. National Science Foundation under the award NSF-DMR 1835848 (changed to NSF-DMR 2015852 on funding moved from UC Davis to ASU). Operation of aero-acoustic levitator at RWTH Aachen is funded by German Research Foundation Grants No Te146/37-2 and No GZ/Inst 222/779-1 FUGG AOBJ:544260.

Institutional Review Board Statement: Not applicable.

Informed Consent Statement: Not applicable.

Data Availability Statement: The data presented in this study are available on request from the corresponding authors.

Acknowledgments: The authors gratefully acknowledge Robert W. Hyers for discussion of experiments and initial evaluation of video data quality, Paul C. Nordine and Richard J. K. Weber for discussion of aero-acoustic levitator operation. Part of this work was presented by S.V.U at the 12th International Workshop on Subsecond Thermophysics (IWSSTP 2019), at DLR Cologne, Germany.

Conflicts of Interest: The authors declare no conflict of interest.

References

1. Luo, A.A. Material design and development: From classical thermodynamics to CALPHAD and ICME approaches. *Calphad* **2015**, *50*, 6–22. [CrossRef]
2. Penilla, E.H.; Devia-Cruz, L.F.; Wieg, A.T.; Martinez-Torres, P.; Cuando-Espitia, N.; Sellappan, P.; Kodera, Y.; Aguilar, G.; Garay, J.E. Ultrafast laser welding of ceramics. *Science* **2019**, *365*, 803–808. [CrossRef]
3. Chen, Z.; Li, Z.; Li, J.; Liu, C.; Lao, C.; Fu, Y.; Liu, C.; Li, Y.; Wang, P.; He, Y. 3D printing of ceramics: A review. *J. Eur. Ceram. Soc.* **2019**, *39*, 661–687. [CrossRef]
4. Simpson, T.W.; Williams, C.B.; Hripko, M. Preparing industry for additive manufacturing and its applications: Summary & recommendations from a National Science Foundation workshop. *Addit. Manuf.* **2017**, *13*, 166–178.
5. Ferrage, L.; Bertrand, G.; Lenormand, P.; Grossin, D.; Ben-Nissan, B. A review of the additive manufacturing (3DP) of bioceramics: Alumina, zirconia (PSZ) and hydroxyapatite. *J. Aust. Ceram. Soc.* **2017**, *53*, 11–20. [CrossRef]
6. Council, N.R. *3D Printing in Space*; The National Academies Press: Washington, DC, USA, 2014; p. 106.
7. Ferrage, L.; Bertrand, G.; Lenormand, P. Dense yttria-stabilized zirconia obtained by direct selective laser sintering. *Addit. Manuf.* **2018**, *21*, 472–478. [CrossRef]
8. McGrath, J.R. *Exploding Wire Research 1774–1963*; NRL Memorandum Report 1698; U.S. Naval Research Laboratory: Washington, DC, USA, 1966.
9. Gallob, R.; Jaeger, H.; Pottlacher, G. A submicrosecond pulse heating system for the investigation of thermophysical properties of metals at high temperatures. *Int. J. Thermophys.* **1986**, *7*, 139–147. [CrossRef]
10. Savvatimskiy, A.I.; Onufriev, S.V.; Valyano, G.E.; Muboyadzhyan, S.A. Thermophysical properties for hafnium carbide (HfC) versus temperature from 2000 to 5000 K (experiment). *J. Mater. Sci.* **2020**, *55*, 13559–13568. [CrossRef]
11. Savvatimskiy, A.I.; Onufriev, S.V.; Muboyajan, S.A.; Tsygankov, P.A. Pulsed Heating of Carbides. *Bull. Russ. Acad. Sci. Phys.* **2018**, *82*, 363–368. [CrossRef]
12. Savvatimskiy, A.I.; Onufriev, S.V.; Muboyadzhyan, S.A. Measurement of ZrC properties up to 5000 K by fast electrical pulse heating method. *J. Mater. Res.* **2017**, *32*, 1287–1294. [CrossRef]
13. Knyazkov, A.M.; Kurbakov, S.D.; Savvatimskiy, A.I.; Sheindlin, M.A.; Yanchuk, V.I. Melting of carbides by electrical pulse heating. *High Temp. High Press.* **2011**, *40*, 349–358.
14. Wroughton, D.M.; Okress, C.E. Magnetic Levitation and Heating of Conductive Materials. U.S. Patent 2,686,864, 17 August 1954.
15. Rhim, W.K.; Chung, S.K.; Barber, D.; Man, K.F.; Gutt, G.; Rulison, A.; Spjut, R.E. An Electrostatic Levitator for High-Temperature Containerless Materials Processing in 1-G. *Rev. Sci. Instrum.* **1993**, *64*, 2961–2970. [CrossRef]
16. Szekely, J.; Schwartz, E.; Hyers, R. Electromagnetic levitation-A useful tool in microgravity research. *JOM* **1995**, *47*, 50–53. [CrossRef]

17. Matson, D.M.; Fair, D.J.; Hyers, R.W.; Rogers, J.R. Contrasting Electrostatic and Electromagnetic Levitation Experimental Results for Transformation Kinetics of Steel Alloys. *Ann. N. Y. Acad. Sci.* **2004**, *1027*, 435–446. [CrossRef] [PubMed]
18. Hyers Robert, W.; Rogers Jan, R. A Review of Electrostatic Levitation for Materials Research. *High Temp. Mater. Process.* **2008**, *27*, 461.
19. Assael, M.J.; Armyra, I.J.; Brillo, J.; Stankus, S.V.; Wu, J.T.; Wakeham, W.A. Reference Data for the Density and Viscosity of Liquid Cadmium, Cobalt, Gallium, Indium, Mercury, Silicon, Thallium, and Zinc. *J. Phys. Chem. Ref. Data* **2012**, *41*, 285. [CrossRef]
20. Kuribayashi, K. Containerless Crystallization of Semiconductors. In *Solidification of Containerless Undercooled Melts*; Herlach, D.M., Matson, D.M., Eds.; Wiley: Weinheim, Germany, 2012.
21. Brillo, J.; Lohoefer, G.; Schmidt-Hohagen, F.; Schneider, S.; Egry, I. Thermophysical property measurements of liquid metals by electromagnetic levitation. *Int. J. Mater. Prod. Technol.* **2006**, *26*, 247–273. [CrossRef]
22. Brillo, J. *Thermophysical Properties of Multicomponent Liquid Alloys*; De Gruyter: Berlin, Germany, 2016.
23. Higuchi, K.; Kimura, K.; Mizuno, A.; Watanabe, M.; Katayama, Y.; Kuribayashi, K. Precise measurement of density and structure of undercooled molten silicon by using synchrotron radiation combined with electromagnetic levitation technique. *Meas. Sci. Technol.* **2005**, *16*, 381–385. [CrossRef]
24. Shiraishi, S.Y.; Ward, R.G. The density of nickel in the superheated and supercooled liquid states. *Can. Metall. Q.* **1964**, *3*, 117–122. [CrossRef]
25. Bradshaw, R.C.; Schmidt, D.P.; Rogers, J.R.; Kelton, K.F.; Hyers, R.W. Machine vision for high-precision volume measurement applied to levitated containerless material processing. *Rev. Sci. Instrum.* **2005**, *76*, 125108. [CrossRef]
26. Adachi, M.; Aoyagi, T.; Mizuno, A.; Watanabe, M.; Kobatake, H.; Fukuyama, H. Precise Density Measurements for Electromagnetically Levitated Liquid Combined with Surface Oscillation Analysis. *Int. J. Thermophys.* **2008**, *29*, 2006–2014. [CrossRef]
27. Watanabe, M.; Adachi, M.; Fukuyama, H. Densities of Fe-Ni melts and thermodynamic correlations. *J. Mater. Sci.* **2016**, *51*, 3303–3310. [CrossRef]
28. Egry, I.; Langen, M.; Lohofer, G.; Earnshaw, J.C. Measurements of thermophysical properties of liquid metals relevant to Marangoni effects. *Philos. Trans. R. Soc. Lond. Ser. A* **1998**, *356*, 845–856. [CrossRef]
29. Fukuyama, H.; Kobatake, H.; Takahashi, K.; Minato, I.; Tsukada, T.; Awaji, S. Development of modulated laser calorimetry using a solid platinum sphere as a reference. *Meas. Sci. Technol.* **2007**, *18*, 2059–2066. [CrossRef]
30. Fukuyama, H.; Watanabe, M.; Adachi, M. Recent studies on thermophysical properties of metallic alloys with PROSPECT: Excess properties to construct a solution model. *High Temp. High Press.* **2020**, *49*, 851. [CrossRef]
31. Ishikawa, T.; Paradis, P.F.; Yoda, S. New sample levitation initiation and imaging techniques for the processing of refractory metals with an electrostatic levitator furnace. *Rev. Sci. Instrum.* **2001**, *72*, 2490–2495. [CrossRef]
32. Paradis, P.F.; Ishikawa, T.; Yu, J.; Yoda, S. Hybrid electrostatic-aerodynamic levitation furnace for the high-temperature processing of oxide materials on the ground. *Rev. Sci. Instrum.* **2001**, *72*, 2811–2815. [CrossRef]
33. Paradis, P.F.; Yu, J.; Ishikawa, T.; Aoyama, T.; Yoda, S.; Weber, J.K.R. Contactless density measurement of superheated and undercooled liquid $Y_3Al_5O_{12}$. *J. Cryst. Growth* **2003**, *249*, 523–530. [CrossRef]
34. Paradis, P.-F.; Yu, J.; Ishikawa, T.; Yoda, S. Contactless Density Measurement of Liquid Nd-Doped 50%CaO–50%Al_2O_3. *J. Am. Ceram. Soc.* **2003**, *86*, 2234–2236. [CrossRef]
35. Tamaru, H.; Koyama, C.; Saruwatari, H.; Nakamura, Y.; Ishikawa, T.; Takada, T. Status of the Electrostatic Levitation Furnace (ELF) in the ISS-KIBO. *Microgravity Sci. Technol.* **2018**, *30*, 643–651. [CrossRef]
36. Ishikawa, T.; Koyama, C.; Saruwatari, H.; Tamaru, H.; Oda, H.; Ohshio, M.; Nakamura, Y.; Watanabe, Y.; Nakata, Y. Density of molten gadolinium oxide measured with the electrostatic levitation furnace in the International Space Station. *High Temp. High Press.* **2020**, *49*, 5–15. [CrossRef]
37. Koyama, C.; Tahara, S.; Kohara, S.; Onodera, Y.; Småbråten, D.R.; Selbach, S.M.; Akola, J.; Ishikawa, T.; Masuno, A.; Mizuno, A.; et al. Very sharp diffraction peak in nonglass-forming liquid with the formation of distorted tetraclusters. *NPG Asia Mater.* **2020**, *12*, 43. [CrossRef]
38. Winborne, D.A.; Nordine, P.C.; Rosner, D.E.; Marley, N.F. Aerodynamic Levitation Technique for Containerless High Temperature Studies on Liquid and Solid Samples. *Metall. Mater. Trans. B* **1976**, *7*, 711–713. [CrossRef]
39. Nordine, P.C.; Atkins, R.M. Aerodynamic levitation of laser-heated solids in gas jets. *Rev. Sci. Instrum.* **1982**, *53*, 1456–1464. [CrossRef]
40. Nordine, P.C.; Weber, J.K.R.; Abadie, J.G. Properties of high-temperature melts using levitation. *Pure Appl. Chem.* **2000**, *72*, 2127–2136. [CrossRef]
41. Benmore, C.J.; Weber, J.K.R. Aerodynamic levitation, supercooled liquids and glass formation. *Adv. Phys. X* **2017**, *2*, 717–736. [CrossRef]
42. Rey, C.A.; Merkley, D. Aero-Acoustic Levitation Device and Method. U.S. Patent 5,096,017, 17 March 1992.
43. Nordine, P.C. Personal communication, 2020.
44. Weber, R.J.K.; Hampton, S.; Merkley, D.S.; Rey, C.A.; Zatarski, M.M.; Nordine, P.C. Aero-acoustic levitation: A method for containerless liquid-phase processing at high temperatures. *Rev. Sci. Instrum.* **1994**, *65*, 456–465. [CrossRef]
45. Sato, H.; Tsukamoto, K.; Kuribayashi, K. Growth of Olivine by Aero-Acoustic Levitation: Reproduction of Meteorite Texture. *J. Jap. Assoc. Cryst. Growth* **1998**, *25*, A155.
46. Nagashio, K.; Hofmeister, W.H.; Gustafson, D.E.; Altgilbers, A.; Bayuzick, R.J.; Kuribayashi, K. Formation of $NdBa_2Cu_3O_{7-\delta}$ amorphous phase by combining aero-acoustic levitation and splat quenching. *J. Mater. Res.* **2001**, *16*, 138–145. [CrossRef]
47. Nagashio, K.; Takamura, Y.; Kuribayashi, K.; Shiohara, Y. Microstructural control of $NdBa_2Cu_3O_{7-\delta}$ superconducting oxide from highly undercooled melt by containerless processing. *J. Cryst. Growth* **1999**, *200*, 118–125. [CrossRef]

48. Nordine, P.C.; Merkley, D.; Sickel, J.; Finkelman, S.; Telle, R.; Kaiser, A.; Prieler, R. A levitation instrument for containerless study of molten materials. *Rev. Sci. Instrum.* **2012**, *83*, 125107/1–125107/14. [CrossRef]
49. Telle, R.; Greffrath, F.; Prieler, R. Direct observation of the liquid miscibility gap in the zirconia-silica system. *J. Eur. Ceram. Soc.* **2015**, *35*, 3995–4004. [CrossRef]
50. Marzo, A.; Barnes, A.; Drinkwater, B. TinyLev: A multi-emitter single-axis acoustic levitator. *Rev. Sci. Instrum.* **2017**, *88*, 085105. [CrossRef]
51. Ushakov, S.V.; Navrotsky, A.; Weber, R.J.K.; Neuefeind, J.C. Structure and Thermal Expansion of YSZ and $La_2Zr_2O_7$ Above 1500 °C from Neutron Diffraction on Levitated Samples. *J. Am. Ceram. Soc.* **2015**, *98*, 3381–3388. [CrossRef]
52. Bendert, J.C.; Gangopadhyay, A.K.; Mauro, N.A.; Kelton, K.F. Volume Expansion Measurements in Metallic Liquids and Their Relation to Fragility and Glass Forming Ability: An Energy Landscape Interpretation. *Phys. Rev. Lett.* **2012**, *109*, 185901. [CrossRef]
53. Community, B.O. Blender-a 3D Modelling and Rendering Package. Available online: http://www.blender.org (accessed on 30 December 2020).
54. Foex, M.; Traverse, P.J. Polymorphism of rare earth sesquioxides at high temperatures. *Bull. Soc. Fr. Mineral. Cristallogr.* **1966**, *89*, 184–205.
55. Foex, M.; Traverse, J.P. Investigations about crystalline transformation in rare earths sesquioxides at high temperatures. *Rev. Int. Hautes Temp. Refract.* **1966**, *3*, 429–453.
56. Zinkevich, M. Thermodynamics of rare earth sesquioxides. *Prog. Mater. Sci.* **2007**, *52*, 597–647. [CrossRef]
57. Ushakov, S.V.; Navrotsky, A. Experimental approaches to the thermodynamics of ceramics above 1500 °C. *J. Am. Ceram. Soc.* **2012**, *95*, 1463–1482. [CrossRef]
58. Cabannes, F.; Vu, T.L.; Coutures, J.P.; Foex, M. Melting point of yttria as a secondary temperature standard. *High Temp. High Press.* **1976**, *8*, 391–396.
59. Granier, B.; Heurtault, S. Density of liquid rare earth sesquioxides. *J. Am. Ceram. Soc.* **1988**, *71*, C466–C468. [CrossRef]
60. Kapush, D.; Ushakov, S.; Navrotsky, A.; Hong, Q.-J.; Liu, H.; van de Walle, A. A combined experimental and theoretical study of enthalpy of phase transition and fusion of yttria above 2000 °C using "drop-n-catch" calorimetry and first-principles calculation. *Acta Mater.* **2017**, *124*, 204–209. [CrossRef]
61. Wang, C.; Zinkevich, M.; Aldinger, F. The zirconia-hafnia system: DTA measurements and thermodynamic calculations. *J. Am. Ceram. Soc.* **2006**, *89*, 3751–3758. [CrossRef]
62. Gallington, L.; Ghadar, Y.; Skinner, L.; Weber, J.; Ushakov, S.; Navrotsky, A.; Vazquez-Mayagoitia, A.; Neuefeind, J.; Stan, M.; Low, J.; et al. The Structure of Liquid and Amorphous Hafnia. *Materials* **2017**, *10*, 1290. [CrossRef] [PubMed]
63. Hong, Q.-J.; Ushakov, S.V.; Kapush, D.; Benmore, C.J.; Weber, R.J.K.; van de Walle, A.; Navrotsky, A. Combined computational and experimental investigation of high temperature thermodynamics and structure of cubic ZrO_2 and HfO_2. *Sci. Rep.* **2018**, *8*, 14962. [CrossRef] [PubMed]
64. Sibieude, F.; Rouanet, A. Effect of cationic substitutions of the type A^{4+}-Ln^{3+} on the polymorphism of lanthanide sesquioxides. Application to the interpretation of the equilibrium diagrams of the zirconium dioxide-lanthanide sesquioxide and thorium dioxide-lanthanide sesquioxide systems. *Colloq. Int. Cent. Nat. Rech. Sci.* **1972**, *205*, 459–468.
65. Kohara, S.; Akola, J.; Patrikeev, L.; Ropo, M.; Ohara, K.; Itou, M.; Fujiwara, A.; Yahiro, J.; Okada, J.T.; Ishikawa, T.; et al. Atomic and electronic structures of an extremely fragile liquid. *Nat. Commun.* **2014**, *5*, 5892. [CrossRef] [PubMed]
66. Kondo, T.; Muta, H.; Kurosaki, K.; Kargl, F.; Yamaji, M.; Furuya, M.; Ohishi, Y. Density and viscosity of liquid ZrO_2 measured by aerodynamic levitation technique. *Heliyon* **2019**, *5*, e02049. [CrossRef]
67. Konings, R.J.M.; Beneš, O.; Kovács, A.; Manara, D.; Sedmidubský, D.; Gorokhov, L.; Iorish, V.S.; Yungman, V.; Shenyavskaya, E.; Osina, E. The Thermodynamic Properties of the f-Elements and their Compounds. Part 2. The Lanthanide and Actinide Oxides. *J. Phys. Chem. Ref. Data* **2014**, *43*, 013101. [CrossRef]
68. Fyhrie, M.; Hong, Q.-J.; Kapush, D.; Ushakov, S.V.; Liu, H.; van de Walle, A.; Navrotsky, A. Energetics of melting of Yb_2O_3 and Lu_2O_3 from drop and catch calorimetry and first principles computations. *J. Chem. Thermodyn.* **2019**, *132*, 405–410. [CrossRef]
69. Granier, B.; Heurtault, S. Method for measurement of the density of liquid refractories. Application to alumina and yttrium oxide. *Rev. Int. Hautes Temp. Refract.* **1983**, *20*, 61–67.
70. Watanabe, H.; Ishii, J.; Wakabayashi, H.; Kumano, T.; Hanssen, L. Spectral Emissivity Measurements. In *Experimental Methods in the Physical Sciences*; Germer, T.A., Zwinkels, J.C., Tsai, B.K., Eds.; Academic Press: Cambridge, MA, USA, 2014; Volume 46, pp. 333–366.
71. Yamada, T.; Noguchi, T. Digital pyrometry in a solar furnace. *Sol. Energy* **1976**, *18*, 533–539. [CrossRef]
72. Noguchi, T.; Kozuka, T. Temperature and emissivity measurement at 0.65 µ with a solar furnace. *Sol. Energy* **1966**, *10*, 125–131. [CrossRef]
73. Krishnan, S.; Weber, J.K.R.; Schiffman, R.A.; Nordine, P.C.; Reed, R.A. Refractive index of liquid aluminum oxide at 0.6328 µm. *J. Am. Ceram. Soc.* **1991**, *74*, 881–883. [CrossRef]
74. Nigara, Y. Measurement of the Optical Constants of Yttrium Oxide. *Jpn. J. Appl. Phys.* **1968**, *7*, 404–408. [CrossRef]
75. Hu, H.; Zhu, C.; Lu, Y.F.; Wu, Y.H.; Liew, T.; Li, M.F.; Cho, B.J.; Choi, W.K.; Yakovlev, N. Physical and electrical characterization of HfO_2 metal-insulator-metal capacitors for Si analog circuit applications. *J. Appl. Phys.* **2003**, *94*, 551–557. [CrossRef]
76. Medenbach, O.; Dettmar, D.; Shannon, R.D.; Fischer, R.X.; Yen, W.M. Refractive index and optical dispersion of rare earth oxides using a small-prism technique. *J. Opt. A Pure Appl. Opt.* **2001**, *3*, 174–177. [CrossRef]
77. Wood, D.L.; Nassau, K. Refractive index of cubic zirconia stabilized with yttria. *Appl. Opt.* **1982**, *21*, 2978–2981. [CrossRef]

78. Stein, A.; Rabinowitz, P.; Kaldor, A. Laser Radiometer. U.S. Patent 4,417,822, 29 January 1983.
79. Felice, R.A. Temperature Determining Device and Process. U.S. Patent 5,772,323, 30 June 2002.
80. Felice, R.A. The spectropyrometer—a practical multi-wavelength pyrometer. *AIP Conf. Proc.* **2003**, *684*, 711–716.
81. Earl, D.D.; Kisner, R.A. Emissivity Independent Optical Pyrometer. U.S. Patent 2,015,124,244, 15 March 2017.
82. Pavlik, A.; Ushakov, S.V.; Navrotsky, A.; Benmore, C.J.; Weber, R.J.K. Structure and thermal expansion of Lu_2O_3 and Yb_2O_3 up to the melting points. *J. Nucl. Mater.* **2017**, *495*, 385–391. [CrossRef]

MDPI
St. Alban-Anlage 66
4052 Basel
Switzerland
Tel. +41 61 683 77 34
Fax +41 61 302 89 18
www.mdpi.com

Materials Editorial Office
E-mail: materials@mdpi.com
www.mdpi.com/journal/materials

www.ingramcontent.com/pod-product-compliance
Lightning Source LLC
LaVergne TN
LVHW070146100526
838202LV00015B/1902